ENDOPHYTE BIOLOGY

Recent Findings from the Kashmir Himalayas

ENDOPHYTE BIOLOGY

Recent Findings from the Kashmir Himalayas

Edited by

Zahoor Ahmed Wani, PhD

Masroor Qadri, PhD

Palak Arora, PhD

Khalid Rehman Hakeem, PhD

AAP | APPLE ACADEMIC PRESS

First edition published 2023

Apple Academic Press Inc.
1265 Goldenrod Circle, NE,
Palm Bay, FL 32905 USA

760 Laurentian Drive, Unit 19,
Burlington, ON L7N 0A4, CANADA

CRC Press
6000 Broken Sound Parkway NW,
Suite 300, Boca Raton, FL 33487-2742 USA

4 Park Square, Milton Park,
Abingdon, Oxon, OX14 4RN UK

Apple Academic Press exclusively co-publishes with CRC Press, an imprint of Taylor & Francis Group, LLC

Library and Archives Canada Cataloguing in Publication

Title: Endophyte biology : recent findings from the Kashmir Himalayas / edited by Zahoor Ahmed Wani, PhD, Masroor Qadri, PhD, Palak Arora, PhD, Khalid Rehman Hakeem, PhD.
Names: Wani, Zahoor Ahmed, editor. | Qadri, Masroor, editor. | Arora, Palak, editor. | Hakeem, Khalid Rehman, editor.
Description: First edition. | Includes bibliographical references and index.
Identifiers: Canadiana (print) 20220200343 | Canadiana (ebook) 20220200378 | ISBN 9781774638323 (hardcover) | ISBN 9781774638330 (softcover) | ISBN 9781003277262 (ebook)
Subjects: LCSH: Endophytes—India—Jammu and Kashmir.
Classification: LCC QR351 .E53 2023 | DDC 579/.17095463—dc23

Library of Congress Cataloging-in-Publication Data

CIP data on file with US Library of Congress

ISBN: 978-1-77463-832-3 (hbk)
ISBN: 978-1-77463-833-0 (pbk)
ISBN: 978-1-00327-726-2 (ebk)

This book is dedicated to Sheikh ul Alam,
Sheikh Noor Ud din Noorani (Nund Reshi).
Nund Reshi was the spiritual leader of Kashmir.
His famous quote about the nature and food security is
"An poshi teli yeli wan poshi"
(Food will suffice only till the forests survive).

About the Editors

Zahoor Ahmed Wani, PhD

Zahoor Ahmed Wani, PhD, is working as Assistant Professor in the Department of Botany at Government Degree College Kishtwar, Jammu & Kashmir, India. His research specialization is in plant microbe interactions with special focus on understanding the basic intricacies of plant-endophyte associations and the ways to modulate them for desired purposes. His work has been published in peer-reviewed quality journals, including *Nature-Scientific Reports, Fungal Biology, and RSC Advances.* He has a book and a book chapter to his credit as well. Dr. Wani is associated with many NGOs as an advisor/technical expert and is working on various issues related to education, environment, and social transformation in rural areas of J&K state, India. Dr. Wani qualified National Eligibility Test securing 47th rank and was awarded a Junior Research Fellowship by the Council of Scientific and Industrial Research to pursue his research career and was awarded a PhD degree in 2017 by the Academy of Scientific and Innovative Research (AcSIR), India.

Masroor Qadri, PhD

Masroor Qadri, PhD, is working as a Postdoctoral Research Associate in the Entomology and Nematology Department, University of Florida, USA. Her research work was on bioprospecting endophytic fungi for bioactive metabolites, characterizing biosynthetic pathway genes and their epigenetic modulation for increased metabolite production in microbes. Dr. Qadri has published her research work in reputed journals and has a book chapter to her credit as well. She was awarded her doctorate in 2015 by Jamia Hamdard University, New Delhi, India. After completing her doctorate, she joined University of Brasilia, Brazil as a Postdoctoral Research Associate in 2015 and worked there for one year before shifting to USA.

Palak Arora, PhD

Palak Arora, PhD, joined CSIR-Indian Institute of Integrative Medicine (CSIR-IIIM), Jammu, as a Junior Research Fellow and was awarded a PhD Degree in 2019 by the Academy of Scientific and Innovative Research (AcSIR), India. She completed her BTech (Biotechnology) at SMVD University in 2013, where she secured first position and was awarded an INSPIRE Fellowship by the Department of Biotechnology, Government of India, to pursue her research career. Dr. Palak worked on culturable endophytes of *Glycyrrhiza glabra* and studied their bioactive potential and functional characterization. Dr. Palak also worked on endohyphal bacterial associations. She has published her research work in leading journals and has a book chapter to her credit. She has received many awards including the INSPIRE award during her short research career.

Khalid Rehman Hakeem, PhD

Khalid Rehman Hakeem, PhD, is Professor at King Abdulaziz University, Jeddah, Saudi Arabia. After completing his doctorate (botany; specialization in plant eco-physiology and molecular biology) from Jamia Hamdard, New Delhi, India, he worked as a lecturer at the University of Kashmir, Srinagar, India. At Universiti Putra Malaysia, Selangor, Malaysia, he was a Postdoctorate Fellow and Fellow Researcher (Associate Professor) for several years. Dr. Hakeem has more than 10 years of teaching and research experience in plant eco-physiology, biotechnology and molecular biology, medicinal plant research, plant-microbe-soil interactions, as well as in environmental studies. He is the recipient of several fellowships at both national and international levels. He has served as a visiting scientist at Jinan University, Guangzhou, China. Currently, he is involved with a number of international research projects with different government organizations. To date, Dr. Hakeem has authored and edited more than 35 books with international publishers. He also has to his credit more than 80 research publications in peer-reviewed international journals and 55 book chapters in edited volumes with international publishers. At present, Dr. Hakeem serves as an editorial board member and reviewer for several high-impact international scientific journals. He is included in the advisory board of Cambridge Scholars Publishing, UK. He is also a fellow of the Plantae group of the American Society of Plant Biologists, member of the World Academy of Sciences, member of the International Society for Development and Sustainability, Japan, and member of the Asian Federation of Biotechnology, Korea.

Contents

Contributors

Vidushi Abrol
CSIR-Indian Institute of Integrative Medicine, Jammu180001, India
Shri Mata Vaishno Devi University, Katra 182320, India; E-mail: vidushiabrol.1285@gmail.com

Palak Arora
302 Sec 1, Channi Himmat, Jammu Tawi, India-180015

Sayeeda Kousar Bhatti
Department of Botany, Shoolini University of Biotechnology & Management Sciences, Solan 173212, Himachal Pradesh, India

Jiya Chanotiya
Department of Biochemistry, College of Basic Sciences and Humanities, GBPUAT, Pantnagar 263145, Uttarakhand, India

Tanvi Gupta
CSIR-Indian Institute of Integrative Medicine, Jammu180001, India
Institute of Clinical Medicine, National Cheng Kung University, Tainan, Taiwan

Khalid Rehman Hakeem
Department of Biological Sciences, Faculty of Science, King Abdulaziz University, Jeddah, Saudi Arabia; E-mail: kur.hakeem@gmail.com

Himanshu Kachroo
Department of Biotechnology, Indian Institute of Technology Roorkee, Roorkee, Uttarakhand, India; E-mail: hkachroo@bt.iitr.ac.in; himanshukachroo357@gmail.com

Madhvi Kanchan
Department of Botany, Punjab University, Chandigarh 160014, India

Sofia Sharief Khan
School of Biotechnology, Shri Mata Vaishno Devi University, Kakryal, Katra 182320, J&K, India

Kusum
Department of Botany, St. Bede's College, Navbahar, Shimla 171002, Himachal Pradesh, India

Aroosa Jan Mattoo
Department of Botany, University of Jammu, J&K 180006, India

Skarma Nonzom
Department of Botany, University of Jammu, J&K 180006, India; E-mail: skarmanunzom@yahoo.com

Niharika Pandey
Department of Biotechnology, M. B. Govt. P. G. College, Haldwani 263139, Uttarakhand, India

Masroor Qadri
Entomology and Nematology Department, Genetics Institute, University of Florida, 1881 Natural Area Drive, Steinmetz Hall, Gainesville, FL 32611-0620, United States; E-mail: qadrimasroor6@gmail.com

Shafaq Rasool
School of Biotechnology, Shri Mata Vaishno Devi University, Kakryal, Katra 182320, J&K, India;
E-mail: shafaq.rasool@smvdu.ac.in

Pushpa Ruwali
Department of Biotechnology, M. B. Govt. P. G. College, Haldwani 263139, Uttarakhand, India

Anand Sagar
Department of Biosciences, Himachal Pradesh University, Shimla171005, Himachal Pradesh, India

Krishika Sambyal
University Institute of Biotechnology, Chandigarh University, Gharuan, Punjab, India

Jaspreet K. Sembi
Department of Botany, Punjab University, Chandigarh 160014, India; E-mail: jaspreet.sembi@pu.ac.in

Rahul Vikram Singh
Academy of Scientific and Innovative Research (AcSIR), Ghaziabad 201002, Uttar Pradesh, India;
E-mail: rahul.negi121@gmail.com

Kranti Thakur
Department of Botany, Shoolini Institute of Life Sciences & Business Management, Solan 173212, Himachal Pradesh, India

Jagdeep Verma
Department of Botany, Government College, Rajgarh 173101, Himachal Pradesh, India

Zahoor Ahmed Wani
Department of Botany, Govt. Degree College Kishtwar, Kishtwar, Jammu & Kashmir, India;
E-mail: zawani1986@gmail.com

Abbreviations

AM	arbuscular mycorrhizae
CBDA	cannabidiolic acid
CMA	cornmeal agar
CSGs	common symbiosis genes
GABA	γ-aminobutyric acid
ITS	internal transcribed spacer
OM	orchid mycorrhizae
OMA	oatmeal agar
OMF	orchid mycorrhizal fungi
PDA	potato dextrose agar
THC	Δ9-tetrahydrocannabinol
THCA	Δ9-tetrahydrocannabinolic acid

Preface

Endophyte biology is an emerging discipline of science with a multitude of applications in ecology, agriculture, and other industries. The origin of endophyte biology dates back to 1866, when De Barry put forth the concept of "endophyte"—any organism living inside the plant tissues. Endophyte biology, due to its ecological and pharmaceutical significance, has created an immense curiosity among the scientific and academic world. This book addresses the scope, applications, and future perspectives of endophyte biology.

The state of Jammu and Kashmir (J&K) is located in the northwestern (NW) Himalayas and is regarded as the biomass of state of India. The whole Himalayan belt is one of the 26 biodiversity hotspots in India, and it has eight critical areas including Ladakh and Kashmir. Despite having a huge diversity of plants, the endophyte research in this part of the world is still in its infancy, and whatever information is available, it is very scattered. Therefore, this book is an attempt to bridge the information gap on endophyte biology pertaining to Kashmir Himalayas.

This book is a compilation of original, latest, and updated information on endophyte biology of Kashmir Himalayas. It covers the definition of endophytes, endophytic diversity of some important plants of Kashmir Himalayas, bioprospection of endophytes for various drug metabolites and sustainable agriculture, etc. This book will serve as a manual for research scholars as it contains the methodologies and techniques involved in endophyte biology research. This book is written in a very lucid and comprehensible language and is supplemented with illustrations, figures, tables, etc. Therefore, it can also be used as a reference book by teachers and students at graduate and undergraduate level in colleges and universities.

CHAPTER 1

Endophytes: An Overview

ZAHOOR AHMED WANI[1*] and KHALID REHMAN HAKEEM[2]

[1]*Department of Botany, Govt. Degree College Kishtwar, Kishtwar, Jammu & Kashmir, India*

[2]*Department of Biological Sciences, Faculty of Science, King Abdulaziz University, Jeddah, Saudi Arabia*

[*]*Corresponding author. E-mail: zawani1986@gmail.com; kur.hakeem@gmail.com*

ABSTRACT

Plant-microbe interactions are ubiquitous and diverse in nature. Microbes colonizing plant tissues without causing any apparent disease symptoms are called endophytes. Diversity and community structure of the endophytic assemblages vary according to the plant genotype, tissue type, plant age, edaphic factors, microclimate conditions, and anthropogenic factors. Endophytic microbial diversity is elucidated by different techniques like culture dependent, metagenomics, etc. Endophytes are classified into various categories clavicipitaceous and non-clavicipitaceous, systemic and non-systemic, etc. based on several criteria like taxonomy, functional diversity, biology, and mode of transmission, evolution, ecological functions, etc.

1.1 INTRODUCTION

Plant–microbe interactions are ubiquitous and diverse in nature. The association of plants with microbes dates back to more than 400 million years ago. This prolonged and close association between two or more different organisms is considered as a major driving force for expansion of biological diversity from genes to ecosystems (Kuo, 2015). Microorganisms usually colonize plant surface before entering the plant. From their entry points, microorganisms systemically colonize different plant parts, from root to shoot, shoot to flowers or fruits, and from fruit to seed. They may also cause localized colonization inside/outside plant organs (Brader et al., 2017). The interaction between members of microbial communities and plant involves cross feeding of metabolites and other signaling molecules that result in the development of a syntrophic metabolism. In syntrophic metabolism, two partners are needed to establish an energetic positive metabolism (Seth and Taga, 2014). Plants constitute vast and diverse niches for endophytic microorganisms that occupy internal tissues of plants without causing any apparent disease symptoms (Fig. 1.1). The complex interplay of diverse array of microbial communities with the host plant affects its ecophysiology such as plant nutrition, growth rate, resistance to biotic and abiotic stress, as well as plant survival and distribution (Wani et al., 2015). Owing to its engagement in diverse heterospecific associations, each plant is considered as a complex community in itself rather than a single organism and the presence of microorganisms inside the plant tissues is considered to be the rule of thumb in ecology (Partida-Martínez and Heil, 2011; Wani et al., 2015). An understanding of the processes and mechanisms involved in plant–microbe interactions is essential to harness the biotechnological potential of plant–endophyte interactions for a range of applications.

1.2 HISTORY AND DEFINITION OF ENDOPHYTES

Endophytism as a natural phenomenon is a question of history, and its origin probably dates back to the existence of plants on the planet earth (Redecker et al., 2000). However, the advancement of endophyte biology as a discipline of science began in the year 1866 when De Barry put forth the concept of "endophyte." The term endophyte (Gr. *endon*, within; *phyton*, plant) was first coined by De Barry to refer "any organism occurring within the plant tissues" (De Barry, 1866). Endophyte is defined in several ways and the definitions are modified as the research advances (see Box 1.1 for various definitions of endophytes). However, most appropriately an endophyte can be defined as an organism that inhabit the plant organization, at least for a part of its life cycle, without producing any apparent disease symptoms in the host plant under normal conditions (Wilson, 1995). Evidence of plant-associated microbes is reported in the fossilized tissues of stem and leaves (Taylor and Taylor, 2000). As a result of these long-held associations, it is possible that some of the endophytic microbes may have devised genetic systems to allow transfer of information between the microbe and the plant partner and *vice versa* (Freeman and Rodriguez, 1993; Stierle et al., 1993). However, endophytes did not receive much attention until the recent recognition of their pharmaceutical and ecological significance (Gunatilaka, 2006). Since then, endophytes have created immense scientific curiosity pertaining to their biology, evolution, ecology, and applications.

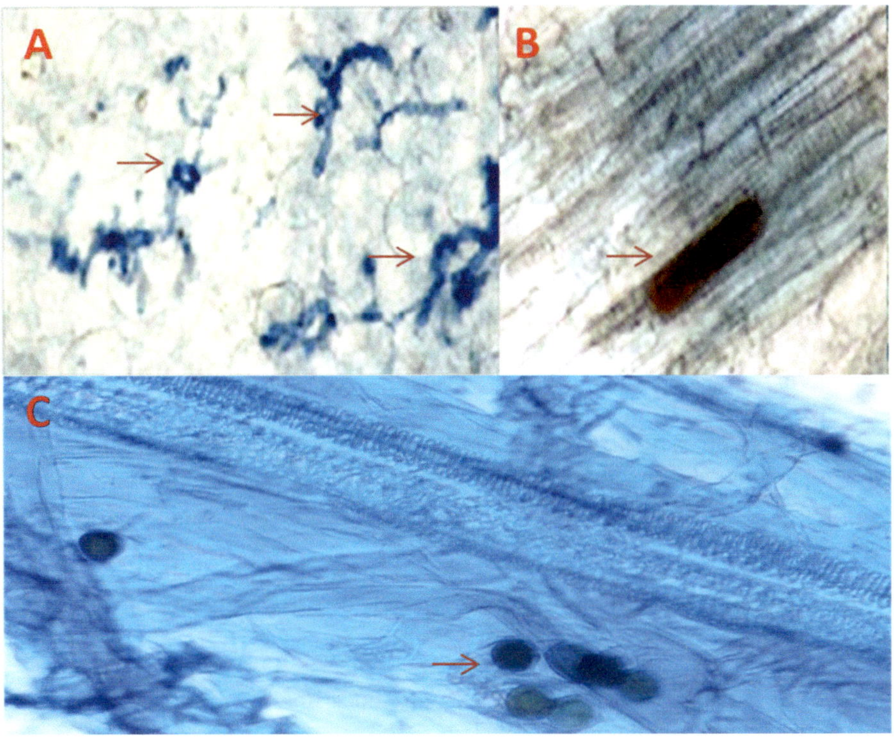

FIGURE 1.1 Microscopic examination of endophytic fungal colonization in plant tissues. (A) Blue-stained structures showing extensive colonization of endophytic fungi, (B) melanized microsclerotia of DSEs inside plant tissues, and (C) chlamydospores-like structures inside plant tissues. (Reprinted with permission from Wani et al. 2016, ©Elsevier)

Contrary to the mycorrhizal symbioses, endophytes do not form cellular interfaces with specialized structures, like arbuscules in the case of arbuscular mycorrhizal fungi. Endophytes have not synchronized their development to the plant partner and the host plant may not get benefit from endophytic colonization, as in the case of commensalism (Brundrett, 2002). The presence of endophyte was first reported by Vogel in 1898, who revealed a mycelium residing in the seed of an annual grass darnel

(*Lolium temulentum*) (Vogel et al., 1898). Later in 1904, Freeman isolated an endophytic fungus from the seeds of *L. temulentum* (Freeman, 1904). Since then, endophytes have been isolated from all the plants, including mosses, bryophytes, pteridophytes, gymnosperms, angiosperms, and even from lichens, studied till date (Arnold and Luzoni, 2007; Porras-Alfaro and Bayman, 2011; Wani et al., 2015, 2016; Qadri et al., 2014; Arora et al., 2017).

Box 1.1. Definitions of Endophyte

- Any organism occurring within plant tissues (De Barry, 1866).
- The microorganisms are able to live inside plants without causing disease symptoms (Tervet and Hollis, 1948).
- All the organisms inhabiting plant organs that at some time in their life can colonize internal plant tissues without causing apparent harm to the host (Petrini, 1991).
- Endophytic bacteria are the population of bacteria that reside within the living organism without doing substantive harm or gaining benefit other than securing residency (Kado, 1992).
- Endophytic bacteria or fungi colonize the host tissue internally, sometimes in high numbers, without damaging the host or eliciting symptoms of plant disease (Quispel, 1992).
- An endophyte is a bacterial or fungal microorganism, which spends the whole or part of its life cycle colonizing inter- and/or intracellularly inside the healthy tissues of the host plant, typically causing no apparent symptoms of disease (Wilson, 1995).
- All bacteria can be detected inside surface-sterilized plant tissues or extracted from inside plants and having no visibly harmful effect on the host plants (Hallmann et al. 1997).
- Microbes colonize living, internal tissues of plants without causing any immediate, overt negative effects (Bacon and White 2000).
- Endophytic bacteria or fungi can be defined as those bacteria or fungi that colonize the internal tissue of the plants showing no external sign of infection or negative effect on their host (Schulz and Boyle 2006).
- Systemic endophytes can be defined as the organisms that inhabit the plant organization, share a symbiotic relationship with the host, and do not produce any visible symptoms of disease at any stage (Wani et al. 2015).
- Transient endophytes, as the organisms that live within the plant tissues at least for part of their life cycle without producing any apparent disease symptoms in plants under normal conditions but turn pathogenic when the host plant is stressed or resource-limited (Wani et al. 2015).

Endophytic biology is pursued in research with multitude of objectives that can be broadly classified into two categories: plant–microbe symbiosis and bioprospecting (Fig. 1.2). For plant–microbe interactions, besides studying the effect of the plant–endophyte symbiosis on the metabolism of the host plant and microbial symbiont, the role of endophytes in adapting the host plants to various biogeographic regions by enhanced stress (biotic and abiotic) tolerance and improved plant productivity is considered. While as under bioprospection, the endophytes are studied for genuine microbial metabolites for medicine, agriculture and industry, potential host metabolites, like taxol and camptothecin, etc., and volatile organic compounds (VOCs) for agriculture, food and aroma industry, and alternate fuels (Wani et al., 2015).

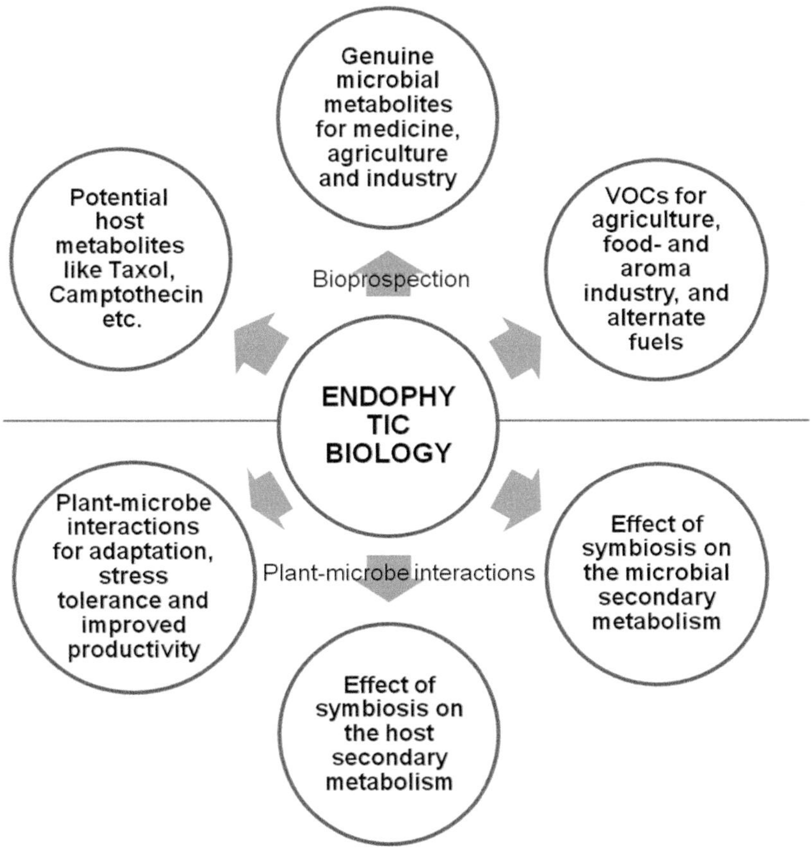

FIGURE 1.2 Endophytic biology is pursued under two broad categories—bioprospection and plant–microbe interactions (Reprinted with permission from Wani et al. 2015, © Springer).

1.3 DIVERSITY AND COLONIZATION OF ENDOPHYTES

Soil microbial communities play an important role in ecosystem functioning and are among the most complex, diverse, and important assemblages in the biosphere. Endophytes are selectively recruited by the plant out of a large pool of soil or rhizospheric microbial communities. The microbes in the soil initially infect the root of host plant, where they enter the plant tissue through cracks formed in lateral root junctions and then colonize by quickly spreading to the intercellular spaces in the root (Rosenblueth and Martínez-Romero, 2006). For example, *Klebsiella* sp. Kp342 aggregate at lateral-root junctions of wheat and alfalfa, before entering the root (Dong et al., 2003). Similarly, *Gluconacetobacter diazotrophicus* and *Herbaspirillum seropedicae* also colonize lateral-root junctions of the host plant (James and Olivares, 1997). There are other portals of entry into the plant, (e.g., wounds caused by microbial or nematode phytopathogens, or the stomata found in leaf tissue), but root cracks are recognized as the main entry points for endophytic colonization (Reinhold-Hurek and Hurek, 1998). The soil microbes that infect plant must be competent root colonizer for successful colonization. Although it is generally assumed that many endophytic communities are the product of a colonization process initiated in the root zone or rhizosphere (Sturz et al., 2000), they may also originate from other sources, such as phyllosphere, anthosphere, or spermosphere (Hallmann et al., 1997). The presence of endophyte in a variety of tissue types within plant suggests its ubiquitous existence in most, if not all, higher plants. Diversity and community structure of the endophytic assemblages and the infection frequency vary according to the plant genotype, tissue type, plant age, edaphic factors, microclimate conditions, and anthropogenic factors (Wani et al., 2015, 2016; Arora et al., 2019). In the case of tree species the distribution of endophytes is affected by the growth stage and canopy height of the plant. The biodiversity of bacterial and fungal endophytes is enormous; however, only a fraction of total endophytic wealth has been subjected to scientific scrutiny and there is a great need to unravel this unexplored and hidden wealth. Various reports on the bacterial and fungal endophytic diversity are given in Tables 1.1 and 1.2, respectively.

1.4 ISOLATION OF ENDOPHYTES

Endophytes have been isolated from all the plants including mosses, bryophytes, pteridophytes, gymnosperms, angiosperms, and lichens (Arnold and Luzoni, 2007; Porras-Alfaro and Bayman, 2011; Wani et al., 2015,

TABLE 1.1 Biodiversity of Bacterial Endophytes Isolated from Different Host Plants.

Plant source	Endophyte diversity	References
27 plant species	47 bacterial species	Mundt and Hinkle (1976)
Saccharum officinarum	*Gluconacetobacter diazotrophicus*	Cavalcante and Döbereiner (1988)
	Burkholderia sp.	Omarjee et al. (2004)
	Pantoea sp.	Loiret et al. (2004)
	32 isolates of endophytic bacteria	Magnani et al. (2010)
Oryza sativa	*Pseudomonas* sp.	You and Zhou (1989)
	Azoarcus sp.	Hurek et al. (1994)
	Herbaspirillum seropedicae	Olivares et al. (1996)
	Burkholderia sp.	Engelhard et al. (2000)
	Serratia sp.	Sandhiya et al. (2005)
	Klebsiella sp.	Rosenblueth et al. (2004)
Citrus sinensis	*Curtobacterium* sp.	Bell et al. (1995)
	Enterobacter sp.	Araújo et al. (2002)
Red clover nodules	15 bacterial species	Sturz et al. (1997)
Phaseolus vulgaris	*Rhizobium etli*	Gutiérrez-Zamora and Martínez-Romero (2001)
Four agronomic crop species (corn, sorghum, soybean, and wheat) and 27 different host species of grasses, forbs, legumes, and wildflowers	853 endophytic strains	Zinniel et al. (2002)
Daucus carota	360 endophytic strains	Surette et al. (2003)
Crocus albiflorus	16 bacterial isolates	Reiter and Sessitsch (2006)
Typha australis	10 diazotrophic bacterial isolates	Jha and Kumar (2007)
Medicinal plants collected from tropical rain forests in Xishuangbanna	2,174 actinobacteria	Qin et al. (2009)
Panax ginseng	13 bacterial genera endophytic bacteria (*Firmicutes*, *Actinobacteria*, α-*Proteobacteria*, and γ-*Proteobacteria)*	Cho et al. (2007) Vendan et al. (2010)
Glycine max	35 endophytic bacteria	Hung et al. (2007)
	1611 bacterial isolates	Dalal and Kulkarni (2013)
Solanum nigrum	77 bacterial isolates	Long et al. (2008)
Piper nigrum	80 bacterial isolates	Aravind et al. (2009)
Medicago sativa	15 bacterial isolates	Stajkovic (2009)
Raphanus sativus	264 bacterial isolates	Seo et al. (2010)
Solanum lycopersicum	72 endophytic bacteria	Yang et al. (2011)
Crocus sativus	54 bacterial isolates	Sharma et al. (2015)

TABLE 1.2 Biodiversity of Fungal Endophytes Isolated from Different Plant Species.

Plant source	Number of endophytic fungal taxa/isolatesa	References
Juniperus communis	114	Petrini and Müller (1979)
Calocedrus decurrens	15	Petrini and Carroll (1981)
Chamaecyparis lawsoniana	18	Petrini and Carroll (1981)
Hordeum vulgare	14	Riesen and Close (1987)
Carpinus caroliniana	155	Bills and Polishook (1991a)
Alnus glutinosa	24	Kowalski and Kehr (1992)
Abies alba	44	Kowalski and Kehr (1992)
Zea mays	23	Fisher et al. (1992)
Oryza sativa	30	Fisher and Petrini (1992)
Eucalyptus globulus	41	Bettucci and Saravay (1993)
Acer macrophyllum	9	Sieber and Dorworth (1994)
Pinus densiflora	9	Hata and Futai (1995)
Picea abies	85	Barklund and Kowalski (1996)
Musa acuminate	24	Brown et al. (1998)
Quercus ilex	149	Collado et al. (2000)
Cuscuta reflexa	45	Suryanarayanan et al. (2000)
Vitis vinifera	17	Moustert et al. (2000)
Vitis vinifera	46	Mostert et al. (2000)
Heisteria concinna and *Ouratea lucens*	418[a]	Arnold et al. (2000)
Tripterygium wilfordii	60	Kumar and Hyde (2004)
81 Thai medicinal plant species	582[a]	Wiyakrutta et al. (2004)
Five medicinal plant species from Western Ghats of India	18	Raviraja (2005)
Three medicinal plants of Similipal Biosphere Reserve India	60	Mohanta et al. (2008)
Nine important medicinal herbs	55	Krishnamurthy et al. (2008)
Azadirachta indica	18	Verma et al. (2007)
Elaeis guineensis	340[a]	Rungjindamai et al. (2008)
Rhododendron fortunei	17	Zhang et al. (2009)
Hevea brasiliensis	58	Gazis and Chaverri (2010)
O. sativa L.	58	Yuan et al. (2010)
Solanum lycopersicum L.	51[a]	Andrade-Linares et al. (2011)
Ericaceae plants	91[a]	Vanó et al. (2011)
Cucumis sativus	18	Waqas et al. (2012)
Sarracenia	12	Glenn and Bodri (2012)
Piper hispidum	21	Orlandelli et al. (2012)
Emblica officinalis	4	Nath et al. (2012)
Glycine max	12	Impullitti and Malvick (2013)
Panax ginseng	38	Wu et al. (2013)

TABLE 1.2 *(Continued)*

Plant source	Number of endophytic fungal taxa/isolatesa	References
Pinus thunbergii	58	Min et al. (2014)
Baccharis trimera	25	Vieira et al. (2014)
Pinus wallichiana	38	Qadri et al. (2014)
Phaseolus vulgaris	42	Parsa et al. (2016)
Catharanthus roseus	7	Pandey et al. (2016)
Crocus sativus	36	Wani et al. (2017)
Z. mays and *O. sativa*	123	Potshangbam et al. (2017)
Glycyrrhiza glabra	38	Arora et al. (2019)

2016; Qadri et al., 2014; Arora et al., 2017). The diversity of endophytic microorganisms in healthy plant tissue is studied using various techniques, like staining techniques for detection of endophytes, cultivation-dependent isolation techniques, and cultivation-independent approaches (Hyde and Soytong, 2008; Qadri et al., 2013, 2014; Whener et al., 2014; Edwards et al., 2015; Wani et al., 2016; Arora et al., 2019). Microbiologists have recognized that as much as 99% of prokaryotic microbial diversity may be uncultivable and suggest that culture-based methods alone underestimate the diversity and misrepresent the taxonomic composition of endophytic communities (Arnold 2007; Hyde and Soytong, 2008). Therefore, the cultivation-independent approaches, such as metagenomics, gradient gel electrophoresis, terminal restriction fragment length polymorphism, and even transcriptomics, are recently used for the detection and identification of different groups of fungi, yeasts, bacteria, and viruses in the plant tissues. Regardless of some limitations, cultivation-based isolation of endophytes is still one of the most exciting fields in endophyte research.

Prior to the isolation of endophytes, the healthy plant material is subjected to surface sterilization to remove any contaminating source attached to the surface of the plant material. Serial washing is often used to remove soil from root tissues, to remove incidental spores from leaf surfaces, and to remove surface contamination in cases where a nontoxic method is desired. This is best accomplished using a large vessel so that the inflowing water vigorously agitates the plant material. An ultrasonic cleaning apparatus removes surface contamination most effectively (Helander et al., 1994). Size of the plant material and surface sterilization procedures vary according to the preferences of the investigator, host plant, and host tissue type sampled (Stone et al., 2012). Surface sterilization of plant material usually entails treating the plant material with a strong oxidant or general disinfectant for a brief period, followed by a sterile rinse to remove residual sterilant. The most commonly used surface sterilant is household chlorine bleach, sodium hypochlorite (NaOCl), usually diluted in water to concentrations of 2–10%. Because the commercial hypochlorite solutions vary in concentration, the percentage hypochlorite or available chlorine as well as the duration of exposure should be specified. Similar oxidant treatments include 3% hydrogen peroxide and 2% potassium permanganate or 0.03% per acetic acid (Torres et al., 2011). Efficacy of surface sterilant is often improved by combining it with a wetting agent, particularly for hydrophobic or densely pubescent leaves. Ethanol (70–95%) is the most commonly used wetting agent; it has limited antibiotic activity and should not be used alone as a surface disinfectant (Schulz et al., 1993). Sometimes surfactants, such as Tween 80, are combined with the sterilant. In some cases, Triton X or Tween 20 is also added to NaOCl solution as surfactant. Other sterilants, not commonly used in endophyte studies, include silver nitrate, mercuric chloride, formalin, and ethylene or propylene oxide. Seeds can be sterilized in a solution of 50% (v/v) Chlorox for 15 min followed by serial washing with sterile distilled water and incubated overnight to allow the endophytic growth (Bacon and White, 1994). For the isolation of endophytic *Streptomyces* strains, the plant material is surface sterilized by exposing them to propylene oxide vapors for 1 hour. Mite infestation can cause significant contamination of the cultures. Mite Infestation can be minimized by autoclaving leftover plant material after plating or by storing unsterilized plant material in sealed plastic bags in a location separate from the sample plates.

Detection and recovery of endophytic microorganisms mainly relies on dissection of plant material into small fragments and subsequent plating of fragments onto a nutrient-rich agar medium (Ezra et al., 2004). The fragment plating method is more common due to its easy handling and high yield of fungal diversity. However, there are some limitations and problems in using the fragment plating method, which must be considered while planning a study (Torres et al., 2011). In addition, the size of tissue fragment used for isolation of endophyte is negatively correlated with the estimated species richness of endophytes (Gamboa et al., 2003). Thus, dividing tissue into smaller sampling units recovers greater species diversity. Conversely, coarsely divided sampling units have greater potential to miss rare or slow-growing endophytic species and to recover mixed genotypes of the same species (Torres et al., 2011). Recently, extinction-to-dilution method is suggested as an alternative method for culturing fungal endophyte communities (Unterseher and Schnittler, 2009). High-throughput culturing (HTC) methods for bacterial cultivation-based extinction-to-dilution have improved strain recovery, especially from slow-growing species or species that are apparently uncultivable (Torres et al., 2011).

1.5 PRESERVATION OF ENDOPHYTES

There are various methods for long-term vouchering and storage of endophytes. Usually, three simple methods are used depending upon the utility of the endophyte culture: frozen glycerol stock, mineral oil stock, and the lyophilization method. Apart from these, endophytes are also stored as nutrient-rich agar slants, frozen barley seed stocks, or room temperature water stocks (Bascom-Slack et al., 2012). Nutrient-rich agar slants and mineral oil stocks are used in the short-term storage of endophytes. However, preservation of endophyte cultures in lyophilized skimmed milk, water storage, or barley seed storage is used in the long-term storage of endophyte. The endophyte cultures stored in lyophilized skimmed milk can be revived even after 10 or 15 years. For Actinobacteria, such as *Streptomyces*, glycerol stocks are usually reliable. It is important to inspect each culture axenically (only one species on a Petri plate) prior to making a permanent stock. Manipulate the culture within the biosafety cabinet to lessen the chance of airborne contamination and constant subculturing is also recommended to reduce the chances of contamination of the endophyte culture.

1.6 IDENTIFICATION OF ENDOPHYTES

Morphological identification of endophyte is based on the color, shape, growth pattern, and microscopic features such as spore shape and arrangement, of the endophytic strain (Fig. 1.3). The use of traditional taxonomic keys to identify the endophytic strain is also recommended. Examples of some useful resources based on reproductive structures of endophytic fungi (mostly filamentous Ascomycota) are listed below. These are often useful for identifying endophytes at the generic level.

Ellis MB, Ellis JP. 1985. *Microfungi on Land Plants, An Identification Handbook.* New York: Macmillan Publishing Company.

Domsch KH, Gams W, Anderson TH. 1993. *Compendium of soil fungi.* Vol. 1, IHWVerlag, Eching.

Barnett HL, Hunter BB. 1998. *Illustrated genera of imperfect fungi.* APS Press, St. Paul, Minnesota.

Hanlin TR. 1998. *Illustrated Genera of Ascomycetes.* APS Press, St. Paul, Minnesota.
Kieffer E, Morelet M. 2000. *The Deuteromycetes: mitosporic fungi classification and generic key.* Science Publishers, Inc, Enfield, NH.
Christopher K, Bruno E. 2003. *Identification of bacterial species.* Pages 103-130, in Tested studies for laboratory teaching, Vol. 24 (MA O'Donnell, Editor). Proceedings of the 24th Workshop/Conference of the Association for Biology Laboratory Education (ABLE).
Kirk PM, Cannon PF, Minter DW, Stalpers JA. 2008. *Ainsworth and Bisby's Dictionary of the Fungi.* (10th edn.), CABI, Wallingford, Oxon, UK.

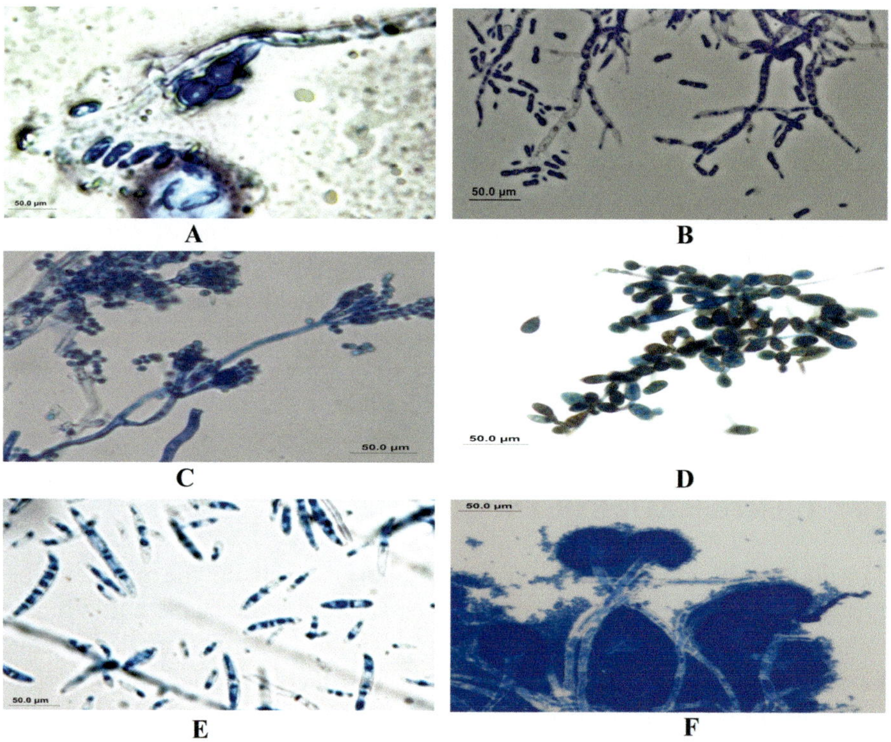

FIGURE 1.3 Microscopic structures of some common endophytes isolated from different plants from Kashmir Himalayas: (A) phialides and conidia of *Phialophora mustea*, (B) phialides and conidia of *Cadophora malorum*, (C) conidiophores of *Penicillium* sp., (D) spores (conidia) of *Alternaria* sp., (E) sickle-shaped conidia of *Fusarium* sp., and (F) conidiophores of *Aspergillus* sp.

There are several online databases that can be accessed to search for fungal names, descriptions, distributions, and up-to-date species lists (Table 1.3).

After identifying a genus, species identification can be made by delving deeper into the literature on the species of concerned genus. Although morphological characterization may be useful in many cases, many endophytes do not readily sporulate in culture conditions thereby limiting the utility of these taxonomic resources. Endophytes that cannot be identified by employing morphological features can be recognized using relevant DNA sequences. Therefore, identification based on molecular approach is gaining more importance. In fungi, the internal transcribed spacer (ITS) region between the conserved flanking regions of the small and large subunit of ribosomal gene is the most frequently sequenced genetic marker of fungi (White et al., 1990; Diaz et al., 2012). In bacteria, 16S ribosomal gene sequence is commonly used for identification (Stackebrandt and Goebel, 1994). Apart from these, few other genetic markers are also used for species identification, like large ribosomal gene, translation elongation factor 1-α and β-tubulin (O'Donnell et al., 2010; Frisvad and Samson, 2004). Among protein-coding genes, the largest subunit of RNA polymerase II (RPB1) is used in fungal barcoding (Cheney et al., 2001; Tanabe et al., 2002). Molecular characterization is used to address research questions relating to systematics, phylogeny, and identification of strains at and even below the species level.

TABLE 1.3 Different Online Fungal Databases with Their Respective Links.

Online database	Link/website
Fungal Databases, Systematic Mycology and Microbiology Laboratory, ARS, USDA	http://nt.ars-grin.gov/fungaldatabases/
MycoBank	http://www.mycobank.org
Index Fungorum	http://www.indexfungorum.org/names/names.asp
Fungal Planet	http://www.fungalplanet.org/index.htm
Bibliography of Systematic Mycology	http://www.indexfungorum.org/BSM/bsm.htm
Tree of Life Web Project	http://tolweb.org/

1.7 CLASSIFICATION OF ENDOPHYTES

Fungal endophytes were categorized into two general groups, namely, clavicipitaceous and nonclavicipitaceous based on their taxonomy, host specificity, evolution, and ecological functions. However, Rodriguez et al. (2009) described four distinct functional groups based on six criteria, host range, tissue(s) colonized, *in planta* colonization pattern, *in planta* biodiversity levels, mechanism of transmission between host generations, and ecological functions. Clavicipitaceous endophytes are referred to as class 1, and nonclavicipitaceous endophytes are further classified into three distinct functional groups as class 2, class 3, and class 4 (Rodriguez et al., 2009). However, endophytes comprise different groups of microorganisms and there is a wide diversity of nonfungal endophytes associated with almost every plant. The endophytic microorganisms can be bacteria, fungi, actinomycetes, or viruses (Bao and Roossinck, 2013). The endophytes express a variety of symbiotic lifestyles ranging from mutualism to parasitism depending on the plant host genotype and/or environmental conditions (Schulz and Boyle 2005; Wani et al., 2015).

Recently, Wani et al. classified endophytes into two general categories, systemic/true endophytes and transient/nonsystemic endophytes, based on seven criteria, including taxonomy, functional diversity, biology, and mode of transmission (Table 1.4) (Wani et al., 2015). The concept of systemic/true endophytes was put forth by Mostert et al., (2000). Systemic endophytes are the organisms that inhabit the plant organization, share a symbiotic relationship with the host, and do not produce any visible symptoms of disease at any stage. However, the transient endophytes are the organisms that live within the plant tissues at least for part of their life cycle without producing any apparent disease symptoms in plants under normal conditions but turn pathogenic when host plant is stressed or resource-limited. Systemic endophytes are cocladogenetic, that is, in different environmental conditions, a given host possesses phylogenetically same endophytes while as transient endophytes vary both in diversity and abundance with change in environment. These endophytes, because of co-evolutionary selection process, share the metabolic and genetic makeup of the host, and are resistant to host metabolites and/or defense mechanism (Christensen et al., 2008; Soliman et al., 2013). Systemic endophytes share a symbiotic relationship with the host plant and when grown under axenic conditions may lose their vitality after subculturing. For example, a camptothecin-producing endophyte, *Fusarium solani* isolated from *Cyanea acuminata*, could indigenously produce the precursors of camptothecin (Kusari et al., 2009). However, a host plant enzyme *strictosidine synthase* absent in the fungus was employed *in planta* for the key step in producing camptothecin (Kusari et al., 2012). This was the main reason for substantial reduction of camptothecin production on subculturing under axenic conditions. The possible reason for this molecular and metabolic cross talk may be horizontal gene transfer between the endophyte and host plant (Kusari and Spiteller, 2012). However, the association of transient endophytes is short lived and seasonal; therefore, they share only physiological cues, and their diversity varies with change in the host's physiological parameters in relation to varying environmental conditions. As systemic endophytes are more likely to be mutualistic with the host plant, their transmission to next generation would be usually vertical, that is, by means of seeds and/or vegetative propagules, while as the transient endophytes are horizontally transmitted *via* spores from one plant to another plant (Saikkonen et al., 1998; Moricca and Ragazzi, 2008).

TABLE 1.4 Classification of Endophytes into Systemic and Nonsystemic Endophytes.

Criteria	True/systemic endophytes	Transient/nonsystemic endophytes
1) Taxonomy	Cocladogenetic species	Varies spatially and temporally
2) Mode of transmission	Usually vertical but in some cases horizontal as well	Horizontal only
3) Lifestyle	Mutualistic	Changes from mutualism to parasitism with change in environment
4) Host defense response	Lack host defense response	Host defense response is active
5) Ecological functions	Beneficial	Beneficial or detrimental depending on the environment, age of the plant, etc.
6) Evolutionary pattern	Co-evolved with the host plant	Association with the host is transient and short lived
7) Diversity	Rare	Rich

(Reprinted with permission from Wani et al. 2015, © Springer).

1.8 CONCLUSION

Endophytes are microbes capable of asymptomatic existence within the plant tissues. They are complex and highly diverse comprising various groups of microorganisms, namely, bacteria, fungi, actinomycetes, or viruses. Endophytes are dynamic in expressing various symbiotic lifestyles which range from mutualism to commensalism to parasitism. However, systemic endophytes are more likely to be mutualistic and stable in comparison to nonsystemic/transient endophytes. Although both morphological and molecular methods are employed for the identification of endophytes, yet molecular methods are more authentic and requisite where traditional methods are not applicable. Endophytes confer biotic and abiotic resistances to host plants and enhance their adaptation in diverse habitats. Bioprospecting of endophytes for known and novel secondary metabolites, especially those of host metabolites, offers numerous applications in agriculture, industry, and medicine.

KEYWORDS

- **endophytes**
- **microbial diversity**
- **horizontal gene transfer**
- **Clavicipitaceous endophytes**
- **systemic endophytes**
- **surface sterilization**

REFERENCES

Arnold, E. A. Understanding the Diversity of Foliar Endophytic Fungi: Progress, Challenges, and Frontiers. *Fungal Biol. Rev.* **2007,** *21*, 51–66.

Arnold, A. E.; Lutzoni, F. Diversity and Host Range of Foliar Fungal Endophytes: Are Tropical Leaves Biodiversity Hotspots? *Ecology* **2007,** *88*, 541–549.

Arora, P.; Wani, Z. A.; Ahmad, T.; Sultan, P.; Gupta, S.; Riyaz-Ul-Hassan, S. Community Structure, Spatial Distribution, Diversity and Functional Characterization of Culturable Endophytic Fungi Associated With *Glycyrrhiza glabra* L. *Fungal Biol.* **2019.** https://doi.org/10.1016/j.funbio.2019.02.003.

Bacon, C. W.; White, J. F. Jr. Stains, Media and Procedures for Analyzing Endophytes. In *Biotechnology of Endophytic Fungi*; Bacon, C. W.; White, J. F. Jr., Eds.; CRC Press Inc.: Boca Raton, FL, 1994; pp 47–56.

Bao, X.; Roossinck, M. J. Multiplexed Interactions: Viruses of Endophytic Fungi. *Adv. Virus Res.* **2013,** *86*, 37–57.

Brader, G.; Compant, S.; Vescio, K.; Mitter, B.; Trognitz, F.; Ma, L. J.; Sessitsch, A. Ecology and Genomic Insights on Plant-Pathogenic and -Nonpathogenic Endophytes. *Annu. Rev. Phytopathol.* **2017,** *55*, 3.1–3.23.

Brundrett, M. C. Co-Evolution of Roots and Mycorrhizas of Land Plants. *New Phytol.* **2002,** *154*, 275–304.

Bascom-Slack, C. A.; Arnold, A. E.; Strobel, S. A. IBI Series Winner. Student-Directed Discovery of the Plant Microbiome and Its Products. *Science* **2012,** *338*, 485–486.

Cheney, S. A.; Lafranchi-Tristem, N. J.; Bourges, D.; Canning, E. U. Relationships of Microsporidian Genera, with Emphasis on the Polysporous Genera, Revealed by Sequences of the Largest Subunit of RNA Polymerase II (RPB1). *J. Eukaryot. Microbiol.* **2001,** *48*, 111–117.

Christensen, M. J.; Bennett, R. J.; Ansari, H. A.; Koga, H.; Johnson, R. D.; Bryan, G. T.; Simpson, J. P.; Koolaard, W. R.; Nickless, E. M.; Voisey, C. R. Epichloë Endophytes Grow by Intercalary Hyphal Extension in Elongating Grass Leaves. *Fungal Genet. Biol.* **2008,** *45*, 84-93.

Diaz, P. L.; Hennell, J. R.; Sucher, N. J. Genomic DNA Extraction and Barcoding of Endophytic Fungi. In *Plant DNA Fingerprinting and Barcoding: Methods and Protocols*; Nikolaus, J., Sucher, et al., Eds.; Methods in Molecular Biology (vol. 862); Springer Science+Business Media, LLC, 2012. DOI 10.1007/978-1-61779-609-8_14.

De Barry, A. Morphologie und physiologia der pilze, flechten, and myxomyceten. Vol. II. In *Hofmeister's Handbook of Physiological Botany*; Leipzig, 1866.

Dong, Y.; Iniguez, A. L.; Triplett, E. W. Quantitative Assessments of the Host Range and Strain Specificity of Endophytic Colonization by *Klebsiella pneumoniae* 342. *Plant Soil* **2003,** *257*, 49–59.

Edwards, J.; Johnson, C.; Santos-Medellín, C.; et al. Structure, Variation, and Assembly of the Root-Associated Microbiomes of Rice. *PNAS USA* **2015,** *112*, E911–E920.

Ezra, D.; Hess, W. H.; Strobel, G. A. New Endophytic Isolates of *M. albus*, A Volatile Antibiotic-Producing Fungus. *Microbiology* **2004,** *150*, 4023–4031.

Freeman, S.; Rodriguez, R. J. Genetic Conversion of a Fungal Pathogen to a Nonpathogenic, Endophytic Mutualist. *Science* **1993,** *260*, 75–78.

Frisvad, J. C.; Samson, R. A. Polyphasic Taxonomy of *Penicillium* Subgenus *Penicillium*—A Guide to Identification of Food and Air-Borne *Terverticillate penicillia* and Their Mycotoxins. *Stud. Mycol.* **2004,** *49*, 1–173.

Gamboa, M. A.; Laureano, S.; Bayman, P. Measuring Diversity of Endophytic Fungi in Leaf Fragments: Does Size Matter? *Mycopathologia* **2003,** *156*, 41–45.

Gunatilaka, A. A. L. Natural Products from Plant-Associated Microorganisms: Distribution, Structural Diversity, Bioactivity, and Implication of Their Occurrence. *J. Nat. Prod.* **2006,** *69*, 509–526.

Hallmann, J.; Quadt-Hallmann, A.; Mahaffee, W. F.; Kloepper, J. W. Bacterial Endophytes in Agricultural Crops. *Can. J. Microbiol.* **1997,** *43*, 895–914.

Helander, M. L.; Sieber, T. N.; Petrini, O.; Neuvonen, S. Endophytic Fungi in Scots Pine Needles: Spatial Variation and Consequences of Simulated Acid Rain. *Can. J. Bot.* **1994,** *72*, 1108–1113.

Heil, M.; Bostock, R. M. Induced Systemic Resistance (ISR) against Pathogens in the Context of Induced Plant Defences. *Ann. Bot.* **2002,** *89*, 503–512.

Hyde, K. D.; Doytong, K. The Fungal Endophyte Dilemma. *Fungal Div.* **2008,** *33*, 163–173.

James, E. K.; Olivares, F. B. Infection and Colonization of Sugarcane and Other Graminaceous Plants by Endophytic Diazotrophs. *Crit. Rev. Plant Sci.* **1997,** *17*, 77–119.

Kuo, C. H. Scrambled and Not-So-Tiny Genomes of Fungal Endosymbionts. *P.N.A.S. U.S.A.* **2015,** *112*, 7622–7623.

Kusari, S.; Zu hlke, S.; Spiteller, M. An Endophytic Fungus from *Camptotheca acuminate* that Produces Camptothecin and Analogues. *J. Nat. Prod.* **2009,** *72*, 2–7.

Kusari, S.; Spiteller, M. Are We Ready for Industrial Production of Bioactive Plant Secondary Metabolites Utilizing Endophytes? *Nat. Prod. Rep.* **2012,** *28*, 1203–1207.

Kusari, S.; Hertweck, C.; Spiteller, M. Chemical Ecology of Endophytic Fungi: Origins of Secondary Metabolites. *Chem. Biol.* **2012,** *19*, 792–798.

Moricca, S.; Ragazzi, A. Fungal Endophytes in Mediterranean Oak Forests: A lesson from Disculaquercina. *Phytopathol.* **2008,** *98*, 380–386.

Mostert, L.; Crous, P. W.; Petrini, O. Endophytic Fungi Associated with Shoots and Leaves of *Vitis vinifera*, with Specific Reference to the *Phomopsis viticola* Complex. Sydowia **2000,** *52*, 46–58.

O'Donnell, K.; et al. Internet-Accessible DNA Sequence Database for Identifying *Fusaria* from Human and Animal Infections. *J. Clin. Microbiol.* **2010,** *48*, 3708–3718.

Peay, K. G.; Kennedy, P. G.; Bruns, T. D. Fungal community ecology: a hybrid beast with a molecular master. *BioScience* **2008,** *58*, 799–810.

Partida-Martínez, L. P.; Heil, M. The Microbe-Free Plant: Fact or Artifact? *Front. Plant Sci.* **2011,** *2*, 100.

Petrini, O. Fungal Endophytes of Tree Leaves. In *Microbial Ecology of Leaves*; Andrews, J. H., Hirano, S. S., Eds.; Springer-Verlag: New York, NY, 1991; pp 179–197.

Porras-Alfaro, A.; Bayman, P. Hidden Fungi, Emergent Properties: Endophytes and Microbiomes. *Annu. Rev. Phytopathol.* **2011,** *49*, 291–315.

Qadri, M.; Johri, S.; Shah, B. A.; Khajuria, A.; Sidiq, T.; Lattoo, S. K.; Abdin, M. Z.; Riyaz-Ul-Hassan, S. Identification and Bioactive Potential of Endophytic Fungi Isolated from Selected Plants of the Western Himalayas. *Springer Plus* **2013,** *2*, 8.

Qadri, M.; Rajput, R.; Abdin, M. Z.; Vishwakarma, R. A.; Riyaz-Ul-Hassan, S. Diversity, Molecular Phylogeny and Bioactive Potential of Fungal Endophytes Associated with the Himalayan Blue Pine (*Pinus wallichiana*). *Microb. Ecol.* **2014,** *67*, 877–887.

Redecker, D.; Kodner, R.; Graham, L. E. Glomalean Fungi from the Ordovician. *Science* **2000,** *289*, 1920–1921.

Reinhold-Hurek, B.; Hurek, T. Interactions of Gramineous Plants with *Azoarcus* spp. and Other Diazotrophs: Identification, Localization, and Perspectives to Study Their Function. *Crit. Rev. Plant Sci.* **1998,** *17*, 29–54.

Rodriguez, R. J.; White, J. F. Jr.; Arnold, A. E.; Redman, R. S. Fungal Endophytes: Diversity and Functional Roles. *New Phytol.* **2009,** *182*, 314–330.

Rosenblueth, M.; Martínez-Romero, E. Bacterial Endophytes and Their Interactions with Hosts. *Mol. Plant Microbe Interact.* **2006,** *19*, 827–837.

Saikkonen, K.; Faeth, S. H.; Helander, M.; Sullivan, T. J. Fungal Endophytes: A Continuum of Interactions with Host Plants. *Annu. Rev. Ecol. Syst.* **1998,** *29*, 319–343.

Schulz, B.; Wanke, U.; Draeger, S.; Aust, H. J. Endophytes from Herbaceous Plants and Shrubs: Effectiveness of Surface Sterilization Methods. *Mycol. Res.* **1993,** *97*, 1447–1450.

Schulz, B.; Boyle, C. The Endophytic Continuum. *Mycol. Res.* **2005,** *109*, 661–687.

Sessitsch, A.; Hardoim, P.; Doring, J.; Weilharter, A.; Krause, A.; Woyke, T.; Mitter, B.; Houberg-lotte, L.; Friedrich, F.; Rahalkar, M.; Hurek, T.; Sarkar, A.; Bodrossy, L.; Van overbeek, L.; Brar, D.; Van elsas, J. D.; Reinhold-Hurek, B. Functional Characteristics of an Endophyte Community Colonizing Rice Roots as Revealed by Metagenomic Analysis. *Mol. Plant Microbe Interact.* **2012,** *25*: 28–36.

Seth, E. C.; Taga, M. E. Nutrient Cross-Feeding in the Microbial World. *Front. Microbiol.* **2014,** *5*, 350.

Soliman, S. S. M.; Trobacher, C. P.; Tsao, R.; Greenwood, J. S.; Raizada, M. N. A Fungal Endophyte Induces Transcription of Genes Encoding a Redundant Fungicide Pathway in Its Host Plant. *BMC Plant Biol.* **2013,** *13*, 93.

Stackebrandt, E.; Goebel, B. M. Taxonomic Note: A Place for DNA-DNA Reassociation and 16S rRNA Sequence Analysis in the Present Species Definition in Bacteriology. *Int. J. Syst. Evol. Microbiol.* **1994,** *44*, 846–849.

Stierle, A.; Strobel, G. A.; Stierle, D. B. Taxol and Taxane Production by *Taxomyces andreanae*, An Endophytic Fungus of Pacific Yew. *Science* **1993,** *260*, 214–216.

Stone, J. K.; Polishook, J. D.; White, J. F. Jr. Endophytic Fungi. In *Biodiversity of Fungi: Inventory and Monitoring Methods*; Mueller, G. M., Bills, G. F., Foster. M. S., Eds.; Elsevier Academic Press: New York, NY, 2012; pp 241–270.

Sturz, A. V.; Christie, B. R.; Nowak, J. Bacterial Endophytes: Potential Role in Developing Sustainable Systems of Crop Production. *Crit. Rev. Plant Sci.* **2000,** *19*: 1–30.

Tanabe, Y.; Watanabe, M. M.; Sugiyama, J. Are Microsporidia Really Related to Fungi?: A Reappraisal based on Additional Gene Sequences from Basal Fungi. *Mycol. Res.* **2002,** *106*, 1380–1391.

Taylor, T. N.; Taylor, E. L. The Rhynie Chert Ecosystem: A Model for Understanding Fungal Interactions. In *Microbial Endophytes*; Bacon, C. W., White, J.F., Eds.; Marcel Decker: New York, NY, 2000; pp 33–45.

Torres, M. S.; Tadych, M.; White, J. F. Jr.; Bills, G. F. Isolation and Identification of Fungal Endophytes. In *Prospects and Applications for Plant-Associated Microbes: A Laboratory Manual Part B: Fungi*; Pirttillä, A. M., Sorvari, S., Eds.; BBi: Finland, 2011; pp 153–164.

Unterseher, M.; Schnittler, M. Dilution-to-Extinction Cultivation of Leaf-Inhabiting Endophytic Fungi in Beech (*Fagus sylvatica* L.) – Different Cultivation Techniques Influence Fungal Biodiversity Assessment. *Mycol. Res.* **2009,** *113*, 645–654.

Vogel, A. Mehl und die anderen Mehlprodukte der Cerealien und Leguminosen. Zeitschrift fu¨ r Nahrungsmittel-Untersuchung, *Hygiene, und Waarenkunde* **1898,** *12*, 25–29.

Wani, Z. A.; Ashraf, N.; Mohi ud din, T.; Riyaz-Ul-Hassan, S. Plant-Endophyte Symbiosis, An Ecological Perspective. *Appl. Microbiol. Biotechnol.* **2015,** *99*, 2955–2965.

Wani, Z. A.; Mirza, D. N.; Arora, P.; Riyaz Ul Hassan, S. Molecular Phylogeny, Diversity, Community Structure, and Plant Growth Promoting Properties of Fungal Endophytes Associated with the Corms of Saffron Plant: An Insight into the Microbiome of *Crocus sativus* Linn. *Fungal Biol.* **2016,** *120*, 1509–1524.

White, T.; Bruns, T.; Lee, S.; et al. Amplification and Direct Sequencing of Fungal Ribosomal RNA Genes for Phylogenetics. In *PCR Protocols: A Guide to Methods and Applications*; Innis, M., Gelfand, D., Sninsky, J., White, T., Eds.;). Academic Press Inc.: New York, NY, 1990.

Wilson, D. Endophytes—The Evolution of a Term and Clarification of Its Use and Definition. *Oikos* **1995,** *73*, 274–276.

CHAPTER 2

Endophytes as a Natural Resource for Bioprospection

ZAHOOR AHMED WANI

Department of Botany, Govt. Degree College Kishtwar, Kishtwar, Jammu & Kashmir, India

E-mail: zawani1986@gmail.com

ABSTRACT

Microorganisms produces huge repertoire of natural products with wide range of applications from pharmaceuticals to agriculture, antimicrobials to biofuel, and ecological adaptation to environmental remediation. Plants harbor diverse array of microbial symbionts known as endophytes, which are known to produce bioactive secondary metabolites. These secondary metabolites mostly induce specific phenotypic function such as crosstalk with associated organisms, chemical warfare/defense, and stress adaptation in the host plant and/or endophyte. Secondary metabolites produced by the endophytic microorganisms include Polyketides, non-ribosomal peptides, terpenes, steroids, alkaloids, phenolic and flavonoids, aliphatic compounds, and volatile organic compounds (VOCs). Some endophytes are known to produce secondary metabolites which mimic the host plant metabolites and one of the probable mechanisms for this is suggested to be the horizontal gene transfer between the host plant and the endophytic microbial partner.

2.1 INTRODUCTION

Microorganisms are known for their potential to produce novel molecules that have found applications in medicine, agriculture, and industry. From the discovery of antibiotics, like penicillin from *Penicillium notatum* by Alexander Fleming in 1928, to other lifesaving drug molecules like Paclitaxel from

Taxomyces andreanae (Stierle et al., 1993), microorganisms have contributed numerous molecules to natural product repositories having the potential to cure human diseases. A specific microorganism can sometime produce 15–20 metabolites with different chemical types, and a particular compound may frequently be produced by multiple microorganisms belonging to different genera, family, class, or phylum (Berdy, 2012). More than 70,000 metabolites of microbial origin are known, and among these, approximately 33,500 are reported to be bioactive. Among these, maximum numbers of bioactive metabolites are derived from microscopic fungi (~15,600) and rest of the metabolites being derived from actinomycetes and bacteria (Berdy, 2012). Microbes are highly diverse in nature as compared to other natural sources; however, microbes are the least explored biological sources. Studies based on estimation of microbial populations have conservatively revealed that only about 1% of bacteria and 5% of fungal species have been identified and explored for human welfare while the rest remain unexplored (Staley et al., 1997). Even the potential of the microorganisms that are isolated in pure culture is further limited by the presence of cryptic/orphan biosynthetic pathways that do not express themselves under standard culture conditions (Bok et al., 2006; Kusari et al., 2014). Challenging the microorganism through various methods and modulating their secondary metabolism has resulted in several new molecules in recent times. Microorganism are also challenged in nature by extreme environmental conditions that prompt them to evolve for new ecological settings, thus activating gene clusters that do not express under the optimum conditions.

Endophytic microbes evolve themselves in the host plant and counter its defense mechanism. Endophytic microbes evolve its biochemistry in response to the host defense and other signal molecules resulting in the production of new molecules by the microbe. It is not therefore surprising that the endophytes are metabolically proficient and their secondary metabolites possess important bioactivities. Each plant represents a unique environment due to its specific chemical constituents, different stages of growth and development that shape the endophytic community. Therefore, it is suggested the vast plant diversity will be harboring huge diversity of endophytic microbes hidden in the plant tissues representing an important resource for human welfare. However, the infinite diversity of endophytic communities harbored in plants in almost all ecological and agricultural niches poses a big challenge in selection of plants for isolating competent/active endophytes (i.e., capable of producing desired bioactive secondary metabolites) worth bioprospection. To overcome this challenge, both curiosity-driven and candidate approach bioprospecting tactics are being used by the researchers for the discovery of potent endophytes with desirable functional traits (Fig. 2.1). A specific rationale for the selection of plant with potent endophytes is proposed to maximize possibility of discovering endophytes with significant bioactive potential. The rationale for the selection of plant for endophyte isolation is as follows:

i) Plants adapted to specialized ecological niche, like harsh climate and/or stressed conditions.
ii) Plants with unique morphology and possessing unusual strategies for subsistence.
iii) Plants having an ethnobotanical importance.
iv) Endemic or endangered plants.
v) Plants growing in areas of abundant biodiversity (biodiversity hotspots).

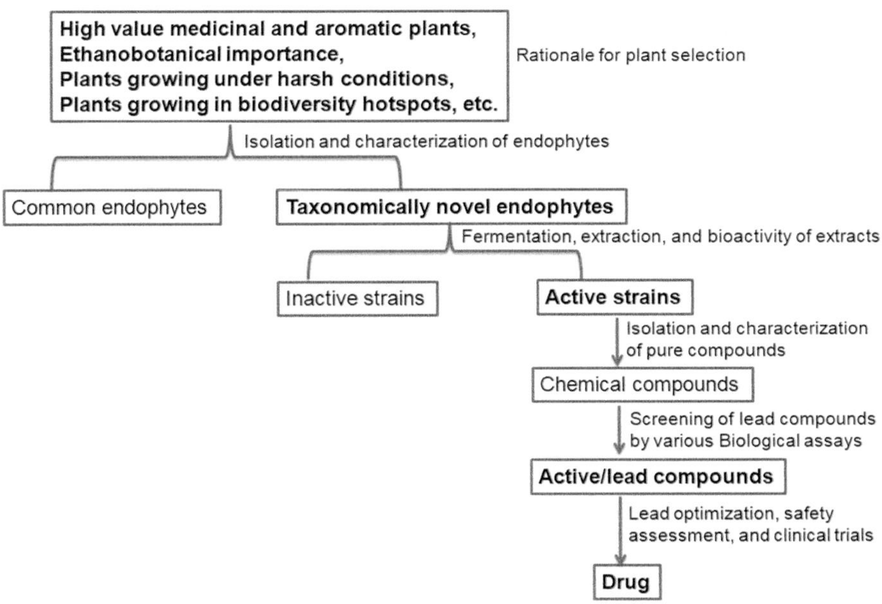

FIGURE 2.1 Bioprospecting strategy utilized to discover novel or competent endophytes with desirable features and the process of drug development from microbial natural products.

For instance, in drug discovery programs, the medicinal plants used in ethnobotanical or folk medicines by rural/tribal people in traditional medicinal preparations are mined for endophytes rather than sampling plants randomly from different populations. These value-added strategies have resulted in the discovery of numerous potent and/or even novel endophytic microbes that might be utilized for bioprospection. Success in discovering new metabolite from microbial natural products depends on growing the given microorganism in conditions appropriate to induce the production of the desired metabolite. Culturing a single strain of microorganism under different conditions may produce substantially different compounds. This has created great interest in the discovery of new secondary metabolites from microorganisms. Thus, it is important to investigate the microorganisms from different ecological niches for the production of novel metabolites. Microorganisms showing bioactivity and taxonomic novelty are selected for further characterization. Microbial extracts prepared using different solvent systems are screened for bioactivities and the bioactive agents will be purified and characterized using different separation and spectroscopic techniques. Small-scale fermentation broth is extracted by the following protocol as shown in the flowchart (Fig. 2.2).

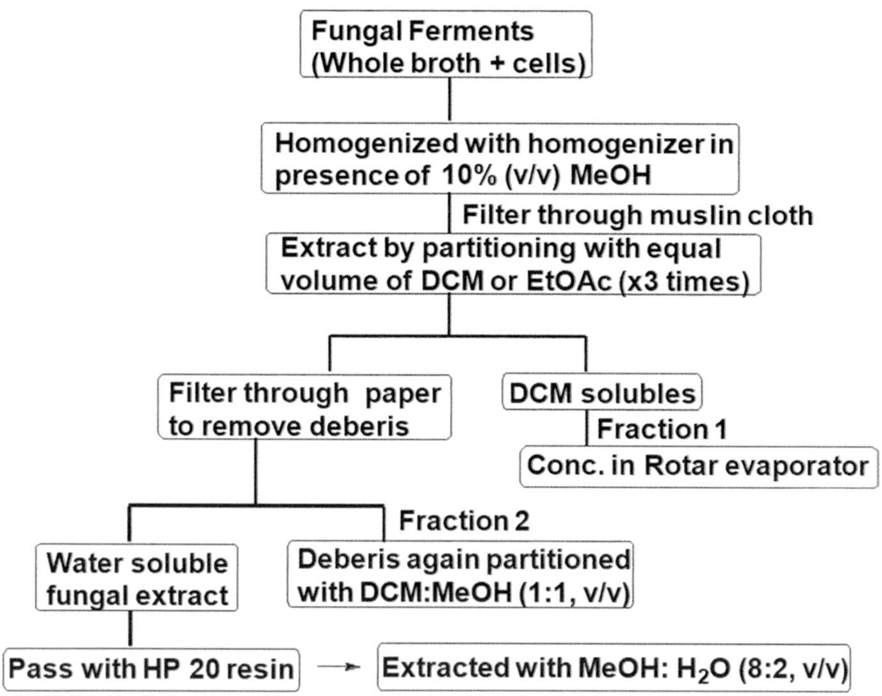

FIGURE 2.2 Scheme of natural product isolation from fungal endophytes, using different solvent systems (MeOH stands for methanol, DCM stands for dichloromethane, EtOAc stands for ethyl acetate).

2.2 SECONDARY METABOLITES OF ENDOPHYTES

Plants engage in multispecies crosstalk with diverse array of microorganisms leading to acquisition of specific and generalized functional traits by each interacting partner. There is a complex and transient metabolic flux across the interacting partners that have a direct bearing on their phenotype and functional traits. Every natural product produced is potentially a functional trait developed by the interacting organisms sharing the same habitat and plays a central role in their interaction and survival (Kusari et al., 2012). The selective pressure on the endophytes exerted by other organisms like competitors, predators, and pathogens induces metabolic shift in the endophytic microbe leading to the production of molecules helping them to overcome the stress conditions (Gloer, 1997).

The term "secondary metabolite" refers to the compounds/natural products derived from primary molecules of the producing organism and has secondary importance to the organism. Secondary metabolites do not play a direct role in growth, development, and reproduction of the producing organism. The production of secondary metabolites in endophytic microorganisms is induced by certain selection pressures, like biotic or abiotic stress, in a given ecological niche. These secondary metabolites mostly induce specific phenotypic function such as crosstalk with associated organisms, chemical warfare/defense, and stress adaptation in the endophyte (Kusari et al., 2014). Secondary metabolites, also known as idiolites, are complex chemical structures with low molecular mass resulting from long enzymatic pathways in living organisms. Secondary metabolites of microorganisms are usually of low molecular mass and are often bioactive. They are produced as families of related compounds. The production of secondary metabolites is often correlated with a specific stage of morphological differentiation of the microorganism (Keller et al., 2005). The endophytes under submerged culture conditions produce secondary metabolites in the stationary phase. Secondary metabolites represent pharmaceutically important class of compounds with wide range of bioactivities like antibiotic, antiparasitic, immunosuppressive, anticancer, and cholesterol-lowering properties (Newman and Cragg, 2012). In addition to their medical relevance to humans, many compounds also play important role within their host organisms as pigments, defense molecules, or virulence factors. Endophytic microorganism produces important classes of secondary metabolites, including polyketides (e.g., rapamycin, lovastatin), nonribosomal peptides (e.g., sirodesmin), terpenes (T-2 toxin), steroids (penicisteroid A), alkaloids (cryptocin), phenolic compounds (colletotric acid), aliphatic compounds (brefeldin A), and peptides (cryptocandin).

Endophytic microorganisms producing bioactive secondary metabolites, including compounds similar to the associated plant metabolites (discussed in next subheading), are important from both academic and industrial perspectives. However, notwithstanding the apparent benefits in their industrial utilization, the commercial production of biologically useful compounds from endophytes has not achieved so far (Kusari and Spiteller, 2011). The production of bioactive natural products from endophytes has so far been made by fermentation of axenic monocultures under standard culture conditions (Scherlach and Hertweck, 2009). The sustained production of desired bioactive secondary metabolite by an endophyte requires constant expression of the biosynthetic genes responsible for the production of the metabolite. However, it has been observed that there is substantial reduction in secondary metabolite production upon repeated subculturing under axenic monoculture condition probably, which is one of the key challenges in achieving their commercial production. It is suggested that the failure in constant expression of the biosynthetic genes required for the desired metabolite and nonexpression of the cryptic biosynthetic gene clusters present in the endophytes under standard culture conditions is responsible for reduction in secondary metabolite production in axenic cultures (Kusari et al., 2014). Cryptic biosynthetic gene clusters are clusters of biosynthetic genes in a microorganism that are not expressed under standard in vitro culture conditions. Thus, even if a microorganism has the potential to biosynthesize a particular secondary metabolite, it will not be produced under standard laboratory conditions. This deactivation of cryptic biosynthetic gene clusters under culture conditions leads to production of a far lesser diversity of secondary metabolites than the actual repertoire of endophyte secondary metabolome. This great challenge is obscuring the true potential of endophytes as microbial factories for industrial production of desired bioactive metabolites.

2.3 ENDOPHYTES AND PLANT METABOLITES

During the last 20 years, it is observed that much of the wealth of microbial biodiversity resides in plant tissues showing interesting ecology, biochemistry, and secondary metabolome (Strobel, 2006). Endophytic microorganisms live in symbiotic relationship with the host plant, growing inside the plant tissues without causing any apparent disease symptoms. The endophytic microorganisms remained a hidden treasure due to their asymptomatic nature until their potential was realized a few decades ago. Since the discovery of Taxol from an endophytic fungus *T. andreanae* inhabiting *Taxus brevifolia,* there has been a paradigm shift in endophytic research and more focus was laid on obtaining plant metabolites from the endophytic partners (Stierle et al., 1993). Since Taxol is fungicide in nature, therefore it was interesting to observe that yew trees producing Taxol harbor fungi that also produce Taxol. This mystery was solved by Soliman and coworkers, where they elucidated the role of a Taxol-producing endophytic fungus in the defense mechanism of the host plant. The endophytic fungus, *Paraconiothyrium* sp. sequesters Taxol in intracellular hydrophobic bodies (HBs) and these fungicide-containing HBs are released by the fungus upon sensing pathogens at the pathogen entry points. The fungicide-laced HBs coalesce to form remarkable extracellular barriers at the pathogen entry points (Soliman et al., 2015). Thus, it makes a novel plant defense mechanism against the phytopathogens. Since then Taxol has been detected in many endophytic strains isolated from different plant species (Mousa and Raizada, 2013; Chen et al., 2016).

As the focus on exploring the plant metabolites from the endophytic microbial partner increases, the molecular mechanisms underlying multi-species chemical crosstalk gained much impetus and one of the probable mechanisms is suggested to be the horizontal gene transfer (HGT) (Stierle et al., 1993; Puri et al., 2006; Kusari et al., 2009). The theory of HGT suggests transfer of the gene clusters, responsible for biosynthesis of secondary metabolites in the host plant, between the plant and its endophytes (Strobel and Daisy, 2003; Tan et al., 2011). Interestingly, recent reports also suggest the existence of independent biosynthetic pathways in the endophytic micro-organisms (Staniek et al., 2009). However, it also seems logical that a micro-organism producing similar metabolites to those of the host plant may have more chances of thriving in the host tissues because of its resistance to the key metabolites. Undoubtedly, endophytic microorganisms are metabolically more proficient than their free-living counterparts and they have the potential to produce exceedingly high number of secondary metabolites, related or unrelated to the host plant. There are several reasons for the increased metabolic activity of endophytes. Firstly, the microorganisms need to evolve to survive in the host tissue thus activating the production of molecules that help the microorganism to evade host plant defense mechanism. Secondly, in order to establish a balanced antagonistic relationship, the endophyte produces several molecules/phytotoxins in host plant (Strobel and Daisy, 2003; Strobel, 2006). Recently, it is proposed that the epigenetic modifications in the endophytes due to host plant metabolites result in turning on some of its otherwise "silent biosynthetic pathways" (Riyaz-Ul-Hassan et al., 2012).

The whole-genome sequencing of microorganisms opened a new area in the field of natural product research and drug discovery programs. Large-scale microbial genome sequencing started from the whole-genome shotgun sequencing of *Haemophilus influenzae*, which demonstrated that microbial genome sequences can be obtained with an unimagined ease and rapidity (Fleischmann et al., 1995). Our understanding of the genetics and enzymology of microbial natural product biosynthesis has also led to the identification and analysis of gene/clusters involved in the biosynthetic pathways in sequenced microbial genomes (Fischbach and Walsh, 2006). *Streptomyces coelicolor* is one of the first sequenced microbes in which it is reported that the gene clusters coding for natural product-like biosynthetic pathways are more than the natural products of the organism known so far (Bentley et al., 2002). Similar observations have been reported in several diverse sequenced microorganisms like *Streptomyces avermitilis* (Ikeda et al., 2003), *Pseudomonas fluorescens* (Paulsen et al., 2005), *Aspergillus* (Bok et al., 2006), *Saccharopolyspora erythraea* (Oliynyk et al., 2007), and *Salinispora tropica* (Udwary et al., 2007). Over the past few years, genome mining of microorganisms for the discovery of new natural products and/or biosynthetic pathways has rapidly advanced as a new field in endophyte biology (Corre and Challis, 2007; Challis, 2008). The above findings strongly support the one-strain-many-compounds (OSMAC) approach, according to which metabolite profile of microorganisms can be altered by varying the growth conditions under standard culture conditions. Therefore, it can be suggested that a large number of potentially useful natural products of microbial origin await discovery (Peric-Concha and Long, 2003). The endophytes thus represent a hidden bioresource, mostly unexplored, keeping in view the plant biodiversity of the world and the fact that each plant investigated is found to harbor endophytic microorganisms.

2.4 ENDOPHYTES AS POTENTIAL DRUG RESOURCE

Endophytic microorganisms are suggested to produce secondary metabolites to overcome the plant defense system and microbial competitors or pathogenic invasion inside the plants. Endophytes are proficient producers of bioactive secondary metabolites due to the constant process of strain development by passing through various stages of plant growth and development, and their ecological functions (Strobel, 2003; Porras-Alfaro and Bayman 2011; Nalli et al., 2015). The ability of these organisms to produce bioactive molecules may be attributed to their evolution over millions of years in diverse ecological niches and natural habitats in which extreme competition for survival is needed (Strobel, 2003; Mousa and Raizada, 2013). Endophytic microorganisms are known to produce various antimicrobial compounds belonging to several structural classes such as alkaloids, peptides, steroids, polyketides, terpenoids, phenols, lignans, cytochalasins, quinines, and flavonoids. Terpenoids and polyketides are the most common antimicrobial compounds recovered from endophytes, while flavonoids and lignans are rare antimicrobial secondary metabolites isolated from endophytes (Kharwar et al., 2011; Mousa and Raizada, 2013). The discovery of billion-dollar anticancer drug, Taxol® (generic name: paclitaxel), from an endophytic fungus (*T. andreanae*) isolated from the Pacific yew tree (*T. brevifolia*) (Stierle et al., 1993), initiated the modern research on endophytes that focuses on discovery of new molecules from microbial sources for drug discovery program. Since this important discovery, the number of patents on potential bioactive metabolites of endophytes has increased dramatically (Strobel, 2006; Wang et al., 2011).

The researchers working on drug discovery from endophytes are mainly focusing on the medicinal plants used by indigenous people in traditional medicinal preparations, or plants used in ethnobotanical or folk medicines, for isolation of endophytes. It is indeed noteworthy that these candid approach strategies have yielded discovery of numerous potent and novel endophytes with great pharmaceutical worth. The various steps involved in bioprospection of endophytes for drug discovery programs are as follows (Fig. 2.1).

i. Isolation, identification, and molecular characterization of endophytes.
ii. Fermentation, extraction, and biological activity of extracts of taxonomically novel endophytes.
iii. Isolation and characterization of bioactive natural products from endophytes selected on the basis of taxonomic novelty and bioactivity profiles.
iv. Exploration of potential strains for novel metabolites through stain development, metabolite remodeling, and activation of silent biosynthetic pathways.
v. Screening of active compounds/lead molecules by performing various biological assays for potential drug development.
vi. Lead optimization studies, safety assessment, and clinical trials of the selected bioactive molecules to be developed as drug.

The endophytic diversity and their intrinsic metabolites are largely unexplored. Endophytes are challenged in nature by extreme environmental and ecological settings resulting in activation of gene clusters or metabolic pathways that do not express under normal conditions; therefore, they are metabolically more active than their free-living counterparts (Strobel and Daisy, 2003; Strobel, 2006; Riyaz-Ul-Hassan et al., 2012). There is an increasing demand for new antimicrobial molecules/drugs due to rampant increase in drug-resistant microbes, life-threatening infections, and recurring infectious diseases (Zhanel et al., 2014). Similarly, microbes are regarded as an unexplored source of cytotoxic agents that have application in cancer therapy (Chen et al., 2016). Thus, endophytes are being widely explored for bioactive natural products with antimicrobial as well as cytotoxic potential. Some of the anticancer molecules obtained from endophytes are camptothecin, vinblastine, torreyanic acid, cytoskyrins, phomoxanthones A & B, photinides A-F, rubrofusarin B, and (+)- epiepoxydon (Isaka et al., 2001; Puri et al., 2005; Chen et al., 2016). Thiodiketopiperazine derivatives reported from *Tilachlidium* sp. were reported to show potential cytotoxic activity against P388 leukemia cells (Feng et al., 2004). Subsequently, these compounds were prepared synthetically and found having a broad range of cytotoxic activity against several cancer cell lines (DeLorbe et al., 2013).

Endophytic microorganisms constitute an important biological repository of molecules for pharmaceutical and agricultural applications (Keller et al., 2005; Strobel, 2006). Natural products have played a lead role in drug discovery programs and it is reported that microbial sources alone have contributed more than 40% of New Chemical Entities (NCEs) from 1981 to 2005 (Clardy and Walsh, 2004; Sieber and Marahiel, 2005). Further, more than 60% of the anticancer and 69% of the FDA-approved antimicrobial drugs are natural products or natural product derivatives (McAlpine et al., 2005). Examples of potential antimicrobial compounds include a unique antimycotic peptide, termed cryptocandin (Strobel et al., 1999) which demonstrated excellent antifungal activity against some important human fungal pathogens—*Candida albicans*, and *Trichophyton* sp. Another antimicrobial compound is a unique tetramic acid known as cryptocin, isolated from an endophytic fungus *Cryptosporiopsis cf. quercina*, having potent activity against a number of plant-pathogenic fungi (Li et al., 2000). Ambuic acid is an antifungal agent isolated from *Pestalotiopsis microspora*. Phomopsichalasin is another novel antimicrobial agent isolated from an endophytic fungus *Phomopsis* sp. (Horn et al., 1995). Antimicrobial polyketides, epicolactone and epicoccolides A and B, isolated from endophytic fungus, *Epicoccum* sp. obtained from *Theobroma cacao,* displayed significant inhibitory effects on the mycelial growth of *Pythium ultimum*, and *Aphanomyces cochlioides*, and *Rhizoctonia solani* (Talontsi et al., 2013). Several of the antimicrobial compounds have been licensed to the pharmaceutical companies for drug development.

Much of the research on endophytes is focused on obtaining important metabolites of the host plant like paclitaxel, camptothecin, podophyllotoxin, etc., initiated by the theory of horizontal transfer of gene clusters from the host plant to the endophytic symbiont. Taxol producing endophytic fungus, *T. andreanae* has a *taxadiene synthase* (txs) gene different than the txs gene of the host plant. This indicates that microorganisms possess independent pathways for the production of such high-value natural products. The bioactive metabolites obtained from endophytes include drug molecules like emodepside and nodulisporic acid A, both approved for use in different pets (Shoop et al., 2001; Altreuther et al., 2011; FDA, 2007; US Patent No. US5399582A). Emodepside is an anthelmintic drug that is effective against a number of gastrointestinal nematodes. It is licensed for use in cats and belongs to the class of drugs known as the octadepsipeptide (Altreuther et al., 2011; FDA, 2007). This drug was isolated from a mycelial cake of the endophytic fungus mycelia sterilia PF1022, found in the leaves of *Camellia japonica* (Terada, 1992). Nodulisporic acid A, isolated from an endophytic *Nodulisporium* sp. ATCC74245, has potent antiflea activity in dogs and lacks overt mammalian toxicity (Shoop et al., 2001). Thus, numerous bioactive molecules that are genuine microbial products have been characterized from the endophytes and many more await discovery (Mousa and Raizada, 2013). Thus, it is very essential to explore microbes for novel lead–drug–molecules to run sustainable drug discovery programs. Also the microbial growth and their manipulation by various approaches, including media engineering, chemical induction, epigenetic modulation, metabolite remodeling, and fermentation technology for scale up, make them suitable for the production of useful natural products (Bok et al., 2006; Knappe et al., 2008; Riyaz-Ul-Hassan et al., 2012). Hence, microbiologists screen every possible source for microorganisms, including extreme environments like ocean beds, geothermal vents, cold desserts, etc., in search of novel strains with promising bioactive potential (Staley et al., 1997).

2.5 ENDOPHYTES AS SOURCE OF BIOFUEL

Throughout modern times the global energy matrix is predominantly dependent on nonrenewable resources, particularly fossil fuels. However, the potential impact of fossils fuels on the environment and global economy has led to a rethinking regarding the global energy scene. The constant increase in the emission levels of carbon dioxide (CO_2) and other heat-trapping gases from different sources is recognized as the main cause of climate change in the present world (UN-GAP report 2014). The increased levels of environmental CO_2 are directly related to the consumption of fossil fuels in different sectors (UN-GAP report 2014). Therefore, finding substitutes for these fossil fuels is a major challenge to the academia and industrial communities. Off late research in alternate resources of renewable energy has received impetus and biofuel has shown to be one of the most promising alternatives. Out of the total biodiesel production, approximately 90% are obtained from plants (Durrett et al., 2008). However, with ever-increasing demand of fuel in modern industrialized world, the production of biodiesel from plants will negatively impact the ecosystem. Therefore, there is a need to look for potential biodiesel precursors in organisms other than plants and microbes are suggested to be the better choice. Microbes do not require land for growth and also do not compete in food production as they accumulate high levels of lipids in their cells. Furthermore, they can be produced on large scale by fermentation technology making them potentially more viable source for biofuel production (Strobel, 2015). One of the classical examples of a microbe producing a liquid fuel (ethanol) is yeast (*Saccharomyces cerevisiae*). Unfortunately, ethanol is not the most desirable fuel since it has low energy content and can foul engines (Strobel, 2014). Thus, this potentially dire situation represents the main impetus for an effort to find and learn more about microbes as a source of hydrocarbons which represents a nonfossil source of liquid fuels.

Many of the fungal endophytes are found to produce volatile organic compounds (VOCs) with applications in agriculture, food processing, aroma and biofuel industry (Strobel et al., 2004; Bitas et al., 2013; Wani et al., 2015). A fungal endophyte, *Muscodor* sp. produces bioactive VOCs having antimycotic potential; thus, *Muscodor* sp. can be used as mycofumigating agent, and also for the preservation of postharvest agricultural produce and decontamination of animal waste (Strobel, 2006; Bitas et al., 2013; Wani et al., 2015). Some others, like *Gliocladium roseum* (reidentified as *Ascocoryne sarcoides*), produce potential biofuel molecules that can be used as alternate fuels if efforts to produce them in large quantities become successful (Strobel et al., 2008; Mallette et al., 2014). The most important aspect of this fungus is its ability to make an extensive series of the acetate esters of straight-chained alkanes, including those of hexyl, heptyl, octyl, and sec-octyl alcohols (Strobel et al., 2008). It is reported that the alkyl alcohol side chains of these acetate esters represent the basic free alkanes of diesel, as they are specifically found in every diesel sample that has been analyzed (Strobel, 2014). Several endophytic fungi are reported to produce 1,8-cineole that can be used in aroma as well as fuel industry (Tomsheck et al., 2010; Riyaz-Ul-Hassan et al., 2012, 2013). Another study investigated endophytic fungi of different tropical plant species for biodiesel precursor production and reported that Xylariaceous fungi possess high concentrations of methyl esters (91%). Therefore, Xylariaceous fungi have the potential of promising source for biofuel production (Santos-Filho et al., 2011).

2.6 ENDOPHYTE BIOPROSPECTION IN INDIA

In India, various groups are working on the bioprospection endophytes and a significant amount of literature is available in this particular field (Table 2.1). However, the main focus has been to isolate promising plant metabolites from the endophytes of the host plant (Puri et al., 2006; Kusari et al., 2009). For example, Podophylotoxin was isolated from an endophytic fungus *Trametes hirsuta,* present in the host plant *Podophyllum hexandrum* (Puri et al. 2006), Podophylotoxin was extracted from another endophytic fungus *Fusarium oxysporum* found in *Juniperus recurva* (Kour et al. 2008), Camptothecin was isolated from an endophytic fungus *Entrophospora infrequens,* harboring the host plant *Nothapodytes foetida* (Puri et al. 2005). Other plant metabolites isolated from endophytes include rohitukine (Kumara et al., 2012), Javinicin (Kharwar et al., 2009), paclitaxel (Sreekanth et al., 2011), and piperine (Verma et al., 2011). In most of the abovementioned studies, the endophytes have been mostly acquired from individual plants sporadically. However, during the last few years, significant effort is being made to study the endophytes from different plants growing in different locations in India for bioprospection (Puri et al., 2006; Shweta et al., 2013; Qadri et al., 2013, 2014, 2015; Wani et al., 2016, 2017, 2018; Arora et al., 2017, 2019).

Kumar and colleagues reported that an endophytic fungus isolated from the plant, *Cedrus deodara,* produces four compounds with potential cytotoxicities against human cancer cell lines (HCT-116, A-549, HEP-1, THP-1, and PC-3). All the compounds are found to induce apoptosis in HL-60 cells, by causing significant microtubule inhibition in HL-60 cells (Kumar et al., 2013). Similarly, three bioactive molecules are isolated from an endophytic fungus *Cryptosporiopsis* sp. inhabiting *Clidemia hirta,* with moderate anticancer and antimicrobial activities (Zilla et al., 2013). An endophytic fungus *Diaporthe phaseolorum* isolated from *Picrorhiza kurroa*

TABLE 2.1 Bioactive Compounds Isolated from Different Endophytic Sources (Bacterial and Fungal Endophytes) in India.

Endophytes	Bioactive compounds	Reference
Streptomyces coelicolor	Munumbicins	Singh and Dubey (2015)
Chloridium sp.	Javanicin	Jalgaonwala et al (2011)
Allamanda cathartica	Munumbicins Phomopsilactone	Nithya and Muthumary (2011)
Cryptosporiopsis quercina	Saadamycin	Dutta et al. (2014)
Cytonaema sp.	Cytonic acids A and B	Bhardwaj and Agrawal (2014)
Fusarium sp.	Xularosides, munumbicins	Jalgaonwala et al. (2011)
Ganoderma boninense	Rapamycin, cyclododecane,	Parthasarathi et al. (2012)
Phomopsis sp.	Munumbicins	Jalgaonwala et al. (2011)
Cinnamomum mollissimum	Saadamycin	Kaul et al. (2012)
Streptomyces hygroscopicus	Coronamycin, rapamycin	Parthasarathi et al. (2012)
Streptomyces sp.	Munumbicins	Kumar et al. (2014)
Entrophospora infrequens	Camptothecin	Puri et al. (2005)
Neurospora crassa	Camptothecin	Rehman et al. (2008)
Trametes hirsute	Podophyllotoxin	Puri et al. (2006)
Fusarium oxysporum	Podophyllotoxin	Kour et al. (2008)
Fusarium proliferatum	Rohitukine	Kumara et al. (2012)
Chloridium sp.	Javinicin	Kharwar et al. (2009)
Fusarium solani	camptothecin	Shweta et al. (2010)
Cryptosporiopsis sp.	®-5-hydroxy-2-methylchroman-4-one	Zilla et al. (2013)
Talaromyces sp.	Ramulosin, epoformin	Kumar et al. (2013)
Diaporthe phaseolorum	Menthol, 3-hydroxypropionic acid	Qadri et al. (2015)
Phialophora mustea	Phialomustins A-D	Nalli et al. (2015)
Phoma sp.	thiodiketopiperazine derivatives	Arora et al. (2016)
Diaporthe terebinthifolii	Diapolic acid A-B	Nalli et al. (2016)
Porostereum sp.	chlorinated aromatic metabolites	Wani et al. (2018)
Muscodor yucatanensis	polyketides	Qadri et al. (2016)
Mortierella alpina	Arachidonic acid	Wani et al. (2017)

produces 3-hydroxypropionic acid (3-HPA) as a major metabolite apart from many VOCs, including menthol, phenylethyl alcohol, (+)-isomenthol, β-phellandrene, β-bisabolene, limonene, 3-pentanone, and 1-pentanol (Qadri et al., 2015). An endophytic fungus *Phialophora mustea* isolated from *Crocus sativus* is reported to produce four new metabolites phialomustins A-D, with antimicrobial and cytotoxic potential (Nalli et al., 2015). *P. mustea* produces several other compounds that remain to be characterized. An endophyte *Muscodor yucatanensis* (Ni30) isolated from *Elleanthus* sp. produces brefeldin A as the major compound in the culture broth (Qadri et al., 2016). This compound is previously reported from *Eupenicillium brefeldianum, Paecilomyces* sp., and *Aspergillus clavatus* (Harri et al., 1963; Wang et al., 2007). Brefeldin A, a polyketide, has antibacterial, antiviral, antinematode, and antifungal activities (Wang et al., 2007; Betina, 1992). It is also widely used in biological research to study protein transport as it blocks protein secretion by causing disassembly of Golgi apparatus (Misumi et al., 1986). The epigenetic modulation of the fungal endophyte (*M. yucatanensis*) by small-molecule epigenetic modifiers, 5-azacytidine, and suberoylanilide hydroxamic acid, results in the production of a different set of volatile organic compounds distinct from the wild type. This may be due to the activation of otherwise silent polyketide synthase (PKS) genes (Qadri et al., 2016) Another fungal endophyte, *Porostereum* sp. isolated from *C. sativus*, produces an array of volatile organic compounds, including few chlorinated aromatic metabolites (CAM) having phytotoxic potential (Wani et al., 2018). An endophyte, *Phoma* sp. isolated from *Glycyrrhiza glabra* produced thiodiketopiperazine derivatives having antimicrobial and cytotoxic potential (Arora et al., 2016). In another study, diapolic acid A-B was characterized from another endophytic fungus *Diaporthe terebinthifolii* isolated from *G. glabra*. These compounds displayed antimicrobial and cytotoxic activity (Nalli et al., 2016). There are reports on using endophytes as growth-promoting agents in various crops/plants (Singh et al., 2013; Wani et al., 2017). The majority of the endophytes isolated from different plant species remain to be explored for bioactive natural products. Considering the enormity of biodiversity in India, concerted efforts are needed to tap the huge repository of endophytic microorganisms for bioprospection.

2.7 CONCLUSION

The industrial production of important molecules from microbial sources may revolutionize the drug market owing to their malleability. Microbes can be grown in very large volumes and they can be regarded as renewable sources of the target molecules. In addition to bringing down the cost of respective drug molecules, such a development may also save the environment by preserving the plants that otherwise are cut down regularly for the extraction of drug molecules. Despite extensive research in this area, no such industrial process has been developed, as the microbes produce these molecules in very low quantities or in some cases stop the production completely after multiple subculturing.

KEYWORDS

- **secondary metabolites**
- **horizontal gene transfer**
- **volatile organic compounds**
- **Polyketides**
- **natural products**
- **stress adaptation**

REFERENCES

Altreuther, G.; Gasda, N.; Adler, K.; Thurieau, H.; Schimmel, A.; Hutchens, D.; Krieger, K. J. Field Evaluations of the Efficacy and Safety of Emodepside Plus Toltrazuril (Procox® Oral Suspension for Dogs) Against Naturally Acquired Nematode and *Isospora* spp. Infections in Dogs. *Parasitol. Res.* **2011,** *190*, 21–28.

Arora, P.; Wani, Z. A.; Nalli, Y.; Ali, A.; Riyaz-Ul-Hassan, S. Antimicrobial Potential of Thiodiketopiperazine Derivatives Produced by *Phoma* sp., an Endophyte of *Glycyrrhiza glabra* Linn. *Microb. Ecol.* **2016.** DOI:10.1007/s00248-016-0805-x.

Arora, P.; Wani, Z. A.; Ahmad, T.; Sultan, P.; Gupta, S.; Riyaz-Ul-Hassan, S. Community Structure, Spatial Distribution, Diversity and Functional Characterization of Culturable Endophytic Fungi Associated with *Glycyrrhiza glabra* L. *Fungal Biol.* **2019.** https://doi.org/10.1016/j.funbio.2019.02.003.

Bentley, S. D.; et al. Complete Genome Sequence of the Model Actinomycete *Streptomyces coelicolor* A3(2). *Nature* **2002,** *417*, 141–147.

Betina, V. Biological Effects of the Antibiotic Brefeldin A (Decumbin, Cyanein, Ascotoxin, Synergisidin): A Retrospective. *Folia Microbiol.* **1992,** *37*, 3–11.

Berdy, J. Thoughts and Facts About Antibiotics: Where We Are Now and Where We Are Heading. *J. Antibiot.* **2012,** *65*, 385–395.

Bitas, V.; Kim, H. S.; Bennett, J. W.; Kang, S. Sniffing on Microbes: Diverse Roles of Microbial Volatile Organic Compounds in Plant Health. *Mol. Plant Microbe Interact* **2013,** *26*, 835–843.
Bok, J. W.; Hoffmeister, D.; Maggio-Hall, L. A.; Renato, M.; Glasner, J. D.; Keller, N. P. Genomic Mining for *Aspergillus* Natural Products. *Chem. Biol.* **2006,** *13*, 31–37.
Challis, G. L. Genome Miming for Novel Natural Product Discovery. *J. Med. Chem.* **2008,** *51*, 2618–2628.
Chen, L.; Zhang, Q. Y.; Jia, M.; Ming, Q. L.; Yue, W.; Rahman, K.; Qin, L. P.; Han, T. Endophytic Fungi With Antitumor Activities: Their Occurrence and Anticancer Compounds. *Crit. Rev. Microbiol.* **2016,** *42*(3), 454–473.
Clardy, J.; Walsh, C. Lessons from Natural Molecules. *Nature* **2004,** *432*, 829–837.
Corre, C.; Challis, G. L. Heavy Tools for Genome Mining. *Chem. Biol.* **2007,** *14*, 7–9.
FDA. Freedom of Information Summary. Original New Animal Drug Application. NADA, (2007) 141-275. Online available at: http://www.fda.gov/downloads/AnimalVeterinary/Products/ApprovedAnimalDrugProducts/FOIADrugSummaries/ucm062332.pdf
DeLorbe, J. E.; Horne, D.; Jove, R.; Mennen, S. M.; Nam, S.; Zhang, F. L.; Overman, L. E. General Approach for Preparing Epidithiodioxopiperazines From Trioxopiperazine Precursors: Enantioselective Total Syntheses of (+)- and (-)-Gliocladine C, (+)-Leptosin D, (+)-T988C, (+)-Bionectin A, and (+)-Gliocladin A. *J. Am. Chem. Soc.* **2013,** *135*, 4117–4128.
Durrett, T. P.; Benning, C.; Ohlrogge, J. Plant Triacylglycerols as Feedstocks for the Production of Biofuels. *Plant J.* **2008,** *54*, 593–607.
Feng, Y.; Blunt, J. W.; Cole, A. L.; Munro, M. H. Novel Cytotoxic Thiodiketopiperazine Derivatives from a *Tilachlidium* sp. *J. Nat. Prod.* **2004,** *67*, 2090–2092.
Fischbach, M. A.; Walsh, C. T. Assembly-Line Enzymology for Polyketide and Nonribosomal Peptide Antibiotics: Logic, Machinery, and Mechanisms. *Chem. Rev.* **2006,** *106*, 3468–3496.
Fleischmann, R. D.; et al. Whole-Genome Random Sequencing and Assembly of *Haemophilus influenzae* Rd. *Science* **1995,** *269*, 496–512.
Gloer, J. B. Applications of Fungal Ecology in the Search for New Bioactive Natural Products. In *The Mycota*. Vol. IV. *Environmental and Microbial Relationships*; Wicklow, D. T., Soderstrom, B. E., Eds.; Springer-Verlag: New York, NY, 1997; pp 249–268.
Harri, E.; Loeffler, W.; Sigg, H. P.; Staehelin, S.; Tamm, C. Uber die isolierung der stoffwechselprodukte aus *Penicellium brefeldianum* Dodge. *Helv Chim Acta* **1963,** *46*, 1235–1243.
Horn, W. S.; Simmonds, M. S. J.; Schwartz, R. E.; Blaney, W. M. Phomopsichalasin, A Novel Antimicrobial Agent from an Endophytic *Phomopsis* sp. *Tetrahedron* **1995,** *51*, 3969–3978.
Ikeda, H.; Ishikawa, J.; Hanamoto, A.; Shinose, M.; Kikuchi, H.; Shiba, T.; Sakaki, Y.; Hattori, M.; Omura, S. Complete Genome Sequence and Comparative Analysis of the Industrial Microorganism *Streptomyces avermitilis*. *Nat. Biotechnol.* **2003,** *21*, 526–531.
Isaka, M.; Jaturapat, A.; Rukseree, K.; Danwisetkanjana, K.; Tanticharoen, M.; Thebtaranonth, Y. Phomoxanthones A and B, Novel Xanthone Dimers from the Endophytic Fungus *Phomopsis* sp. *J. Nat. Prod.* **2001,** *64*, 1015–1018.
Keller, N. P.; Turner, G.; Bennett, J. W. Fungal Secondary Metabolism-from Biochemistry to Genomics. *Nat. Rev. Microbiol.* **2005,** *3*, 937–947.
Kharwar, R. N.; Verma, V. C.; Kumar, A.; Gond, S. K.; Harper, J. K.; Hess, W. M.; Lobkovosky, E.; Ma, C.; Ren, Y.; Strobel, G. A. Javanicin, An Antibacterial Naphthaquinone from an Endophytic Fungus of Neem, *Chloridium* sp. *Curr. Microbiol.* **2009,** *58*, 233–238.

Kharwar, R. N.; Mishra, A.; Gond, S. K.; Stierle, A.; Stierle, D. Anticancer Compounds Derived from Fungal Endophytes: Their Importance and Future Challenges. *Nat. Prod. Rep.* **2011,** *28*, 1208–1228.

Knappe, T. A.; Linne, U.; Zirah, S.; Rebuffat, S.; Xie, X.; Marahiel, M. A. Isolation and Structural Characterization of Capistruin, a Lasso Peptide Predicted from the Genome Sequence of *Burkholderia thailandensis* E264. *JACS* **2008,** *13*, 11446–11454.

Kour, A.; Shawl, A. S.; Rehman, S.; Sultan, P.; Qazi, P. H.; Suden, P.; Khajuria, R. K.; Verma, V. Isolation and Identification of an Endophytic Strain of *Fusarium oxysporum* Producing Podophyllotoxin from *Juniperus recurva*. World J. Microbiol. Biotechnol. **2008,** *24*, 1115–1121.

Kusari, S.; Zühlke, S.; Spiteller, M. An Endophytic Fungus from *Camptotheca acuminata* that Produces Camptothecin and Analogues. *J. Nat. Prod.* **2009,** *72*, 2–7.

Kusari, S.; Spiteller, M. Are We Ready for Industrial Production of Bioactive Plant Secondary Metabolites Utilizing Endophytes? *Nat. Prod. Rep.* **2011,** *28*, 1203–1207.

Kusari, S.; Hertweck, C.; Spiteller, M. Chemical Ecology of Endophytic Fungi: Origins of Secondary Metabolites. *Chem. Biol.* **2012,** *19*, 792–798.

Kusari, S.; Singh, S.; Jayabaskaran, C. Biotechnological Potential of Plant-Associated Endophytic Fungi: Hope Versus Hype. *Trends Biotechnol*. **2014,** *32*, 297–303.

Kumar, M.; Qadri, M.; Sharma, P. R.; et al. Tubulin Inhibitors from an Endophytic Fungus Isolated from *Cedrus deodara*. *J. Nat. Prod.* **2013,** *76*, 194−199.

Kumara, P. M.; Zuehlke, S.; Priti, V.; Ramesha, B. T.; et al. *Fusarium proliferatum,* an Endophytic Fungus from *Dysoxylum binectariferum* Hook.f, Produces Rohitukine, a Chromane Alkaloid Possessing Anti-Cancer Activity. *Antonie Van Leeuwenhoek*. **2012,** *101*(2), 323–329.

Li, J. Y.; Strobel, G.; Harper, J.; Lobkovsky, E.; Clardy, J. Cryptocin, a Potent Tetramic Acid Antimycotic from the Endophytic Fungus *Cryptosporiopsis cf. quercina*. *Org. Lett.* **2000,** *2*, 767–770.

Mallette, N.; Pankratz, E.; Busse, S.; Strobel, G. A.; Carlson, R.; Peyton, B. Evaluation of Cellulose as a Substrate for Hydrocarbon-Fuel Production by *Ascocoryne sarcoides* (NRRL 50072). *J. Sustain. Bioener. Syst.* **2014,** *4*, 33–49.

McAlpine, J. B.; Bachmann, B. O.; Piraee, M.; Tremblay, S.; Alarco, A. M.; Zazopoulos, E.; Farnet, C. M. Microbial Genomics as a Guide to Drug Discovery, Structural Elucidation: ECO02301, a Novel Antifungal Agent, as an Example. *J. Nat. Prod.* **2005,** *68*, 493–496.

Misumi, I.; Misumi, Y.; Miki, K.; Takntsuki, A.; Tamura, G.; lkehara, Y. Novel Blockade by Brefeldin A of Intracellular Transport of Secretory Proteins in Cultured Rat Hepatocytes. *J. Biol. Chem.* **1986,** *261*, 11398–11403.

Mousa, W. K.; Raizada, M. N. The Diversity of Anti-Microbial Secondary Metabolites Produced by Fungal Endophytes: An Interdisciplinary Perspective. *Front Microbiol.* **2013,** *4*, 65.

Nalli, Y.; Mirza, DN.; Wani, Z. A.; Wadhwa, B.; Mallik, F. A.; Raina, C.; Chaubey, A.; Riyaz-Ul-Hassan, S.; Ali, A. Phialomustin A-D, New Antimicrobial and Cytotoxic Metabolites from an Endophytic Fungus, *Phialophora mustea. RSC Adv.* **2015,** *5*, 95307–95312.

Nalli, Y.; Arora, P.; Wadhwa, B.; Malik, F. A.; Vishwakarma, R. A.; Riyaz-Ul-Hassan, S.; Ali, A. Diapolic Acid A-B from an Endophytic Fungus, *Diaporthe terebinthifolii* Depicting Anti-Microbial and Cytotoxic Activity. *J. Antbiot.* **2016.** doi:10.1038/ja.2016.109.

Newman, D. J.; Cragg, G. M. Natural Products as Sources of New Drugs Over the 30 Years From 1981 to 2010. *J. Nat. Prod.* **2012,** *75*, 311–335.

Oliynyk, M.; Samborsky, M.; Lester, J. B.; Mironenko, T.; Scott, N.; Dickens, S.; Haydock, S. F.; Leadlay, P. F. Complete Genome Sequence of the Erythromycin-Producing Bacterium *Saccharopolyspora erythraea* NRRL2338. *Nat. Biotechnol.* **2007,** *25*, 447–453.

Paulsen, I. T.; et al. Complete Genome Sequence of the Plant Commensal *Pseudomonas fluorescens* Pf-5. *Nat. Biotechnol.* **2005,** *23*, 873–878.
Peric-Concha, N.; Long, P. F. Mining the Microbial Metablome: A New Frontier for Natural Product Lead Discovery. *Drug Dis. Today* **2003,** *8*, 1078–1084.
Porras-Alfaro, A.; Bayman, P. Hidden Fungi, Emergent Properties: Endophytes and Microbiomes. *Annu. Rev. Phytopathol.* **2011,** *49*, 291–315.
Puri, S. C.; Verma, V.; Amna, T.; Qazi, G. N.; Spiteller, M. An Endophytic Fungus from *Nothapodytes foetida* that Produces Camptothecin. *J. Nat. Prod.* **2005,** *68*, 1717–1719.
Puri, S. C.; Nazir, A.; Chawla, R.; Arora, R.; et al. The Endophytic Fungus *Trametes hirsuta* as a Novel Alternative Source of Podophyllotoxin and Related Aryl Tetralin Lignans. *J. Biotechnol.* **2006,** *122*(4), 494–510.
Qadri, M.; Johri, S.; Shah, B. A.; Khajuria, A.; Sidiq, T.; Lattoo, S. K.; Abdin, M. Z.; Riyaz-Ul-Hassan, S. Identification and Bioactive Potential of Endophytic Fungi Isolated from Selected Plants of the Western Himalayas. *SpringerPlus* **2013,** *2*, 8.
Qadri, M.; Rajput, R.; Abdin, M. Z.; Vishwakarma, R. A.; Riyaz-Ul-Hassan, S. Diversity, Molecular Phylogeny and Bioactive Potential of Fungal Endophytes Associated With the Himalayan blue pine (*Pinus wallichiana*). *Microb. Ecol.* **2014,** *67*, 877–887.
Qadri, M.; Nalli, Y.; Jain, S. K.; Ali, A.; Strobel, G. A.; Vishwakarma, R. A.; Riyaz-Ul-Hassan, S. An Insight into the Secondary Metabolism of *Muscodor yucatanensis*: Small Molecule Epigenetic Modifiers Induce Expression of Secondary Metabolism Related Genes and Production of New Metabolites in the Endophyte. *Microb. Ecol.* **2016,** *73*, 954–965.
Riyaz-Ul-Hassan, S.; Strobel, G. A.; Booth, E.; Knighton, B.; Floerchinger, C.; Sears, J. Modulation of Volatile Organic Compound Formation in the Mycodiesel Producing Endophyte- *Hypoxylon* sp. C1-4. *Microbiol* **2012,** *158*, 464–473.
Riyaz-Ul-Hassan, S.; Strobel, G.; Geary, B.; Sears, J. An Endophytic *Nodulisporium* sp. from Central America Producing Volatile Organic Compounds with Both Biological and Fuel Potential. *J. Microbiol. Biotechnol.* **2013,** *23*, 29–35.
Santos-Filho, F. C.; Fill, T. P.; Nakamura, J.; Monteiro, M. R.; Rodrigues-Fo, E. Endophytic Fungi as a Source of Biofuel Precursors. *J. Microbiol. Biotechnol.* **2011,** *21*(7), 728–733.
Scherlach, K.; Hertweck, C. Triggering Cryptic Natural Product Biosynthesis in Microorganisms. *Org. Biomol. Chem.* **2009,** *7*, 1753–1760.
Shoop, W. L.; Gregory, L. M.; Zakson-Aiken, M.; Michael, B. F.; Haines. H W.; Ondeyka, J. G.; Meinke, P. T.; Schmatz, D. M. Systemic Efficacy of Nodulisporic Acid Against Fleas on Dogs. *J. Parasitol.* **2001,** *87*, 419–423.
Shweta, S.; Bindu, J. H.; Raghu, J.; Suma, H. K.; Manjunatha, B. L.; Kumara, P. M.; Ravikanth, G.; Nataraja, K. N.; Ganeshaiah, K. N.; Uma Shaanker, R. Isolation of Endophytic Bacteria Producing the Anti-Cancer Alkaloid Camptothecine from *Miquelia dentata* Bedd. (Icacinaceae). *Phytomed* **2013,** *20*(10), 913–917.
Sieber, S. A.; Marahiel, M. A. Molecular Mechanisms Underlying Nonribosomal Peptide Synthesis: Approaches to New Antibiotics. *Chem. Rev.* **2005,** *105*, 715–738.
Singh, R. K.; Malik, N.; Singh, S. Improved Nutrient Use Efficiency Increases Plant Growth of Rice with the Use of IAA-Overproducing Strains of Endophytic *Burkholderia cepacia* strain RRE25. *Microb. Ecol.* **2013,** *66*(2), 375–384.
Soliman, S. S.; Greenwood, J. S.; Bombarely, A.; Mueller, L. A.; Tsao, R.; Mosser, D. D.; Raizada, M. N. An Endophyte Constructs Fungicide-Containing Extracellular Barriers for Its Host Plant. *Curr. Biol.* **2015,** *25*(19), 2570–2576.

Sreekanth, D.; Sushim, G. K.; Syed, A.; Khan, B. M.; Ahmad, A. Molecular and Morphological Characterization of a Taxol-Producing Endophytic Fungus, *Gliocladium* sp., from *Taxus baccata. Mycobiol* **2011,** *39*(3), 151–157.

Staley, J. T.; Castenholz, R. W.; Colwell, R. R.; Holt, J. G.; Kane, M. D.; Pace, N. R.; Saylers, A. A.; Tiedje, J. M. *The Microbial World: Foundation of the Biosphere*; American Academy of Microbiology: Washington, DC, 1997, pp 32.

Staniek, A.; Woerdenbag, H. J.; Kayser, O. *Taxomyces andreanae:* A Presumed Paclitaxel Producer Demystified? *Planta Med.* **2009,** *75*, 1561–1566.

Stierle, A.; Strobel, G. A.; Stierle, D. B. Taxol and Taxane Production by *Taxomyces andreanae*, an Endophytic Fungus of Pacific Yew. *Science* **1993,** *260*, 214–216.

Strobel, G. A.; Miller, R. V.; Martinez-Miller, C.; Condron, M. M.; Teplow, D. B.; Hess, W. M. Cryptocandin, A Potent Antimycotic from the Endophytic Fungus *Cryptosporiopsis cf. quercina*. *Microbiol.* **1999,** *145*, 1919–1926.

Strobel, G. A. Endophytes as Sources of Bioactive Products. *Microb. Infect.* **2003,** *5*, 535–544.

Strobel, G.; Daisy, B.; Castillo, U.; Harper, J. Natural Products from Endophytic Microorganisms. *J. Nat. Prod.* **2004,** *67*(2), 257–268.

Strobel, G. Harnessing Endophytes for Industrial Microbiology. *Curr. Opin. Microbiol.* **2006,** *9*, 240–244.

Strobel, G. A.; Daisy, B. Bioprospecting for Microbial Endophytes and Their Natural Products. *Microbiol. Mol. Biol. Rev.* **2003,** *67*, 491–502.

Strobel, G. A.; Knighton, B.; Kluck, K.; Ren, Y.; Livinghouse, T.; Griffin, M.; Spakowicz, D.; Sears, J. The Production of Myco-Diesel Hydrocarbons and their Derivatives by the Endophytic Fungus *Gliocladium roseum* (NRRL 50072). *Microbiol.* **2008,** *154*, 3319–3328.

Strobel, G. A. The Story of Mycodiesel. *Curr. Opin. Microbiol.* **2014,** *19*, 52–58.

Strobel, G. A. Bioprospecting—Fuels from Fungi. *Biotechnol. Lett.* **2015.** DOI 10.1007/s10529-015-1773-9

Talontsi, F. M.; Dittrich, B.; Schüffler, A.; Sun, H.; et al. Epicoccolides: Antimicrobial and Antifungal Polyketides from an Endophytic Fungus *Epicoccum* sp. Associated with *Theobroma cacao*. *Eur. J. Org. Chem.* **2013,** *15*, 3174–3180.

Tan, Y. Y.; Spiering, M. J.; Scott, V.; Lane, G. A.; Christensen, M. J.; Schmid, J. In Planta Regulation of Extension of an Endophytic Fungus and Maintenance of High Metabolic Rates in its Mycelium in the Absence of Apical Extension. *Appl. Environ. Microbiol.* **2011,** *67*, 5377–5383.

Terada, M. Neuropharmacological Mechanism of Action of PF1022A, An Antinematode Anthelmintic with a New Structure of Cyclic Depsipeptide, on *Angiostrongylus cantonensis* and Isolated Frog Rectus. *Jpn. J. Parasitol.* **1992,** *41*, 108–117.

Tomsheck, A.; Strobel, G. A.; Booth, E.; Geary, B.; Spakowicz, D.; Knighton, B.; Floerchinger, C.; Sears, J.; Liarzi, O.; Ezra, D. *Hypoxylon* sp. an Endophyte of *Persea indica*, Producing 1,8-Cineole and Other Bioactive Volatiles with Fuel Potential. *Microb. Ecol.* **2010,** *60*, 903–914.

Udwary, D. W.; Zeigler, L.; Asolkar, R. N.; Singan, V.; Lapidus, A.; Fenical, W.; Jensen, P. R.; Moore, B. S. Genome Sequencing Reveals Complex Secondary Metabolome in the Marine Actinomycete *Salinispora tropica*. *Proc. Natl. Acad. Sci. U. S. A.* **2007,** *104*, 10376–10381.

Verma, V. C.; Lobkovsky, E.; Gange, A. C.; Singh, S. K.; Prakash, S. Piperine Production by Endophytic Fungus *Periconia* sp. Isolated from *Piper longum* L. *J. Antibiot.* **2011,** *64*, 427–431.

Wang, F.; Jiao, R.; Cheng, A.; Tan, S.; Song, Y. Antimicrobial Potentials of Endophytic Fungi Residing in *Quercus variabilis* and Brefeldin A Obtained from *Cladosporium* sp. *World J. Microbiol. Biotechnol.* **2007,** *23*, 79–83.

Wang, L. W.; Zhang, Y. L.; Lin, F. C.; Hu, Y. Z.; Zhang, C. L. Natural Products with Antitumor Activity from Endophytic Fungi. *Mini Rev. Med. Chem.* **2011,** *11*, 1056–1074.

Wani, Z. A.; Mirza, D. N.; Arora, P.; Riyaz Ul Hassan, S. Molecular Phylogeny, Diversity, Community Structure, and Plant Growth Promoting Properties of Fungal Endophytes Associated with the Corms of Saffron Plant: An Insight Into the Microbiome of *Crocus sativus* Linn. *Fungal Biol.* **2016,** *120*, 1509–1524.

Wani, Z. A.; Kumar, A.; Sultan, P.; Bindu, K.; Riyaz-Ul-Hassan, S.; Ashraf, N. *Mortierella alpina* (CS10E4), An Oleaginous Fungal Endophyte of *Crocus sativus* L. Enhances Apocarotenoid Biosynthesis and Stress Tolerance in Host Plant. *Sci. Rep.* **2017,** *7*(1), 8598.

Wani, Z. A.; Ahmad, T.; Nalli, Y.; Ali, A.; Singh, A. P.; Vishwakarma, R. A.; Ashraf, N.; Riyaz-Ul-Hassan, S. *Porostereum* sp., Associated with Saffron (*Crocus sativus* L.), is a Latent Pathogen Capable of Producing Phytotoxic Chlorinated Aromatic Compounds. *Curr. Microbiol.* **2018.** doi: 10.1007/s00284-018-1461-9.

Zhanel, G. G.; Chung, P.; Adam, H.; Zelenitsky, S.; et al. Ceftolozane/Tazobactam: A Novel Cephalosporin/β-Lactamase Inhibitor Combination with Activity Against Multidrug-Resistant Gram-Negative Bacilli. *Drugs* **2014,** *74*, 31–51.

Zilla, M. K.; Qadri, M.; Pathania, A. S.; et al. Bioactive Metabolites from an Endophytic *Cryptosporiopsis* sp. Inhabiting *Clidemia hirta*. *Phytochemistry* **2013,** *95*, 291–297.

CHAPTER 3

Plant–Endophyte Symbiosis

ZAHOOR AHMED WANI[1*] and KHALID REHMAN HAKEEM[2]

[1]*Department of Botany, Govt. Degree College Kishtwar, Kishtwar, Jammu & Kashmir, India*

[2]*Department of Biological Sciences, Faculty of Science, King Abdulaziz University, Jeddah, Saudi Arabia*

[*]*Corresponding author. E-mail: zawani1986@gmail.com*

ABSTRACT

Endophytism is the phenomenon of mutualistic association of a plant with a microorganism wherein the microbe lives within the tissues of the plant without causing any symptoms of disease. In addition to being a treasured biological resource, endophytes play diverse indispensable functions in nature for plant growth, development, stress tolerance, and adaptation. Our understanding of endophytism and its ecological aspects are overtly limited, and we have only recently started to appreciate its essence. Endophytes may impact plant biology through the production of diverse chemical entities including, but not limited to, plant growth hormones and by modulating the gene expression of defense and other secondary metabolic pathways of the host. Studies have shown differential recruitment of endophytes in endophytic populations of plants growing in the same locations, indicating host specificity and that endophytes evolve in a coordinated fashion with the host plants. Endophytic technology can be employed for the efficient production of agricultural and economically important plants and plant products. The rational application of endophytes to manipulate the microbiota, intimately associated with plants, can help in enhancement of production of agricultural produce, increased production of key metabolites in medicinal and aromatic plants, as well as adaption to new bio-geographic regions through tolerance to various biotic and abiotic conditions.

3.1 INTRODUCTION

In 1994, Richard Jefferson put forth an idea that the evolutionary selection unit is not a single organism but a macro-organism together with its entire associated microbial consortia. He further concluded that "agriculture can only progress sustainably when balanced hologenetic combinations or holo-alleles are present" (Jefferson, 1994). Although plants are sessile organisms, they are involved in intensive mutualistic associations with other organisms like microbes that help them interact with the surrounding environment. These mutualistic microbes called endophytes are diversely localized in the internal tissues of plant without causing any apparent disease symptoms. Endophytes include neutrals, commensals and/or beneficial micro-organisms, dormant saprobes, and pathogens during the latent phase of their life cycle. Endophytes being intimately associated with the plant exert a significant influence on many plant parameters like growth and development, metabolism and nutrient acquisition, tolerance to biotic and abiotic stress, as well as plant survival and distribution (Wani et al., 2015). Owing to its engagement in diverse heterospecific associations, each plant is considered as a complex community in itself, rather than a single organism. The multispecies crosstalk with these micro-organisms leads to the selection of specific and generalized functional traits by each interacting partner (Kusari et al., 2014). Endophytes are considered to originate in the outside environment and enter the plant through stomata, lenticels, wounds, areas of emergence of lateral roots and germinating radicals. They later colonize the plant tissues by quickly spreading to the intercellular spaces inside the host plant (Chi et al., 2005). It is generally assumed that many endophytic communities are the product of a colonizing process initiated in the root zone or rhizosphere (Sturz et al., 2000). However, they may also originate from other sources, such as the phyllosphere, the anthosphere, or the spermosphere (Hallmann et al., 1997). It has been proposed that cellulolytic (endoglucanase and polygalacturonase) and pectinolytic (pectate lyase) enzymes produced by endophytes are involved in the infection process (Baldrian and Valaskova, 2008; Compant et al., 2005). The cell wall of a plant cell is mainly composed of polysaccharides (pectic substances, hemicelluloses, and cellulose) and glycoproteins. Therefore, the microbial cellulases and protease play a significant role in cell wall lysis that occurs during plant–microbe interactions (Baldrian and Valaskova, 2008). Many endophytic fungi produce auxin in axenic cultures. The physiological role of auxin in fungi is not well understood; it is presumed that auxin plays an important role in colonization of roots (Hilbert et al., 2012). The basis of chemical crosstalk between plant and associated microbes lies in certain communication molecules that are responsible for plant–microbe interactions. However, sophisticated communication systems have been created during the course of evolution by which the plant influences the behavior of micro-organisms in the root environment to its own favor.

3.2 DYNAMICS OF PLANT–ENDOPHYTE INTERACTION

According to the theory of "Balanced antagonism," the endophyte–host interactions (endophytism) exhibit a paradigm shift from mutualism to antagonism. This paradigm shift indicates phenotypic plasticity of the plant–endophyte interaction that depends on the biocommunication between endophyte and host plant, and certain environmental conditions (Schulz and Boyle, 2005; Wani et al., 2015). Most endophytes are tested on a single or a few plant species and even if they show no deleterious effects on these plants, they may exhibit pathogenicity on other plants. Furthermore, pathogenicity depends on a number of environmental parameters and biotic interactions. For instance, fluorescent pseudomonad showing plant-beneficial effects can cause disease on leather leaf ferns under specific conditions (Kloepper et al., 2013). In the case of epichloe-endophyte association, the onset of flowering in some species of the host plant induces the fungal endophyte to switch from mutualistic asexual life cycle to antagonistic sexual life cycle (Schardl et al., 2004). It is suggested that some of the fungal endophytes have presumably evolved from plant pathogenic fungi, as evidenced by some root endophytic fungi that require host cell death for proliferation during the formation of mutualistic symbiosis with the host plant. They remain asymptomatic for many years and only become parasitic when their hosts are stressed (Fig. 3.1) (Wani et al., 2015). In general, a variety of microbes may enter and become transient endophytes, and those consistently found inside the host tissues for long periods of time may eventually share the physiological and genetic makeup of the host and thus serve as candidate symbionts or true endophytes (Wani et al., 2015). The composition of plant microbiome is affected by various host-driven factors, including the plant genotype, and by other external factors such as edaphic factors, climatic factors, and agricultural practices (Wani et al., 2015; Mitter et al., 2016). The microbial component of healthy seeds, tubers, or corms used to propagate a plant appears to be inherited between plant generations and is likely to represent an important resource for the microbiome buildup. The soil- and root-derived micro-organisms arrive later and have to compete with the already established microflora inside the plant (Hardoim et al., 2015; Mitter et al., 2016). This has resulted in the evolution of sophisticated biocommunication systems through which the plant influences the behavior of micro-organisms in the root environment to its own favor. Recently, it has been reported that the biotrophic lifestyle of endophyte during the colonization of host plant is an important feature of the plant–endophyte interactions, as it implies a strong genetic and metabolic relief to both the interacting partners. Also, some endophytes produce molecular signals like reactive oxygen species (ROS), which result in switching of endophytism to either antagonism or mutualism in response to some environmental cues (White and Torres, 2010). Thus, endophytes in the earlier stage are detected as minor pathogens which over a period of time co-evolve with the host plant into benign or mutualistic symbionts with varying degrees of dependence, depending on the cost–benefit analysis of the plant–endophyte association (Schulz and Boyle, 2005; Conn et al., 2008). However, the response of long-term association between an endophyte and its host may be mutualistic or antagonistic in nature, depending on nutrient availability to the endophyte and metabolic status of the host plant (Lahrmann et al., 2013; Wani et al., 2015).

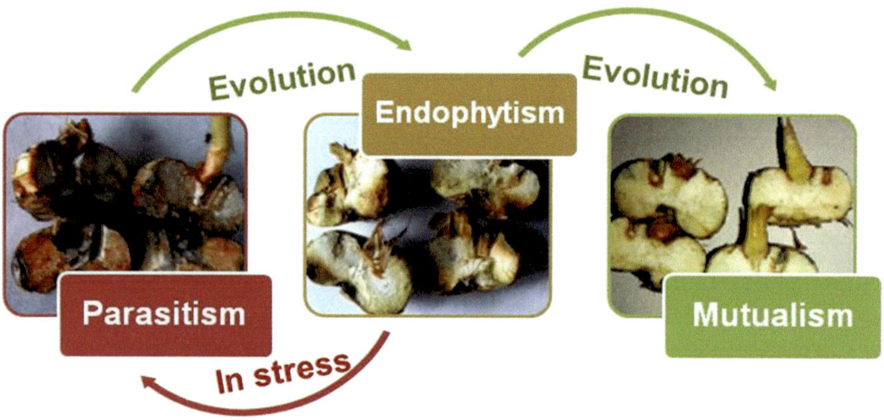

FIGURE 3.1 Evolutionary progression of the host–endophyte relationship.

The other unique aspect of endophytism is the multitrophic association between different endophytic groups and the host plants. Recent studies on "fungal–bacterial" associations, wherein the bacteria reside within the endophytic fungal hyphae, unfolded a novel chapter in microbial ecology. These associations are more common and important than previously thought. Moreover, many of these associations are central to agriculture, forestry, and bioremediation. The variegated crosstalks between, different endophytes within a plant, the endophyte and its endosymbiont, and the host plant under different biotic and abiotic selection pressure collectively shape the outcome of this multitrophic symbiosis. One of the most extensively studied multitrophic symbioses is the endohyphal association of *Burkholderia* with *Rhizopus* sp., central to this association is a characteristic phytotoxin rhizoxin produced by the endosymbiont rather than the phytopathogen causing rice seedling blight (Partida-Martinez and Hertweck, 2005). In the case of some arbuscular mycorrhizal fungi, endohyphal bacteria are reported to facilitate phosphate acquisition and transport in plants (Ruiz-Lozano and Bonfante, 1999). The thermotolerance ability in *Dichanthelium lanuginosum* provided by the endophytic fungal symbiont (*Curvularia protuberata*) is attributed to a double-stranded virus harbored by the fungal endophyte (Rodriguez and Roossinck, 2012). These findings illustrate that this tripartite association between plant–fungus–endohyphal symbiont is a complex interaction which broadens the scope of plant–endophyte association. Therefore, in order to understand the complexity of endophytism, future endophyte research should focus on multitrophic association models with cost–benefit analysis of communications between different interacting partners.

3.3 RECRUITMENT OF ENDOPHYTES BY HOST PLANTS

Plants growing in different geographical regions are confronted with different environmental challenges. The environmental cues in combinatorial effect with host genotype shape the composition of endophytic microbes harbored by the host plant (Wani et al., 2015). The diversity of endophytes associated with plants shows temporal as well as spatial variation (Vega et al., 2010). For instance, studies showed that endophytes may increase in incidence, diversity, and host breadth as a function of latitude. Furthermore, endophyte communities from higher latitudes are characterized by relatively few fungal species representing several classes of Ascomycota, while tropical endophyte assemblages are dominated by a small number of classes but a very large number of different endophytic species (Arnold and Lutzoni, 2007). The variations in the endophytic communities can also be attributed to plant age, plant source, tissue type, and time of sampling (Kobayashi and Palumbo, 2000). Vendan et al., (2010) showed that the age of the plant could largely influence the variation in the endophytic community of ginseng plants. Similarly, variation in endophytic diversity might be a function of the different maturation stages specific to each plant, which might influence the different types and amounts of root exudates (Ferreira et al., 2008).

Interestingly plants growing in similar environmental conditions do not harbor similar endophytes. It was observed that the endophytes isolated from cottonwood were altogether different from the endophytes of willow, even though both the tree species were growing at the same site within a meter of distance from each other (Doty et al., 2009). Similarly, the endophytic diversity of Saffron and *Glycyrrhiza* growing under similar environmental conditions in Kashmir Valley were different from each other (Wani et al., 2016; Arora et al., 2019). The Saffron microbiome was dominated by dark septate endophytes (DSEs) with an isolation frequency of more than 30%, particularly *Phialophora mustea* and *Cadophora malorum* being the most dominant endophytes. Molecular phylogeny assigned these DSEs into a single clad, indicating a strong effect of the host genotype on the selective recruitment of the DSEs (Wani et al., 2016). This differential recruitment of endophytes has been noted in other studies of endophytic populations from plants growing in the same location. The diversity of endophytes within four clones of poplar harbored four distinct endophytic populations; supporting the hypothesis that plant genotype plays an important role in selective recruitment of endophytes (Ulrich et al., 2008). In an important study involving metagenomic analysis of root-associated microbiome in rice plant cultivated under controlled condition as well as field condition. It is reported that the composition of the microbial consortia varies in different root-associated compartments, namely, endosphere (root interior), rhizoplane (root surface), and rhizosphere (soil close to the root surface). Under controlled condition, microbiome composition varies with soil source and plant genotype, whereas in field condition, geographical location and cultivation practices are responsible for the variation in microbiome composition. This differential recruitment of microbes across the rhizocompartments is a result of active selection of microbial consortia at different steps and each step involves various molecular signals (general plant metabolites, cell wall components or membrane proteins, small-molecule hormones particularly jasmonic acid, salicylic acid, and ethylene) released by the plant. These results suggest that the core microbiota are recruited from very diverse microbial surroundings, narrowing down both the most relevant community members and pointing to the host detriments controlling the mechanism of assembly of microbes (Lebeis, 2014; Edwards et al., 2014).

The aforementioned findings suggest that the endophytic community of a plant is determined by the combinatorial effect of the plant genotype and the environment, consistent with a co-evolutionary process whereby the endophytes may have evolved in a coordinated fashion with the host plant (Wani et al., 2015). There is the evidence for multiple horizontal transfers of genes between the symbionts as an important ecological event that conferred a selective advantage on the interacting partners (Saikkonen et al., 2004). However, the interactions between plant and symbiotic microbial genomes (i.e., intergenomic epistasis, or genotype (G)×genotype (G) interactions) can have an important effect on the rate and direction of co-evolutionary selection of interacting partners in endophytism (Wade, 2007). Thus, it is hypothesized that differential recruitment of endophytes in plants is a result of co-evolutionary selection process determined by intergenomic interactions of between the interacting partners with environmental conditions acting as a catalyst in this evolutionary selection process (Wani et al., 2015). However, the genetic principles governing the differential recruitment of endophytes by a specific host in a particular environment are poorly understood and need to be deliberated in future.

3.4 MECHANISM OF ACTION OF ENDOPHYTES IN THE HOST PLANT

Endophytism is a complex ecological and fascinating cost–benefit association, in which numerous fair-trade partners such as symbionts or mutualists engage in co-operative and social interactions with other micro-organisms such as antagonists and pathogens (West et al., 2006; Werner et al., 2014). Off late researchers have endeavored to elucidate the molecular dialogues involved in the establishment of plant–endophyte association and responses thereof, but very limited data is available as of now (Sherameti et al., 2008; Straub et al., 2013; Wani et al., 2017, 2018). One of the challenges is the complex relationship between the plant and the endophyte and the other is difficulty in imitating the living condition of endophytes in vitro as well as studying the mechanisms *in planta*. There are various factors that influence plant–endophyte interactions, but the host plant response to endophyte infection is mainly mediated by the host genotype, microbial genotype, resource availability, and environmental cues (Fig. 3.2) (Wani et al., 2015). It is reported that the endophytic response in plants is largely primed by the plant genotype, endophyte species, and the endophyte strain involved (Gundel et al., 2012; Qawasmeh et al., 2012). As reviewed by Wani et al., 2015, broadly there are two basic mechanisms, through which endophytes affect their responses in host plants:

3.4.1 BY PRODUCING DIVERSE CHEMICAL ENTITIES/MESSENGERS

Endophytes influence the plant growth and metabolism by producing an array of biocommunication signals/signaling molecules such as phytohormones, secondary metabolites, phytoanticipins, ROS, phytoalexins, volatile organic compounds (VOCs), toxicants, antibiotics, peptaibols, fatty acids, and siderophores. It is reported that endophytes induce the root growth in plants, and the most common mechanism that endophytes invoke to stimulate root growth is through secretion of phytohormones (particularly Indole acetic acid) within the plant (Khan et al., 2012; Waqas et al., 2014; Wani et al., 2017). Similarly, endophytes are reported to elicit the biosynthesis of plant metabolites by inducing phytohormone production in the host plant. For example, an oleaginous fungal endophyte *Mortierella alpina* significantly enhances the secondary metabolite content, particularly the specialized metabolites like crocin and safranal in *Crocus sativus*. It is suggested that the elevated levels of crocin and safranal in *C. sativus* might be due to some phytohormone, like indole acetic acid (IAA) and jasmonic acid (JA) (Wani et al., 2017). Similarly, abscisic acid is reported to be required for flavonoid and catharanthine accumulation, elicited by the fungal endophyte *Sphaeropsis* sp. in *Ginkgo biloba* (Hao et al., 2010). A fungal endophyte (*Piriformospora indica*) helps rice plant (*Oryza sativa*) to tolerate root herbivory through changes in gibberellins and jasmonate signaling mechanism (Cosme et al., 2016). There are various reports that indicate signaling crosstalk between salicylic acid and JA results in a complementary action in mediating endophyte-induced secondary metabolite accumulation in plants (Wang et al., 2015). Host–endophyte interactions are also known to generate ROS stimulating antioxidant production in their host plant, which in turn is responsible for protecting the plant from oxidative stress (Tanaka et al., 2006; White and Torres, 2010). In *Lolium perenne* the colonization of endophyte (*Neotyphodium lolii)* significantly influenced the phenolic content and antioxidant activity of the host plant. However, the effect varied depending on type of the endophytic strain (Qawasmeh et al., 2012). A fungal endophyte (*M. alpina)* significantly increased total carotenoids, phenolic, and flavonoid content in the endophyte-treated *Crocus* plants. It is reported that the phenols, flavonoids, and carotenoids help in preventing plants from oxidative stress (Baba et al., 2015). In addition to this, some endophytes manipulate the host plant metabolism by modifying the nutrient uptake and nutrient homeostasis (Sherameti et al., 2005; Singh et al., 2013). A basidiomycete *Porostereum* sp. CSE26 isolated from *C. sativus* produced a mixture of 15 VOCs, including two chlorinated anisyl metabolites (CAM), that is, 3-chloro-4-methoxybenzaldehyde and 2,3-dichlorophenyl isothiocyanate. The CAM molecules produced by the *Porostereum* sp. were found to have phytotoxic property against *Arabidopsis thaliana* (Wani et al., 2018).

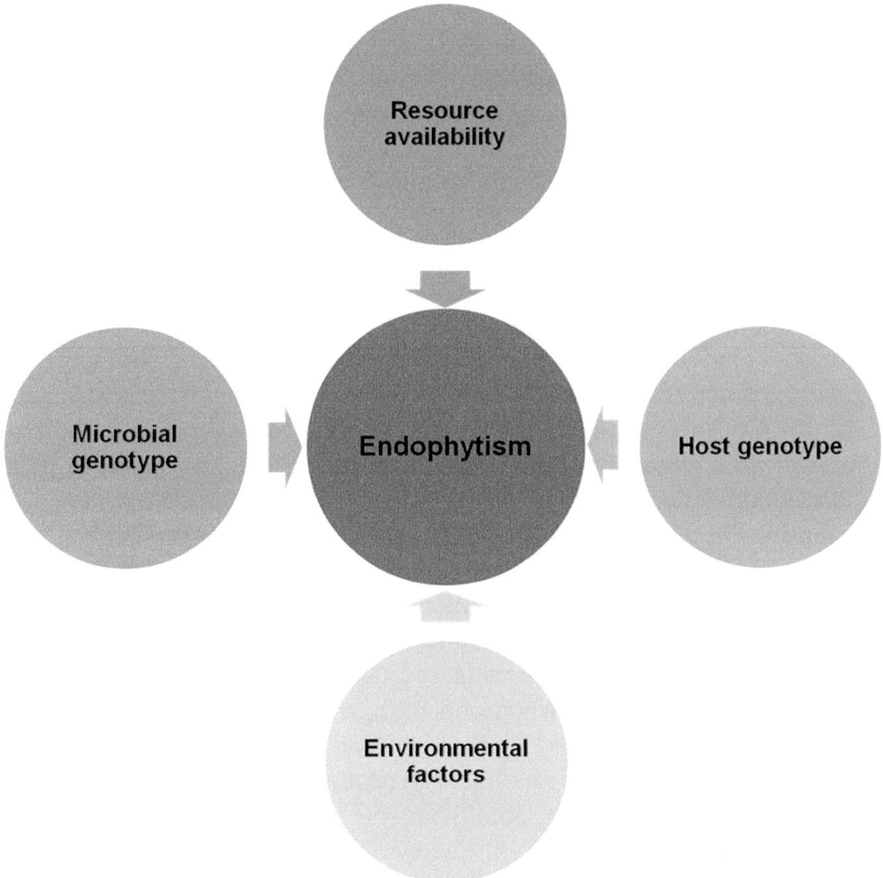

FIGURE 3.2 Endophytism is determined by the combinatorial effect of intergenomic interactions between the host and microbial symbiont genotype, in consistent with the environmental factors and resource availability.

3.4.2 BY ALTERING/INDUCING PLANTS' METABOLIC AND DEFENSE PATHWAYS

Plant can detect the signal molecules released by endophyte through chemo-perception systems. This triggers a cascade of signal transduction giving rise to a series of plant defense responses akin to plant–pathogen interaction, leading to a noticeable change in plant metabolic state (Qawasmeh et al., 2012; Wani et al., 2017). For example, fungal endophyte *M. alpina* shifts the metabolic flux toward enhanced production of apocarotenoids by modulating the expression of key pathway genes in *C. sativus*. The endophyte produces IAA and releases arachidonic acid (AA) which in turn induces the production of jasmonic acid in the host plant (Wani et al., 2017). Both IAA and JA are reported to have positive influence on the expression of apocarotenoid biosynthetic pathway genes in *Crocus* (Ashraf et al., 2015). Further, the endophyte enhanced tolerance to corm rot disease in the host plant by releasing arachidonic acid, which acts as conserved defense signal and activates JA-induced defense signaling in endophyte-treated *Crocus* plants (Wani et al., 2017). In another study, two fungal endophytes (*Curvularia* sp. and *Choanephora infundibulifera*) of *Catharanthus roseus* were found to enhance vindoline content in the host plant by modulating structural and regulatory genes related to terpenoid indole alkaloid biosynthesis (Pandey et al., 2016). An endophytic actinobacterium *Pseudonocardia* sp. isolated from *Artemisia annua* induced artemisinin production in *Artemisia* plant byupregulating the expression of cytochrome P450 monooxygenase and cytochrome P450 oxidoreductase genes involved in artemisinin biosynthesis (Li et al., 2012). Elicitors from the endophytic fungus *Trichoderma atroviride* stimulated the biosynthesis of tanshinone in the host plant by increasing the expression of genes related to tanshinone biosynthesis (Ming et al., 2013). Studies of plant gene expression in response to endophytic colonization reveal that the genes for carbon and nitrogen metabolism, plant growth, and plant defense are induced by the endophyte (Elvira-Recuenco and Van Vuurde, 2000). The endophytic rhizobacteria and actinobacteria are reported to induce disease resistance by stimulating the systemic defense pathways in the host plant (Conn et al., 2008). The systemic acquired resistance (SAR) pathway in *Arabidopsis* is normally activated by biotrophic pathogens either as a part of the hypersensitive response or as a symptom of disease. The jasmonic acid/ethylene (JA/ET) pathway is triggered by infection with necrotrophic pathogens (Durrant and Dong, 2004; Glazebrook, 2005). The activation of plant defense genes by endophytic actinobacteria in the absence of a pathogen reveals that the endophytes are detected as "minor" pathogens that do not trigger a full resistance response on their own; this may result in more effective priming of the defense response (Conn et al., 2008). Recently, it was reported that an endophytic bacterium *Bacillus thuringiensis* isolated from *Pteridium aquilinum* induced defense response against *Rhizoctonia solani* in cucumber plants. The possible mode of action is reported to be the induction of pathogenesis-related (PR) proteins and defense-related enzymes by the endophyte against pathogen in host plant (Seo et al., 2012). Despite the great efforts put in to understand the mechanistic aspect of endophytism, this discipline of science is still in its infancy. A complete comprehension of this complex ecological phenomenon can only be obtained through systems biology approach by the integration of the "omics" technologies, such as metagenomics, metabolomics, or transcriptomics together with ecogenomics (Wani et al., 2015).

3.5 ENDOPHYTES AN ECOLOGICAL BARGAIN TO PLANTS

Symbiotic plant–fungal interactions are of widespread interest to ecological research as they influence important ecosystem processes, including plant productivity, plant diversity, and plant–pathogen interactions (Wani et al., 2015), as exemplified by the association of endophytic systemic clavicipitaceous fungi with grasses exerting beneficial effects on host plants, through increased resistance to biotic and abiotic stresses, which are of great ecological significance (Kuldau and Bacon, 2008). There are various factors that influence the plant–endophyte interactions; however, the response of plant to endophyte infection is mainly mediated by the host genotype, endophytic strain, resource availability, and environmental factors (Fig. 3.2) (Wani et al., 2015). For example, the endophyte interactions in tall fescue develops low osmotic potential primarily in young meristematic and elongating leaves, which enable the plant to remain stable during drought stress (Elmi and West, 1995). Similarly, thermotolerance and salt tolerance are observed in certain plants colonized with endophyte (Redman et al., 2002; Waller et al., 2005). The higher colonization of DSEs in the corms of *C. sativus* indicates an ecological significance, as it is reported that the melanized hyphae, typical for DSEs, are considered important for the host to survive in stress conditions (Wani et al., 2016). The pigmentation in DSEs is due to the presence of cell wall melanin, which can trap and eliminate oxygen radicals generated during abiotic stress (Richier et al., 2005).

Fungal endophyte colonization significantly affects both the primary and secondary metabolisms of the host plant. There is a need for wider metabolic studies beyond alkaloid accumulation to understand the dynamic functional aspects of this association. It is reported that, a shift in C to N ratio and secondary metabolite production due to endophyte colonization is likely to have impacts on herbivore and pathogen responses to grasses infected with *Neotyphodium* sp. (Rasmussen et al., 2008). The endophytes may produce a range of different types of metabolites that not only play a role in defense and competition but are required for specific interaction and communication with the host plant. (Brader et al., 2014). Further, metagenomic studies in rice found endophytic root bacteria contain several groups of genes involved in motility, plant polymer degradation, iron acquisition (e.g., siderophores), quorum-sensing, and detoxification of ROS, indicating control over those pathways is important for colonization by root microbiome (Sessitsch et al., 2012). Also, the phenotype and functional traits of most plants in nature are the products of multitrophic interactions of plant with other organisms, mainly micro-organisms, sharing the same habitat and resulting in complex and transient metabolic flux across the interacting partners essential for their survival (Kusari et al., 2014). It is also reported that the positive effects of endophyte on plant performance depend on genetic variation in the host and/or endophyte, and on nutrient availability (Cheplick, 2007; Gundel et al., 2012). This link between resource availability and beneficial or neutral versus detrimental effects on plant performance suggests a metabolic cost of the endophyte to the host plant.

The other important aspect of ecological implication of endophyte is in phytoremediation process either directly through degradation and/or accumulation of environmental pollutants or indirectly by promoting the growth of plants having phytoremediation ability (Stepniewska and Kuzniar, 2013). For example, plants inoculated with genetically engineered endophytes were more tolerant to toluene, and they also minimized the transpiration of toluene to the atmosphere (Newman and Reynolds, 2005). Thus, fungal symbionts might be a drain (net cost) on plant metabolism or might upregulate metabolism, but endophyte hosting plants are reported to have increased tolerance to drought, heat, metal toxicity, low pH, and high salinity, thereby invoking an ecological significance to the plants (Wani et al., 2015). Such multiplexed interactions are clearly indispensable for the perfect functioning of ecosystem. Therefore, it can be hypothesized that the plant–endophyte interface is an important "ecological marketplace," in which numerous fair-trade partners such as symbionts or mutualists engage in co-operative and social interactions with other micro-organisms such as antagonists and pathogens (West et al., 2006; Werner et al., 2014).

3.6 ENDOPHYTES AS GATEWAY FOR SUSTAINABLE AGRICULTURE

The complex association of endophyte with the host plant is of great ecological significance owing to their compatibility, ease of reinfection, and pattern of colonization (Backman and Sikora, 2008; Sikora et al., 2010). Whenever we think of a microbial infection in plants, symptoms of diseases or detrimental effects come to our mind, but this is not true in the case of endophytes. However, research work in this aspect of plant–microbe interactions is in infancy and the molecular mechanism to understand this unique relationship is yet to be explored. Interests are often dictated by more immediate socioeconomic impulses because microbes are responsible for many plant diseases that cause substantial economic loss in agriculture. These harmful effects are often manifested directly through pathogen-mediated damage to the plant and a consequent reduction in plant vigor and yield or quality of crops. However, there is a diverse community of micro-organisms (endophytes) that interact positively with plants in agricultural system in relation to their nutrition and ability to resist biotic and abiotic stress and have the potential to be manipulated such that the benefits of their positive effects are harnessed (Wani et al., 2015).

Endophytes are especially interesting for integrative pest management as innovative biological control agents (BCAs) (Li et al., 2012). An important advantage of endophytes as BCAs over the conventional BCAs is that they can be applied directly to seeds or seedlings, thereby avoiding treatment to large quantities of soil or large numbers of already established plants. Recently, an *Enterobacter* sp. is reported as a potent biocontrol agent against *Verticillium dahliae* Kleb, which is the causative agent of verticillium wilt of cotton (Li et al., 2012). Few fungal endophytes are already being produced on large scale as commercial BCAs, for example *Trichoderma harzianum, Paecilomyces lilicinus, Beauveria bassiana*, and *Fusarium oxysporum* (Mendoza and Sikora, 2009; Sikora et al., 2010). However, so far single micro-organisms have been used as BCAs and the use of multiple organisms in a consortium imitating the complexity of associations within the plant system has just begun to be explored. Endophytes can be genetically engineered and these engineered endophytes have the potential to provide an alternative to plant transgenic technology by conferring plants a new pathway to benefit from foreign genes (Li et al., 2007). For example, an endophyte *Leifsonia xyli* subsp. *cynodontis,* a xylem-inhabiting bacterium, was genetically modified with a gene from *B. thuringiensis*. This gene is responsible for producing δ-endotoxin that is active against insects in nature, especially Lepidoptera and Coleoptera. When the genetically engineered endophyte is inoculated in the plant, it secretes the toxin inside the plant tissues protecting it against attacks from the target insects (Selim et al., 2012; Saikkonen et al., 2013).

World is witnessing an unprecedented ecological damage done by synthetic agrochemicals, endophytes continue to serve as selective and safe alternative. Many endophytes show antimicrobial activities, and they are known to impart resistance to the host against a range of microbial infections. Another benefit of these endophytes to the host is the production of growth hormones. Phytohormone production by endophytes is probably the best-studied mechanism of plant growth promotion, leading to morphological and architectural changes in the host plant, thus contributing to the overall growth and development of the plant. Recently, it was reported that an endophyte *M. alpina* showed a significant improvement in many morphological and physiological traits in endophyte-treated *Crocus* plants, including total biomass, size of corms, stigma biomass, number of apical sprouting buds, and number of adventitious roots. The endophyte also shifted metabolic flux toward enhanced production of apocarotenoids by modulating the expression of key pathway genes in the host plant. Further, *M. alpina* enhanced tolerance to corm rot disease by releasing arachidonic acid that acts as conserved defense signal and induces jasmonic acid production in endophyte-treated *Crocus* corms (Wani et al., 2017). Some endophytes are found to help the host plant in nitrogen acquisition, either by tapping atmospheric nitrogen directly (Sherameti et al., 2005) or by translocating the insect-derived nitrogen indirectly (Behie et al., 2012), thereby play a larger role in nitrogen cycling. One of the most potential functions of endophytes is the facilitation of nutrient uptake. Some endophytes are reported to mobilize phosphorous uptake in plants (Yadav et al., 2010), while others are found to impact the growth and development of the plant by producing phytohormones (Khan et al., 2012; Waqas et al., 2014). Many of the fungal endophytes have been found to produce antimycotic VOCs. VOCs produced by micro-organisms are regarded important infochemicals in the biosphere that influence the dynamics of the ecosystem (Wheatley, 2002). Microbial species produce consistent and reproducible VOC profiles under standard culture conditions. Several of these endophytes may find applications in agriculture, aroma industry, food processing, and as potential biofuel molecules (Strobel et al., 2008; Riyaz-Ul-Hassan et al., 2013). Endophytes like *Muscodor* spp. produce bioactive VOCs that inhibit or kill important plant pathogens and they can be used for mycofumigation, postharvest preservation of agricultural produce, and decontamination of animal waste (Strobel, 2006; Bitas et al., 2013). It seems reasonable that the VOC-producing micro-organisms may be preferentially establishing symbiotic associations with higher plants as they contribute to the host defense mechanism by inhibiting the plant pathogens.

3.7 CONCLUSION

Endophytes establish intimate association with the plant. Being present inside the plant tissues, they impact the development of the host plant significantly.

Plant–endophyte interaction is determined by the co-evolution of interacting partners together to impart essential benefits to each other. Recent studies on plant–endophyte symbiosis involving "-omics" technologies in a systems biology approach have started providing insights into different facets of plant–endophyte interaction, including the dynamics of multispecies symbiotic network involved. Also, the greater utilization of microbes of endophytic origin in agricultural system may significantly reduce the use of inorganic fertilizers, herbicides, and pesticides. Thus, endophyte technology holds the key for a potential gateway to sustainable agriculture development in future course of time.

KEYWORDS

- **endophytism**
- **mutualistic association**
- **phytoremediation**
- **plant adaptation**
- **stress tolerance**
- **dark septate endophytes**
- **differential recruitment**

REFERENCES

Arnold, A. E. Understanding the Diversity of Foliar Fungal Endophytes: Progress, Challenges, and Frontiers. *Fungal Biol. Rev.* **2007,** *21*, 51–66.

Arnold, A. E.; Lutzoni, F. Diversity and Host Range of Foliar Fungal Endophytes: Are Tropical Leaves Biodiversity Hotspots? *Ecology* **2007,** *88*, 541–549.

Arora, P.; Wani, Z. A.; Ahmad, T.; Sultan, P.; Gupta, S.; Riyaz-Ul-Hassan, S. Community Structure, Spatial Distribution, Diversity and Functional Characterization of Culturable Endophytic Fungi Associated with *Glycyrrhiza glabra* L. *Fungal Biol.* **2019.** https://doi.org/10.1016/j.funbio.2019.02.003.

Ashraf, N.; Jain, D.; Vishwakarma, R. A. Identification, Cloning and Characterization of an Ultrapetala Transcription Factor *CsULT1* from *Crocus*, A Novel Regulator of Apocarotenoid Biosynthesis. *BMC Plant Biol.* **2015,** *15*, 25.

Baba, S. A.; Malik, A. H.; Wani, Z. A.; Sumji, T.; Shah, Z.; Ashraf, N. Phytochemical Analysis and Antioxidant Activity of Different Tissue Types of *Crocus sativus* and Oxidative Stress Alleviating Potential of Saffron Extract in Plants, Bacteria, and Yeast. *S. Afr. J. Bot.* **2015,** *99*, 80–87.

Backman, P. A.; Sikora, R. A. Endophytes: An Emerging Tool for Biological Control. *Biol. Control.* **2008,** *46*, 1–3.

Baldrian, P.; Valášková, V. Degradation of Cellulose by Basidiomycetous Fungi. *FEMS Microbiol. Rev.* **2008,** *32*, 501–521.

Bao, X.; Roossinck, M. J. Multiplexed Interactions: Viruses of Endophytic Fungi. *Adv. Virus Res.* **2013,** *86*, 37–57.

Behie, S. W.; Zelisko, P. M.; Bidochka, M. J. Endophytic Insect-Parasitic Fungi Translocate Nitrogen Directly from Insects to Plants. *Science* **2012,** *336*, 1576–1577.

Bitas, V.; Kim, H. S.; Bennett, J. W.; Kang, S. Sniffing on Microbes: Diverse Roles of Microbial Volatile Organic Compounds in Plant Health. *Mol. Plant Microbe Interact.* **2013,** *26*, 835–843.

Botella, L.; Díez, J. J. Phylogenic Diversity of Fungal Endophytes in Spanish Stands of *Pinus halepensis. Fungal Divers* **2011,** *47*, 9–18.

Brader, G.; Compant, S.; Mitter, B.; Trognitz, F.; Sessitsch, A. Metabolic Potential of Endophytic Bacteria. *Curr. Opin. Biotechnol.* **2014,** *27*, 30–37.

Cheplick, G. P. Costs of Fungal Endophyte Infection in *Lolium perenne* Genotypes from Eurasia and North Africa Under Extreme Resource Limitation. *Environ. Exp. Bot.* **2007,** *60*, 202–210.

Chi, F.; Shen, S.; Cheng, H.; Jing, Y.; Yanni, Y. G.; Dazzo, F. B. Ascending Migration of Endophytic Rhizobia, from Roots to Leaves, Inside Rice Plants and Assessment of Benefits to Rice Growth Physiology. *Appl. Environ. Microbiol.* **2005,** *71*, 7271–7278.

Christensen, M. J.; Bennett, R. J.; Ansari, H. A.; Koga, H.; Johnson, R. D.; Bryan, G. T.; Simpson, J. P.; Koolaard, W. R.; Nickless, E. M.; Voisey, C. R. Epichloë Endophytes Grow by Intercalary Hyphal Extension in Elongating Grass Leaves. *Fungal Genet Biol.* **2008,** *45*, 84–93.

Compant, S.; Duffy, B.; Nowak, J.; Clement, C.; Barka, E. A. Use of Plant Growth-Promoting Bacteria for Biocontrol of Plant Diseases: Principles, Mechanisms of Action, and Future Prospects. *Appl. Environ. Microbiol.* **2005,** *71*, 4951–4959.

Conn, V. M.; Walker, A. R.; Franco, C. M. Endophytic Actinobacteria Induces Defense Pathways in *Arabidopsis thaliana. Mol. Plant Microbe Interact.* **2008,** *21*, 208–218.

Doty, S. L.; Oakley, B.; Xin, G.; Kang, J. W.; Singleton, G.; Khan, Z.; Vajzovic, A.; Staley, J. T. Diazotrophic Endophytes of Native Black Cottonwood and Willow. *Symbiosis* **2009,** *47*, 23–33.

Durrant, W. E.; Dong, X. Systemic Acquired Resistance. *Annu. Rev. Phytopathol.* **2004,** *42*, 185–209.

Edwards, J.; Johnson, C.; Santos-Medellín, C.; Lurie, E.; Podishetty, N. K.; Bhatnagar, S.; Eisen, J. A.; Sundaresan, V. Structure, Variation, and Assembly of the Root-Associated Microbiomes of Rice. *Proc. Natl. Acad. Sci. U.S.A.* **2014.** doi: 10.1073/pnas.1414592112

Ek-Ramos, M. J.; Zhou, W.; Valencia, C. U.; Antwl, J. B.; Kalns, L. L.; Morgan, G. D.; Kerns, D. L.; Sword, G. A. Spatial and Temporal Variation in Fungal Endophyte Communities Isolated from Cultivated Cotton (*Gossypium hirsutum*). *PLoS ONE* **2013,** *8*, e66049.

Elmi, A. A.; West, C. P. Endophyte Infection Effects on Stomatal Conductance, Osmotic Adjustment and Drought Recovery of Tall Fescue. *New Phytol.* **1995,** *131*, 61–67.

Elvira-Recuenco, M.; Van Vuurde, J. W. L. Natural Incidence of Endophytic Bacteria in Pea Cultivars Under Field Conditions. *Can. J. Microbiol.* **2000,** *46*, 1036–1041.

Fahey, J. W.; Dimock, M. B.; Tomasino, S. F.; Taylor, J. M.; Carlson, P. S. Genetically Engineered Endophytes as Biocontrol Agents: A Case Study in Industry. *Microb. Ecol. Leaves* **1991,** 401–411.

Ferreira, A.; Quecine, M. C.; Lacava, P. T.; Oda, S.; Azevedo, J. L.; Araújo, W. L. Diversity of Endophytic Bacteria from Eucalyptus Species Seeds and Colonization of Seedlings by *Pantoea agglomerans*. *FEMS Microbiol. Lett.* **2008,** *287*(1), 8–14.
Freeman, S.; Rodriguez, R. J. Genetic Conversion of a Fungal Pathogen to a Nonpathogenic, Endophytic Mutualist. *Science* **1993,** *260*, 75–78.
Friesen, M. L. Widespread Fitness Alignment in the Legume–Rhizobium Symbiosis. *New Phytol.* **2012,** *194*, 1096–1111.
Glazebrook, J. Contrasting Mechanisms of Defense against Biotrophic and Necrotrophic Pathogens. *Annu. Rev. Phytopathol.* **2005,** *43*, 205–227.
Gundel, P. E.; Martinez-Ghersa, M. A.; Omacini, M.; Cuyeu, R.; Pagano, E.; Rios, R.; Ghersa, C. M. Mutualism Effectiveness and Vertical Transmission of Symbiotic Fungal Endophytes in Response to Host Genetic Background. *Evol. Appl.* **2012,** *5*, 838–884.
Hallmann, J.; Quadt-Hallmann, A.; Mahaffee, W. F.; Kloepper, J. W. Bacterial Endophytes in Agricultural Crops. *Can. J. Microbiol.* **1997,** *43*, 895–914.
Hao, G. P.; Du, X. H.; Zhao, F. X.; Ji, H. W. Fungal Endophytes-Induced Abscisic Acid is Required for Flavonoid Accumulation in Suspension Cells of *Ginkgo biloba*. *Biotechnol. Lett.* **2010,** *32*: 305–314.
Hardoim, P. R.; Van Overbeek, L. S.; Berg, G.; Pirttilä, A. M.; Compant, S.; Campisano, A.; Döring, M.; Sessitsch, A. The Hidden World Within Plants: Ecological and Evolutionary Considerations for Defining Functioning of Microbial Endophytes. *Microbiol. Mol. Biol. Rev.* **2015,** *79*, 293–320.
Heil, M.; Bostock, R. M. Induced Systemic Resistance (ISR) Against Pathogens in the Context of Induced Plant Defences. *Ann. Bot.* **2002,** *89*, 503–512.
Herrera, J.; Khidir, H. H.; Eudy, D. M.; Porras-Alfaro, A.; Natvig, D. O.; Sinsabaugh, R. L. Shifting Fungal Endophyte Communities Colonize *Bouteloua gracilis*: Effect of Host Tissue and Geographical Distribution. *Mycologia* **2010,** *102*, 1012–1026.
Hilbert, M.; Voll, L. M.; Ding, Y.; Hofmann, J.; Sharma, M.; Zuccaro, A. Indole Derivative Production by the Root Endophyte *Piriformospora indica* is Not Required for Growth Promotion but for Biotrophic Colonization of Barley Roots. *New Phytol.* **2012,** *196*, 520–534.
Jefferson, R. The Hologenome. Agriculture, Environment and the Developing World: A Future of PCR. Cold Spring Harbor, NY: Cold Spring Harbor Laboratory Press, 1994.
Khan, A. L.; Hamayun, M.; Kang, S. M.; Kim, Y. H.; Jung, H. Y.; Lee, J. H.; Lee, I. J. Endophytic Fungal Association via Gibberellins and Indole Acetic Acid can Improve Plant Growth Under Abiotic Stress: An Example of *Paecilomyces formosus* LHL10. *BMC Microbiol.* **2012,** *12*, 3.
Kloepper, J. W.; McInroy, J. A.; Liu, K.; Hu, C. H. Symptoms of Fern Distortion Syndrome Resulting from Inoculation with Opportunistic Endophytic Fluorescent *Pseudomonas* spp. *PLoS One* **2013,** *8*, e58531.
Kobayashi, D. Y.; Palumbo, J. D. Bacterial Endophytes and Their Effects on Plants and Uses in Agriculture. *Microb. Endophyt.* **2000,** *19*, 199–233.
Kuldau, G.; Bacon, C. Clavicipitaceous Endophytes: Their Ability to Enhance Resistance of Grasses to Multiple Stresses. *Biol. Control.* **2008,** *46*, 57–71.
Kusari, S.; Hertweck, C.; Spiteller, M. Chemical Ecology of Endophytic Fungi: Origins of Secondary Metabolites. *Chem. Biol.* **2012,** *19*, 792–798.
Kusari, S.; Singh, S.; Jayabaskaran, C. Biotechnological Potential of Plant-Associated Endophytic Fungi: Hope Versus Hype. *Trends Biotechnol.* **2014,** *32*, 297–303.

Kusari, S.; Zu hlke, S.; Spiteller, M. An Endophytic Fungus from *Camptotheca acuminate* that Produces Camptothecin and Analogues. *J. Natl. Prod.* **2009,** *72*, 2–7.

Lahrmann, U.; Dinga, Y.; Banhara, A.; Rath, M.; Hajirezaeid, M. R.; Döhlemanna, S.; Wirénd, N. V.; Parniskeb, M.; Zuccaroa, A. Host-Related Metabolic Cues Affect Colonization Strategies of a Root Endophyte. *Proc. Natl. Acad. Sci. U.S.A.* **2013,** *110*, 13965–13970.

Lebeis, S. L. The Potential for Give and Take in Plant–Microbiome Relationships. *Front. Plant Sci.* **2014,** *5*, 287.

Li, C. H.; Shi, L.; Han, Q.; Hu, H. L.; Zhao, M. W.; Tang, C. M.; Li, S. P. Biocontrol of *verticillium* Wilt and Colonization of Cotton Plants by an Endophytic Bacterial Isolate. *J. Appl. Microbiol.* **2012,** *113*, 641–651.

Li, T. Y.; Zeng, H. L.; Ping, Y.; Lin, H.; Fan, X. L.; Guo, Z. G.; Zhang, C. F. Construction of a Stable Expression Vector for *Leifsoniaxyli* subsp. *cynodontis* and its Application in Studying the Effect of the Bacterium as an Endophytic Bacterium in Rice. *FEMS Microbiol. Lett.* **2007,** *267*, 176–183.

Márquez, L. M.; Redman, R. S.; Rodriguez, R. J.; Roossinck, M. J. A Virus in a Fungus in a Plant—Three Way Symbiosis Required for Thermal Tolerance. *Science* **2007,** *315*, 513–515.

Mendoza, A. R.; Sikora, R. A. Biological Control of *Radopholus similis* by Co Application of the Mutualistic Endophyte *Fusarium oxysporum* Strain 162, the Egg Pathogen *Paecilomyces lilacinuss* Train 251 and the Antagonistic Bacteria *Bacillus firmus*. *Bio. Control.* **2009,** *54*, 263–272.

Ming, Q.; Su, C.; Zheng, C.; Jia, M.; Zhang, Q.; Zhang, H.; Rahman, K.; Han, T.; Qin, L. Elicitors from the Endophytic Fungus *Trichoderma atroviride* Promote *Salvia miltiorrhiza* Hairy Root Growth and Tanshinone Biosynthesis. *J. Exp. Bot.* **2013,** *64*, 5687–5694.

Mitter, B.; Pfaffenbichler, N.; Sessitsch, A. Plant–Microbe Partnerships in 2020. *Microb. Biotechnol.* **2016.** DOI: 10.1111/1751-7915.12382.

Mostert, L.; Crous, P. W.; Petrini, O. Endophytic Fungi Associated With Shoots and Leaves of *Vitis vinifera*, with Specific Reference to the *Phomopsis viticola* Complex. *Sydowia* **2000,** *52*, 46–58.

Newman, L. A.; Reynolds, C. M. Bacteria and Phytoremediation: New Uses for Endophytic Bacteria in Plants. *Trends Biotechnol.* **2005,** *23*, 6–8.

Pandey, S. S.; et al. Fungal Endophytes of *Catharanthus roseus* Enhance Vindoline Content by Modulating Structural and Regulatory Genes Related to Terpenoid Indole Alkaloid Biosynthesis. *Sci. Rep.* **2016,** *6*, 26583.

Partida-Martinez, L. P.; Hertweck, C. Pathogenic Fungus Harbours Endosymbiotic Bacteria for Toxin Production. *Nature* **2005,** *437*, 884–888.

Partida-Martínez, L. P.; Heil, M. The Microbe-Free Plant: Fact or Artifact? *Front. Plant Sci.* **2011,** *2*, 100.

Petrini, O. Fungal Endophytes of Tree Leaves. In: *Microbial Ecology of Leaves*; Andrews, J. H., Hirano, S. S., Eds.; Springer-Verlag: New York, NY, 1991; pp 179–197.

Qawasmeh, A.; Objed, H. K.; Raman, A.; Wheatley, W. Influence of Fungal Endophyte Infection on Phenolic Content and Antioxidant Activity in Grasses: Interaction between *Lolium perenne* and Different Strains of *Neotyphodiumlolii*. *J. Agric. Food Chem.* **2012,** *60*, 3381–3388.

Rasmussen, S.; Parsons, A. J.; Popay, A.; Xue, H.; Newman, J. A. Plant-Endophyte-Herbivore Interactions: More than Just Alkaloids? *Plant Signal. Behav.* **2008,** *3*, 974–977.

Redecker, D.; Kodner, R.; Graham, L. E. Glomalean Fungi from the Ordovician. *Science* **2000,** *289*, 1920–1921.

Redman, R. S.; Sheehan, K. B.; Stout, R. G.; Rodriguez, R. J.; Henson, J. M. Thermotolerance Conferred to Plant Host and Fungal Endophyte During Mutualistic Symbiosis. *Science* **2002,** *298*, 1581.

Richier, S.; Furla, P.; Plantivaux, A.; Merle, P. L.; Allemand, D. Symbiosis-Induced Adaptation to Oxidative Stress. *J. Exp. Biol.* **2005,** *208*, 277–285.

Riyaz-Ul-Hassan, S.; Strobel, G.; Geary, B.; Sears, J. An Endophytic *Nodulisporium* sp. from Central America Producing Volatile Organic Compounds with Both Biological and Fuel Potential. *J. Microbiol. Biotechnol.* **2013,** *23*, 29–35.

Rodriguez, R. J.; Henson, J.; Van Volkenburgh, E.; Hoy, M.; Wright, L.; Beckwith, F.; Kim, Y.; Redman, R. S. Stress Tolerance in Plants via Habitat-Adapted Symbiosis. *ISME J.* **2008,** *2*, 404–416.

Rodriguez, R. J.; Roossinck, M. Viruses, Fungi and Plants: Cross-Kingdom Communication and Mutualism. *Biocomm. Fungi.* **2012.** DOI 10.1007/978-94-007-4264-2_14

Ruiz-Lozano, J. M.; Bonfante, P. Identification of a Putative P-Transporter Operon in the Genome of a *Burkholderia* Strain Living Inside the Arbuscular Mycorrhizal Fungus *Gigaspora margarita. J. Bacteriol.* **1999,** *181*(13), 4106–4109.

Saikkonen, K.; Wäli, P.; Helander, M.; Faeth, S. H. Evolution of Endophyte-Plant Symbioses. *Trends Plant. Sci.* **2004,** *9*, 275–280.

Schardl, C. L.; Leuchtmann, A.; Spiering, M. J. Symbioses of Grasses with Seedborne Fungal Endophytes. *Annu. Rev. Plant Biol.* **2004,** *55*, 315–340.

Schulz, B.; Boyle, C. The Endophytic Continuum. *Mycol. Res.* **2005,** *109*, 661–687.

Selim, K. A.; El-Beih, A. A.; AbdEl-Rahman, T. M.; El-Diwany, A. I. Biology of Endophytic Fungi. *Curr. Res. Environ. Appl. Mycol.* **2012,** *2*, 31–82.

Seo, D. J.; Nguyen, D. M.; Song, Y. S.; Jung, W. J. Induction of Defense Response against *Rhizoctonia solani* in Cucumber Plants by Endophytic Bacterium *Bacillus thuringiensis* GS1. *J. Microbiol. Biotechnol.* **2012,** *22*, 407–415.

Sessitsch, A.; Hardoim, P.; Doring, J.; Weilharter, A.; et al. Functional Characteristics of an Endophyte Community Colonizing Rice Roots as Revealed by Metagenomic Analysis. *Mol. Plant Microbe Interact.* **2012,** *25*, 28–36.

Sherameti, I.; Shahollari, B.; Venus, Y.; Altschmied, L.; Varma, A.; Oelmuller, R. The Endophytic Fungus *Piriformospora indica* Stimulates the Expression of Nitrate Reductase and The Starch-Degrading Enzyme Glucan-Water Dikinase in Tobacco and *Arabidopsis* roots through a Homeodomain Transcription Factor Which Binds to a Conserved Motif in Their Promoters. *J. Biol. Chem.* **2005,** *280*, 26241–26247.

Sherameti, I.; Tripathi, S.; Varma, A.; Oelmüller, R. The Root-Colonizing Endophyte *Pirifomospora indica* Confers Drought Tolerance in *Arabidopsis* by Stimulating the Expression of Drought Stress–Related Genes in Leaves. *Mol. Plant Microbe Interact.* **2008,** *21*, 799–800.

Shoop, W. L.; Gregory, L. M.; Zakson-Aiken, M.; Michael, B. F.; Haines, H. W.; Ondeyka, J. G.; Meinke, P. T.; Schmatz, D. M. Systemic Efficacy of Nodulisporic Acid Against Fleas on Dogs. *J. Parasitol.* **2001,** *87*, 419–423.

Sikora, R. A.; zumFelde, A.; Mendoza, A.; Menjivar, R.; Pocasangre, L. *In planta* Suppressiveness to Nematodes and Long-Term Root Health Stability Through Biological Enhancement–Do We Need a Cocktail? *Acta Hort.* **2010,** *879*, 553–560.

Singh, R. K.; Malik, N.; Singh, S. Improved Nutrient Use Efficiency Increases Plant Growth of Rice with the Use of IAA-Overproducing Strains of Endophytic *Burkholderia cepacia* Strain RRE25. *Microb. Ecol.* **2013,** *66*, 375–384.

Stepniewska, Z.; Kuzniar, A. Endophytic Microorganisms—Promising Applications in Bioremediation of Greenhouse Gases. *Appl. Microbiol. Biotechnol.* **2013,** *97*, 9589–9596.

Straub, D.; Rothballer, M.; Hartmann, A.; Ludewig, U. The Genome of the Endophytic Bacterium *H. frisingense* GSF30T Identifies Diverse Strategies in the *Herbaspirillum* Genus to Interact With Plants. *Front Microbiol.* **2013,** *4*, 168.

Strobel, G. *Muscodor albus* and Its Biological Promise. *J. Ind. Microbiol. Biotechnol.* **2006,** *33*, 514–522.

Strobel, G. A.; Knighton, B.; Kluck, K.; Ren, Y.; Livinghouse, T.; Griffen, M.; Daniel Spakowicz, D.; Sears, J. The Production of Myco-Diesel Hydrocarbons and Their Derivatives by the Endophytic Fungus *Gliocladium roseum* (NRRL 50072). *Microbiol.* **2008,** *154*, 3319–3328.

Sturz, A. V.; Christie, B. R.; Nowak, J. Bacterial Endophytes: Potential Role in Developing Sustainable Systems of Crop Production. *Crit. Rev. Plant Sci.* **2000,** *19*, 1–30.

Tanaka, A.; Christensen, M. J.; Takemoto, D.; Park, P.; Scott, B. Reactive Oxygen Species Play a Role in Regulating a Fungus-Perennial Ryegrass Mutualistic Interaction. *Plant Cell* **2006,** *18*, 1052–1066.

Ulrich, K.; Ulrich, A.; Ewald, D. Diversity of Endophytic Bacterial Communities in Poplar Grown Under Field Conditions. *FEMS Microbiol. Ecol.* **2008,** *63*, 169–180.

Vega, F. E.; Simpkins, A.; Aime, M. C.; Posada, F.; Peterson, S. W.; Rehner, S. A.; Infante, F.; Castillo, A.; Arnold, A. E. Fungal Endophyte Diversity in Coffee Plants from Colombia, Hawai'i, Mexico and Puerto Rico. *Fungal Ecol.* **2010,** *3*, 122–138.

Vendan, R. T.; Yu, Y. J.; Lee, S. H.; Rhee, Y. H. Diversity of Endophytic Bacteria in Ginseng and Their Potential for Plant Growth Promotion. *J. Microbiol.* **2010,** *48*, 559–565.

Wade, M. J. The Co-Evolutionary Genetics of Ecological Interactions. *Nat. Rev. Genet.* **2007,** *8*, 185–195.

Waller, F.; Achatz, B.; Baltruschat, H.; Fodor, J.; Becker, K.; Fischer, M.; Heier, T.; Huckelhoven, R.; Neumann, C.; von Wettstein, D.; Franken, P.; Kogel, K. H. The Endophytic Fungus *Piriformospora indica* Reprograms Barley to Salt-Stress Tolerance, Disease Resistance, and Higher Yield. *Proc. Natl. Acad. Sci. U.S.A.* **2005,** *102*, 13386–13391.

Wang, X. M.; Yang, B.; Ren, C. G.; Wang, H. W.; Wang, J. Y.; Dai, C. C. Involvement of Abscisic Acid and Salicylic Acid in Signal Cascade Regulating Bacterial Endophyte-Induced Volatile Oil Biosynthesis in Plantlets of *Atractylodes lancea. Physiol. Plant.* **2015,** *153*, 30–42.

Wani, Z. A.; Ashraf, N.; Mohi ud din, T.; Riyaz-Ul-Hassan, S. Plant-Endophyte Symbiosis, An Ecological Perspective. *Appl. Microbiol. Biotechnol.* **2015,** *99*, 2955–2965.

Wani, Z. A.; Mirza, D. N.; Arora, P.; Riyaz Ul Hassan, S. Molecular Phylogeny, Diversity, Community Structure, and Plant Growth Promoting Properties of Fungal Endophytes Associated with the Corms of Saffron Plant: An Insight Into the Microbiome of *Crocus sativus* Linn. *Fungal Biol.* **2016,** *120*, 1509–1524.

Wani, Z. A.; Kumar, A.; Sultan, P.; Bindu, K.; Riyaz-Ul-Hassan, S.; Ashraf, N. *Mortierella alpina* (CS10E4), An Oleaginous Fungal Endophyte of *Crocus sativus* L. Enhances Apocarotenoid Biosynthesis and Stress Tolerance in Host Plant. *Sci. Rep.* **2017,** *7*(1), 8598.

Wani, Z. A.; Ahmad, T.; Nalli, Y.; Ali, A.; Singh, A. P.; Vishwakarma, R. A.; Ashraf, N.; Riyaz-Ul-Hassan, S. *Porostereum* sp., Associated with Saffron (*Crocus sativus* L.), is a Latent Pathogen Capable of Producing Phytotoxic Chlorinated Aromatic Compounds. *Curr. Microbiol.* **2018.** doi: 10.1007/s00284-018-1461-9.

Waqas, M.; Khan, A. L.; Lee, I. J. Bioactive Chemical Constituents Produced by Endophytes and Effects on Rice Plant Growth. *J. Plant Interact.* **2014,** *9*, 478–487.

West, S. A.; et al. Social Evolution Theory for Microorganisms. *Nat. Rev. Microbiol.* **2006,** *4*, 597–607.

Werner, G. D. A.; et al. Evolution of Microbial Markets. *Proc. Natl. Acad. Sci. U.S.A.* **2014,** *111*, 1237–1244.

Wheatley, R. E. The Consequences of Volatile Organic Compound Mediated Bacterial and Fungal Interactions. *Ant. Leeuwen.* **2002,** *81,* 357–364.

White, J. F. Jr; Torres, M. S. Is Plant Endophyte-Mediated Defensive Mutualism the Result of Oxidative Stress Protection? *Physiol. Plant.* **2010,** *138*, 440–446.

Yadav, V.; Kumar, M.; Deep, D. K.; Kumar, H.; Sharma, R.; Tripathi, T.; Tuteja, N.; Saxena, A. K.; Johri, A. K. A Phosphate Transporter from the Root Endophytic Fungus *Piriformosporaindica* Plays a Role in Phosphate Transport to the Host Plant. *J. Biol. Chem.* **2010,** *285*, 26532–26544.

CHAPTER 4

Exploring the Endophytic Fungi of Himalayan *Pinus* sp.

ZAHOOR AHMED WANI[1*] and MASROOR QADRI[2]

[1]*Department of Botany, Govt. Degree College Kishtwar, Kishtwar, Jammu & Kashmir, India*

[2]*Entomology and Nematology Department, Genetics Institute, University of Florida, Gainesville, FL, United States*

[*]*Corresponding author. E-mail: zawani1986@gmail.com*

ABSTRACT

Himalayan forest cover is predominantly constituted of *Pinus* species belonging to Pinaceae family. *Pinus* has a great ethnobotanical importance and with its seeds, bark, and roots used as traditional medicine in various parts of northern India. Coniferous trees are suggested to be the reservoirs of enormous microbial wealth which is yet to be characterized and explored for human welfare. There are various studies on exploring the diversity and interaction of microbes (endophytes) associated with the above- and belowground part of coniferous plants. Himalayan blue pine faces several constraints in its successful regeneration in the field due to consistent pathogen attacks at primary stages of plant establishment posing a serious threat to forest nurseries. The ecological and evolutionary dynamics of natural and agro-ecosystems that shape the microbial communities in these ecosystems are very complex. Therefore, a concerted effort required to understand the microbial flux in natural and agro-ecosystems, the association of the endophytes with the tree species growing in western Himalayas, and their role in establishment of the plants under varied ecological settings.

4.1 INTRODUCTION

India is one of the twelve mega biodiversity centers, with its forest cover predominantly covered by coniferous trees (Aggarwal, 2014). *Pinaceae* is the largest family among the conifers and Pinus is the most abundant genera of this family, with over 110 species worldwide (Richardson et al., 2007). However, only five species (*Pinus roxburghii, Pinus wallichiana, Pinus kesiya, Pinus gerardiana, and Pinus merkusii)* are native to India (Gamble, 1902). Out of these five species, *P. roxburghii*, *P. wallichiana*, and *P. gerardiana* are dominant in Himalayan ranges, whereas *P. kesiya* and *P. merkusii* are found in Assam and Indo-Burma region (Sharma et al., 2018).

P. roxburghii is a large tree with crown spreading several meters, and it is commonly known as chir pine (Fig. 4.1). *P. roxburghii* is predominantly found in northwestern Himalayas at a height of 500–2500 m above mean sea level (Troup, 1921). It is known as an important resin and timber yielding tree. *P. gerardiana* commonly known as chilgoza pine is native to the northwestern Himalayas. It grows at altitudes ranging from 2000 to 3350 m above mean sea level (Troup, 1921). The branches are slightly ascending with needle-like leaves arranged in clusters of three. Male cones are long while as female cones are oblong ovoid with thick woody scales bearing cylindrical seeds with rudimentary wings (Fig. 4.1) (Farjon, 1984). *P. gerardiana* is well known for its edible seeds that are locally known as Chilgoza or Neja or Neje.

P. wallichiana is a coniferous evergreen tree native to northwestern Himalayan range and is found at altitudes ranging from 1500 to 3400 m above mean sea level (Troup, 1921). *P. wallichiana* is commonly known as blue pine and locally known as Kail. In Jammu and Kashmir, the area under forest cover is 20,230 km^2, with conifers alone covering approximately 40.87% of the forest cover, of which 9.73% is occupied by *P. wallichiana* (Anonymous, 2008). *P. wallichiana* is monoecious plant and produces a large number of winged pollen grains that are dispersed by wind. This tree grows to a height of 50 m with a straight trunk and short, down curved branches (Fig. 4.1). *P. wallichiana* is a large tree with regularly spaced down curved branches in whorls forming a pyramidal architecture and the leaves are needle-like arranged in clusters of five (Fig. 4.1). The bark on young trees is smooth, developing fissures with age. The male cones are present on lower branches and are often in dense clusters on younger twigs. The female cones are in groups, erect when young but later turn pendant. *P. wallichiana* usually grows under xerophytic conditions and prefers dry or moist soil, acidic or neutral soil and can tolerate drought. The plant can tolerate strong winds but not maritime exposure.

FIGURE 4.1 Morphology of Himalayan *Pinus* species. (A) *P. wallichiana* (blue pine), (B) *P. gerardiana* (chilgoza pine), and (C) *P. roxburghii* (chir pine).

4.2 PHYTOCHEMISTRY AND PHARMACOLOGY OF HIMALAYAN *PINUS* SPECIES

Pinus species has a great ethnobotanical importance and is widely used in various parts of northern India. *Pinus species* are known to be rich source of alkaloids and phenolics, including terpenoids, flavonoids, lycopenes, tannins, and xanthones (Beri 1970). The secondary metabolites are known for their medicinal properties ranging from anticancer, anti-inflammatory, antioxidant, and antimicrobial property. The wood oil obtained from *P. roxburghii* is aromatic and carminative in nature. It is used as nerve tonic, expectorant, and a remedy for dermatological problems and worm infestations. The wood oil and bark paste is applied to burns, scalds, and ulcers. The volatile component of pine is known as turpentine oil or oleoresin, which is obtained by tapping the pine tree. Oleoresin upon distillation process forms rosin. Rosin is also known as colophony, which is used in paper sizing, adhesives, enamels, solvents, plasticizers, paints and varnishes, antiseptic, and other commercial products (Hussain et al., 2010; Sharma et al., 2018). Turpentine oil is included in the Indian Pharmaceutical Codex as *Oleum terebinthinae* (Bajracharya, 1979). Boiled resin, locally known as *khaida* or *leesa*, is used to heal foot cracks. The carbon collected from the burnt resinous wood (*doi*) of *P. roxburghii* mixed with mustard oil is made into a paste, locally known as Kajal, which is applied on eyelids to keep the eyes clean and attractive (Singh et al., 1990).

P. gerardiana is well known for its edible seeds known as Chilgoza. The nuts are considered to be rich source of proteins, carbohydrates, fibers, minerals besides higher oil content. The oil from nuts is of very good quality and is free of cholesterol. They are rich sources of fatty acids like stearic acid, linoleic acid, linolenic acid, oleic acid, arachidic acid, and palmitic acid (Thakur et al., 2015).

P. wallichiana is widely used for timber, which is valued only next to the wood of *Cedrus deodara*. It is one of the most dominant tree species in the forest ecosystem of the Kashmir Himalayas and has immense medicinal, commercial, and ecological significance (Dar et al., 2012; Sharma et al., 2018). *P. wallichiana* is exploited mainly as a timber source; however, it is good source of oleoresin as well. The oleoresin is used for the production of turpentine oil, rosin, needle oil, and camphor. The essential oil of this plant comprises β-pinene, α-pinene, myrcene, camphene, limonene, α-phellandrene, trans-caryophyllene, α-humulene, α-cadinol, and α-bisabolol. In addition, it also contains undecane, dodecane, tridecane, abietic acid, and isopimaric acid. The essential oil is reported to possess potential anticancer and antioxidant properties (Coppen et al., 1988; Dar et al., 2012). Further, a dark-brown resinous substance locally known as killum/kellum is extracted from this species. Killum is traditionally used by the farmers of Kashmir to protect their skin from insect bites, skin-cracks, and infections while working long hours in waterlogged rice fields.

4.3 PLANT–MICROBE INTERACTION IN *PINUS*

Plants engage with a diverse array of microorganisms present in various tissues of plants, including phyllosphere, rhizosphere, and endosphere. These multitrophic interactions between plant and its microbiome are highly complex and dynamic, having important implications on plant community structure and functioning of ecosystem (Wani et al., 2015; Rua et al., 2016). Conifers, including different species of *Pinus,* harbor diverse group of microorganisms as endophytes. The endophytic community varies with species as well as geographical distribution of plants. Coniferous trees are suggested to be the reservoirs of enormous microbial wealth which is yet to be characterized and explored for human welfare (Hoffman and Arnold, 2008). Various studies have focused on exploring the diversity and interaction of microbes associated with the above- and belowground part of coniferous plants. Conifers interact with a diverse array of microorganisms, including mutualistic mycorrhizal fungi (Smith and Read, 2008), pathogenic microbes (Fogel, 1988), and foliar endophytes (Qadri et al., 2013, 2014; Oono et al., 2014; Carrell and Frank, 2014, 2015; Rua et al., 2016). The relationship between conifers and their microbial partners has a great ecological significance as conifers grow in impoverished acidic soils at high altitudes, and it is potentially facilitated by their interaction with various microbes (Richardson, 2000; Carrell and Frank, 2014).

The belowground mutualistic relationship between coniferous roots and mycorrhizal fungi is an important component of conifer's ecological niche. The two interacting partners are believed to have co-evolved and diversified approximately 200 million years ago forming a functionally obligate mutualistic association (Smith and Read, 2008; Tedersoo et al., 2010). This mutualistic relationship is characterized by the exchange of minerals between the interacting partners, such as carbon (C) in the form of carbohydrates from the plants with nitrogen (N), phosphorus (P), and micronutrients from the microbial partner (Smith and Read, 2008). Coniferous plants are reported to be primarily associated with ectomycorrhizal (ECM) fungi. ECM fungi are an extremely diverse group that contains more than 5000 species from the fungal order *Agaricales* alone (Ryberg and Matheny, 2012). ECM fungi are of immense ecological importance for conifers as they are reported to improve the plant growth and confer resistance against biotic and abiotic stress conditions in conifers (Smith and Read, 2008).

The microbial association in plants varies with plant age, tissue type, and plant health. In conifers, the microbial associations vary between phyllosphere and rhizosphere. The resident microbes in phyllosphere are subjected to varied temperature, moisture content, and radiation throughout the day and night. These factors indirectly affect the phyllosphere microbiome, which varies through changes in plant metabolism (Turner et al., 2013). In phyllosphere, the needles and bark of conifers are heavily colonized by fungal and bacterial endophytes (Pirttilä and Wäli, 2009; Rua et al., 2016). The microbial colonization varies with needle age and position of buds, such that needles and buds tend to differ in their endophytic microbial communities. Bowman and Arnold while investigating the association of ECM and foliar endophytic (FE) fungi with *Pinus ponderosa* reported that the abundance and diversity of ECM fungi were similar across sites, unlike FE fungi where the diversity peaked in mid-to-high elevation. However, the community composition and distribution of the most common ECM fungi differed with elevation, while as there was no such variation in the distribution of FE fungi (Bowman and Arnold, 2018). In another study, Ponpandian and colleagues isolated 1622 culturable bacterial endophytes from four *Pinus* species (*Pinus densiflora*, *Pinus koraiensis*, *Pinus rigida*, and *Pinus thunbergii*) in

Korea. Molecular phylogeny based on the acquisition of 16S ribosomal gene sequence grouped the bacterial endophytes into 215 operational taxonomic units (OTUs) encompassing 68 different genera (Ponpandian et al., 2019). In conifers, fungal endophytes are restricted to discrete portion of tissues in the needles where they remain in a slow growing (latent) state (Suske and Acker, 1986; Deckert et al., 2001). It is assumed that the growth and proliferation of these latent endophytes are triggered by injury or natural senescence of the needle. The transmission of fungal endophytes in *Pinus* is hypothesized to occur horizontally, because there is lack of information on endophytes in seeds and young leaves in *Pinus* (Ganley and Newcombe, 2006).

Himalayan blue pine faces several constraints in its successful regeneration in the field. The plants are often exposed to consistent pathogen attacks, causing root rot and wilt diseases, at primary stages of plant establishment posing a serious threat to forest nurseries (Pinto et al., 2006; Lilja et al., 2010). Root rot disease is a serious problem in pine seedlings worldwide and serious losses due to this disease are reported from Canada, United States, and many European countries (Enabak et al., 1990; Greifenghagen et al., 1991; Pinto et al., 2006). In *Pinus,* root rot disease is caused by numerous fungal pathogens that include species of *Fusarium, Rhizoctonia, Pythium, Macrophomina*, and *Cylindrocladium* (Huang and Kuhlman, 1990; Ahanger et al., 2011; Dar et al., 2011). These fungi invade the root epidermal cells, grow intercellularly by decomposing cell wall constituents and persist by metabolizing cell contents. Dar and colleagues reported that some fungal agents, like *Trichoderma harzianum, Trichoderma viride, Pisolithus tinctorius*, and *Laccaria laccata,* significantly inhibited the growth of fungal pathogens causing root rot. These fungal agents can effectively mitigate root rot disease in blue pine and can be efficiently exploited as potential biocontrol agents in integrated disease management module (Dar et al., 2011). Pine wilt disease (PWD) is another most destructive disease of *Pinus* trees causing immense environmental damage and economic loss around the world worth tens of million dollars (Tóth, 2011). The only known causal agent of the disease is pinewood nematode (PWN) *Bursaphelenchus xylophilus* (Nickle et al. 1981). It is reported that various bacterial endophytes having nematicidal activity against the pinewood nematode have been isolated from different *Pinus* species and they can be used as potential biocontrol agents against pinewood nematode (Liu et al., 2019; Ponpandian et al., 2019). These reports suggest potential use of bacterial endophytes from pine trees as alternative biocontrol agents against pinewood nematode. The use of biological organisms (particularly endophytes) is gaining momentum as innovative biological control agents (BCAs) in integrated disease management module (Wani et al., 2015). World is currently witnessing unprecedented ecological change caused by synthetic agrochemical inputs into ecosystems; microbial activities may provide selective and safe alternatives to agrochemical inputs for integrated pest and disease management in plants (Wani et al., 2015).

4.4 ENDOPHYTES ASSOCIATED WITH HIMALAYAN PINES

As described in the previous section, the coniferous tree, including species of *Pinus*, harbors huge diversity of endophytic microorganisms. Bhardwaj and colleagues while investigating the diversity of fungal endophytes associated with *P. roxburghii* from Garhwal forests in India isolated 17 endophytic fungi from the spikes of *P. roxburghii.* The dominant fungal endophytes of *P. roxburghii* included *Penicillium frequentaus, Thielaviopsis basicola, Geotrichium albida,* and *Alternaria alternata,* with an isolation frequency 41.1%, 29.40%, 11.76%, and 5.88%, respectively (Bhardwaj et al., 2014). The diversity and bioactivity of fungal endophytes associated with the Himalayan blue pine (*P. wallichiana*), particularly from the Kashmir Valley, was studied by Qadri et al. (2014). Qadri and colleagues isolated 130 endophytic fungi from the stem and needles of *P. wallichiana.* Molecular phylogenetic studies based on ITS1-5.8S-ITS2 ribosomal gene sequence analyses assigned these endophytes into 52 ITS genotypes, spreading over 39 different genera (Table 4.1). Most of the ITS genotypes showed more than 99% sequence similarity with the known fungal taxa; however, 15 fungal endophytes showed less than 99% sequence similarity with the known fungal taxa (Table 4.1). The 15 fungal endophytes with less than 99% sequence similarity may represent novel fungal lineages, which needs to be studied further. Most of the fungi isolated from *P. wallichiana* correspond with those recorded from other *Pinus* species. However, the fungal species *Anthostomella conorum*, *Microdiplodia* spp., *Therrya fuckelii*, *Thielavia subthermophila, Tricharina hiemalis*, and *Tritirachium oryzae* were isolated for the first time from *Pinus*. The dominant endophytes in Himalayan blue pine were *Alternaria* sp., *Pestalotiopsis* sp., *Preussia* sp., and *Sclerostagonospora* sp. (Qadri et al., 2014).

The diversity of endophytic microbes varies with age and tissue type. In *P. wallichiana* the needles harbor higher and more diverse endophytic fungi as compared to stem, with a colonization frequency of 66.9% and 53.7%, respectively. Most of the isolated fungi belong to *Ascomycota*; however only one isolate, that is, *T. oryzae*, belongs to *Basidiomycota*. Most of the endophytes belong to the fungal class Dothideomycetes followed

TABLE 4.1 The Table Presents the 52 Different Fungal Endophytes (Their GenBank Accession Numbers and Sequence Similarity) Isolated from *Pinsu wallichiana.*

ITS genotype	Molecular identification (GenBank accession no.)	Sequence similarity (%)
ITS1	*Alternaria brassicae* (KF735044)	99
ITS2	*Sclerostagonospora* sp.(KF735042)	99
ITS3	*Anthostomella sp.* (KF734993)	94
ITS4	*Tricharina sp.* (KF734996)	96
ITS5	*Paraconiothyrium brasiliense* (KF735025)	99
ITS6	*Coniothyrium sp.* (KF734989)	93
ITS7	*Epicoccum nigrum* (KF735002)	99
ITS8	*Cladosporium sp.* (KF735000)	95
ITS9	*Rachicladosporium sp.* (KF734990)	96
ITS10	*Leptosphaeria* sp. (KF734991)	91
ITS11	*Microdiplodia sp.* (KF734984)	98
ITS12	*Aspergillus sp.* (KF963269)	97
ITS13	*Thielavia subthermophila* (KF734997)	99
ITS14	*Sordaria humana* (KF735011)	99
ITS15	*Lecythophora* sp. (KF735049)	96
ITS16	*Rosellinia* sp. (KF735050)	96
ITS17	*Coniochaeta sp.* (KF735054)	97
ITS18	*Phomopsis* sp. (KF963270)	99
ITS19	*Trichoderma harzianum* (KF734994)	99
ITS20	*Pezizomycetes* sp. (KF735056)	99
ITS21	*Neurospora dictyophora* (KF735041)	99
ITS22	*Pestalotiopsis besseyi* (KF734985)	100
ITS23	*Phoma herbarum* (KF734988)	99
ITS24	*Lophiostoma corticola* (KF734987)	99
ITS25	*Nigrospora* sp. (KF735047)	99
ITS26	*Phoma aliena* (KF735035)	99
ITS27	*Truncatella betulae* (KF735004)	99
ITS28	*Therrya sp.* (KF735005)	93
ITS29	*Lophodermium macci* (KF735007)	99
ITS30	*Pestalotiopsis citrine* (KF735037)	99
ITS19	*T. harzianum* (KF734994)	99
ITS20	*Pezizomycetes* sp. (KF735056)	99

TABLE 4.1 *(Continued)*

ITS genotype	Molecular identification (GenBank accession no.)	Sequence similarity (%)
ITS21	*N. dictyophora* (KF735041)	99
ITS22	*P. besseyi* (KF734985)	100
ITS23	*P. herbarum* (KF734988)	99
ITS24	*L. corticola* (KF734987)	99
ITS25	*Nigrospora* sp. (KF735047)	99
ITS26	*P. aliena* (KF735035)	99
ITS27	*T. betulae* (KF735004)	99
ITS28	*Therrya sp.* (KF735005)	93
ITS29	*L. macci* (KF735007)	99
ITS30	*Pestalotiopsis citrine* (KF735037)	99
ITS31	*Fusarium larvarum* (KF735019)	100
ITS32	*P. aliena* (KF735014)	99
ITS33	*Truncatella spadicea* (KF735017)	99
ITS34	*Xylaria* sp. (KF735016)	99
ITS35	*Sporormiella* sp. (KF735029)	99
ITS 36	*Cadophora sp.* (KF935231)	66
ITS 37	*Pezizomycetes* sp. (KF735056)	99
ITS 38	*Lophodermium pini* (KF735039)	99
ITS 39	*Cochliobolus australiensis* (KF735030)	99
ITS 40	*Alternaria* sp. (KF735026)	100
ITS 41	*Coniochaeta* sp. (KF735009)	99
ITS 42	*Preussia* sp. (KF735031)	99
ITS 43	*Preussia intermedia* (KF735013)	100
ITS 44	*Phoma* sp. (KF735028)	99
ITS 45	*Fimetariella rabenhorstii* (KF735023)	99
ITS 46	*Phoma glomerata* (KF735046)	99
ITS 47	*S. humana* (KF735010)	99
ITS 48	*Alternaria porri* (KF735003)	99
ITS 49	*Penicillium restrictum* (KF735001)	99
ITS 50	*Tritirachium oryzae* (KF735034)	99
ITS 51	*Pseudoplectania sp.* (KF734992)	89
ITS 52	*Geopyxis sp.* (KF735056)	97

by Sordariomycetes, Eurotiomycetes, Leotiomycetes, and Pezizomycetes. The relative frequency of different classes of fungal endophytes isolated from foliar and stem tissues varies considerably (Fig. 4.2). Most of the foliar and stem endophytes belong to Dothideomycetes represented by the orders Pleosporales, Capnodiales, and Botryosphaeriales. Most of the needle endophytes belong to Sordariomycetes represented by the orders Diaporthales, Hypocreales, Sordariales, Trichosphaeriales, and Xylariales. All the endophytes belonging to the fungal class Pezizomycetes were isolated from foliar tissues only and none was recovered from the stem tissues. The only basidiomycetous endophytes strain, *T. oryzae*, was isolated from stem tissues only (Qadri et al., 2014).

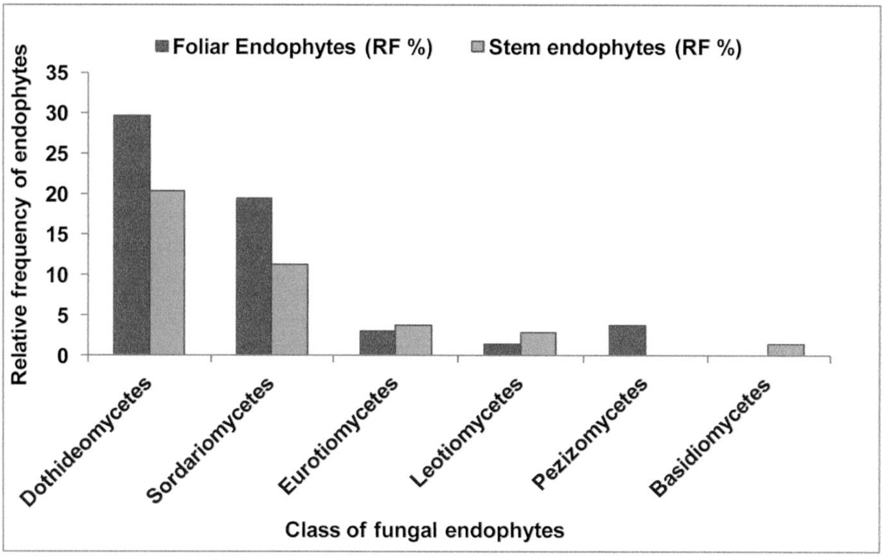

FIGURE 4.2 Relative frequencies of different class of fungal endophytes isolated from foliar and stem tissues of *Pinus wallichiana*.

There is differential colonization of endophytes in the stem and needle tissues of *P. wallichiana*. The probable reason for this differential colonization of endophytes may be the distinct microenvironments of these tissues types, which influence on shaping their microbiota differently. It is reported that many endophytes isolated from the needle tissues were not recovered from the stem tissues and vice versa. Thus, tissue specificity is evident among the endophytes of the Himalayan blue pine. The endophytic fungal genera restricted to stem tissues included *Cladosporium* sp., *Rachicladosporium* sp., *Leptosphaeria* sp.,

Microdiplodia sp., *Neurospora* sp., *Lophiostoma* sp., *Therrya* sp., *Lophodermium* sp., *Glomerella* sp., *Penicillium* sp., *Tritirachium* sp., and *Phomopsis* sp, whereas those that were restricted to needle tissues included *Anthostomella* sp., *Tricharina* sp., *Paraconiothyrium* sp., *Coniothyrium* sp., *Thielavia* sp., *Sordaria* sp., *Lecythophora* sp., *Rosellinia* sp., *Pseudoplectania* sp., *Nigrospora* sp., *Fusarium* sp., *Xylaria* sp., *Sporormiella* sp., *Cadophora* sp., *Geopyxis* sp., *Cochliobolus* sp., and *Fimetariella* sp. The endophytic fungal genera common to both stem and needles tissues included *Alternaria* sp., *Sclerostagonospora* sp., *Epicoccum* sp., *Preussia* sp., *Coniochaeta* sp., *Trichoderma* sp., *Pestalotiopsis* sp., *Phoma* sp., *Truncatella* sp., and *Aspergillus* sp. (Fig. 4.3) (Qadri et al., 2014). These common endophytes may be the systemic endophytes of Himalayan blue pine and the ecological implications of these fungal endophytes in the host plant must be investigated for a better understanding of the plant–endophyte relationship in Himalayan blue pine.

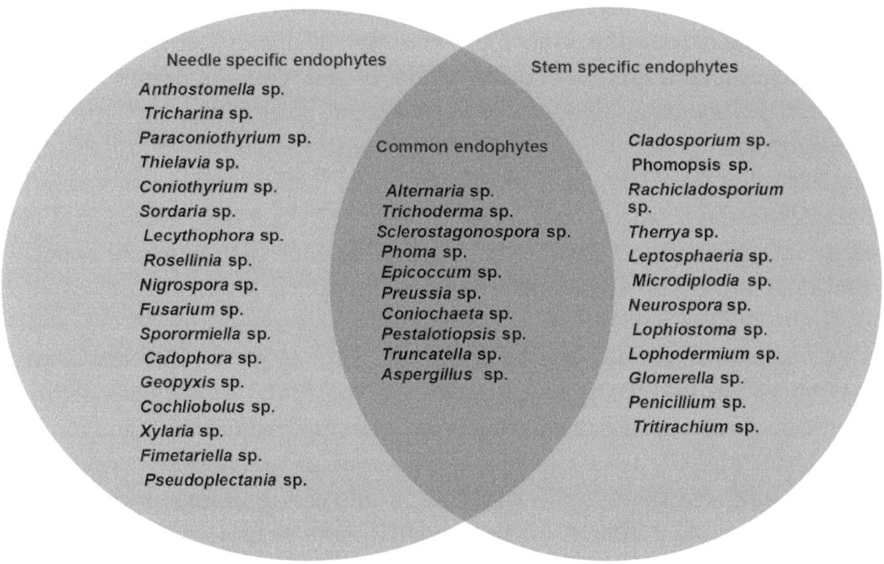

FIGURE 4.3 Venn diagram showing differential recruitment/colonization of fungal endophytes in foliar and stem tissues of *Pinus wallichiana*.

4.5 BIOACTIVE POTENTIAL OF ENDOPHYTES OF *P. WALLICHIANA*

Endophytes produce a number of bioactive secondary metabolites with potential applications in agriculture, pharma, and cosmetics industries

(Strobel and Daisy, 2003; Jalgaonwala et al., 2011; Godstime et al., 2014). These secondary metabolites have been categorized into various functional groups such as alkaloids, benzopyranones, chinones, flavonoids, phenolics, quinones, steroids, saponins, tannins, terpenoids, tetralones, xanthones, etc. The production of metabolites by the endophytes is affected by various factors, such as the climatic conditions in which the host plant is growing, season of sample collection, and geographical location (Shukla et al., 2014). However, the extraction of metabolites from natural microbial sources has now become more rapid, efficient, and convenient due to various synthetic processes developed over the past few years (Hussain et al., 2012). It is suggested that there is a direct correlation between the production of bioactive metabolite by the endophyte and coevolution of the microbe with its host plant. This coevolution may have incorporated genetic information from higher plants to the microbial partner leading to a sustainable symbiotic relationship with some added advantage to both the interacting partners (Strobel, 2003). Further, endophytic microbes are proficient producers of bioactive metabolites and drug-like molecules. This may be due to their evolution over millions of years in diverse ecological niches and natural habitats.

In *P. wallichiana*, eight endophytic fungal strains had ITS ribosomal gene sequence similarity of less than 95% with the known organisms. It is suggested that these may represent new fungal lineages and requires further characterization (Qadri et al., 2014). However, it is hypothesized that new microbial strains may have unique chemical characteristics and are expected to produce novel secondary metabolites with bioactive potential. Therefore, novel microbial strains must be explored for their natural products, as 40% of new chemical entities (NCEs) reported from 1981 to 2005 are derived from microbial sources. It is suggested that microbial sources are the most diverse but least explored natural sources as compared to plants and animals for human welfare. It is reported that microbial populations explored so far constitute only about 1% of bacteria and 5% of fungi, however, the rest remains to be explored for their contribution to biological applications (Staley et al., 1997).

Qadri and colleagues studied the bioactive potential of fungal endophytes isolated from *P. wallichiana.* They reported the extracts of 22 fungal endophytes showed significant antimicrobial potential against one or more plant and human pathogens. Five endophytic strains, *Coniothyrium carteri*, *T. subthermophila*, *Truncatella betulae*, *Cochliobolus australiensis*, and *T. oryzae*, were highly active against *Candida albicans*. The extracts of *T. oryzae* and *Coniochaeta gigantospora* displayed broad spectrum antimicrobial activities. Three endophytic strains, namely, *T. oryzae*, *Truncatella spadicea*, and *Fusarium larvarum* showed prominent antagonistic activity against a panel of fungal phytopathogens (Qadri et al., 2014). The above findings suggest that *P. wallichiana* harbors a rich diversity of fungal endophytes with potential antimicrobial activities. However, in this study, sampling was done from a single site and the endophytic diversity may vary with latitudinal gradient. Therefore, there is need to study the endophytic diversity of Himalayan blue pine and other conifers growing at the same site as well as along the latitudinal gradient in the northwestern Himalayas, and the comparative assessment of endophytes of conifers from other biogeographic regions is also required.

4.6 CONCLUSION

There is a huge diversity of fungal endophytes associated with coniferous trees; however, very little information is available about the importance of this biological wealth vis-à-vis the host plant. The ecological and evolutionary dynamics of natural and agro-ecosystems that shape the microbial communities in these ecosystems are very complex. Therefore, a concerted effort required to understand the microbial flux in natural and agro-ecosystems, the association of the endophytes with the tree species growing in western Himalayas, and their role in establishment of the plants under varied ecological settings. There is also an urgent need for comparative assessment of endophytic communities harbored by various *Pinus* species growing all over the world. It is also important to tap these endophytes and conserve them ex situ for bioprospecting various bioactive metabolites of varied importance.

KEYWORDS

- **Himalayan blue pine**
- **conifers**
- **agro-ecosystems**
- **antagonistic activity**
- **ITS ribosomal gene sequence**

REFERENCES

Aggarwal, S. G. Antibacterial Activity of Plants of Himalayan Region. *World J. Pharm. Pharm. Sci.* **2014,** *3* (5), 581–587.

Anonymous. *A Digest of Forest Statistics*; Statistics Division, Department of Forests, Jammu and Kashmir Government: Srinagar, 2008; p 55.

Bajracharya, M. B. *Ayurvedic Medicinal Plants and General Treatments*; Jore Ganesh Press Pvt Ltd: Kathmandu, 1979; pp 78–85.

Beri, R. M. Chemical Constituents of the Bark of *Pinus roxburghii* Sargent. *Indian J. Chem.* **1970,** *8*, 469–470.

Bhardwaj, A.; Sharma, D.; Agrawal, P. K. Isolation and Characterization of Endophytic Fungi from Spikes of *Pinus rouxburghii* Growing in Himalayan Region. *World J. Pharm. Res.* **2014,** *3* (9), 568–579.

Bowman, E. A.; Arnold, A. E. Distributions of Ectomycorrhizal and Foliar Endophytic Fungal Communities Associated with *Pinus ponderosa* along a Spatially Constrained Elevation Gradient. *Am. J. Bot.* **2018,** *105* (4), 687–699.

Carrell, A. A.; Frank, A. C. *Pinus flexilis* and *Piceae engelmannii* Share a Simple and Consistent Needle Endophyte Microbiota with a Potential Role in Nitrogen Fixation. *Front. Microbiol.* **2014,** *5*, 333.

Carrell, A. A.; Frank, A. C. Bacterial Endophyte Communities in the Foliage of Coast Redwood and Giant Sequoia. *Front. Microbiol.* **2015,** *6*, e01008.

Coppen, J. J. W.; Robinson, J. M.; Kaushal, A. N. Composition of Xylem Resin from *Pinus wallichiana* and *P. roxburghii*. *Phytochemistry* **1988,** *27*, 2873–2875.

Dar, G. H.; Beig, M. A.; Ahanger, F. A.; Ganai, N. A.; Ahanger, M. A. Management of Root Rot Caused by *Rhizoctonia solani* and *Fusarium oxysporum* in Blue Pine (*Pinus wallichiana*) through Use of Fungal Antagonists. *Asian J. Plant. Pathol.* **2011.** DOI:10.3923/ajppaj.2011.

Dar, M. Y.; Shah, W. A.; Mubashir, S.; Rather, M. A. Chromatographic Analysis, Anti-Proliferative and Radical Scavenging Activity of *Pinus wallichiana* Essential Oil Growing in High Altitude Areas of Kashmir, India. *Phytomedicine* **2012,** *19*, 1228–1233.

Deckert, R. J.; Melville, L. H.; Peterson, L. Structural Features of a *Lophodermium* Endophyte during the Cryptic Life-Cycle Phase in the Foliage of *Pinus strobus*. *Mycol. Res.* **2001,** *105*, 991–997.

Farjon, A. *Pines: Drawings and Descriptions of the Genus Pinus*; Antiquarian Book Sellers Association of America: New York, 1984.

Fogel, R. Interactions among Soil Biota in Coniferous Ecosystems. *Agric. Ecosyst. Environ.* **1988,** *24*, 69–85.

Gamble, J. S. *A Manual of Indian Trees*; Sampson Low, Marston and Co.: London, 1902.

Ganley, R. J.; Newcombe, G. Fungal Endophytes in Seeds and Needles of *Pinus monticola*. *Mycol. Res.* **2006,** *110*, 318–327.

Godstime, O. C.; Enwa, F. O.; Augustina, J. O.; Christopher, E. O. Mechanisms of Antimicrobial Actions of Phytochemicals against Enteric Pathogens—A Review. *J. Pharm. Chem. Biol. Sci.* **2014,** *2*, 77–85.

Hoffman, M. T.; Arnold, A. E. Geographic Locality and Host Identity Shape Fungal Endophyte Communities in Cupressaceous Trees. *Mycol. Res.* **2008,** *112*, 331–344.

Hussain, K. F.; Nisar, M.; Majeed, A.; Nawaz, K.; Bhatti, K. H. Ethnomedicinal Survey for Important Plants of Jalalpur Jattan, District Gujrat, Punjab, Pakistan. *Ethnobot. Leaflets* **2010,** *14*, 807–825.

Hussain, M. S.; Fareed, S.; Ansari, S.; Rahman, M. A.; Ahmad, I. Z.; Saeed, M. Current Approaches toward Production of Secondary Plant Metabolites. *J. Pharm. Bioallied Sci.* **2012,** *4*, 10–20.

Jalgaonwala, R. E.; Mohite, B. V.; Mahajan, R. T. Natural Products from Plant Associated Endophytic Fungi. *J. Microbiol. Biotechnol. Res.* **2011,** *1*, 21–32.

Little, E. L.; Critchfield, W. B. *Subdivisions of the Genus Pinus*; USDA Forest Service Miscellaneous Publication, 1969.

Liu, Y.; Ponpandian, L. N.; Kim, H.; Jeon, J.; Hwang, B. S.; Lee, S. K.; Park, S. C.; Bae, H. Distribution and Diversity of Bacterial Endophytes from Four *Pinus* Species and Their Efficacy as Biocontrol Agents for Devastating Pine Wood Nematodes. *Sci. Rep.* **2019,** *9*, 12461.

Oono, R.; Lutzoni, F.; Arnold, A. E.; Kaye, L.; Ren, J. M. U.; May, G.; Carbone, I. Genetic Variation in Horizontally Transmitted Fungal Endophytes of Pine Needles Reveals Population Structure in Cryptic Species. *Am. J. Bot.* **2014,** *101*, 1362–1374.

Pirttilä, A. M.; Wäli, P. Conifer endophytes. In *Defensive Mutualism in Microbial Symbiosis*; White, J. F. Jr., Torres, M. S., Eds.; CRC Press: Boca Raton, FL, 2009; pp 235–246.

Ponpandian, L. N.; Rim, S. O.; Shanmugam, G.; Jeon, J.; Park, Y. H.; Lee, S. K.; Bae, H. Phylogenetic Characterization of Bacterial Endophytes from Four *Pinus* Species and Their Nematicidal Activity against the Pine Wood Nematode. *Sci. Rep.* **2019,** *9*, 12457.

Qadri, M.; Johri, S.; Shah, B. A.; Khajuria, A.; Sidiq, T.; Lattoo, S. K.; Abdin, M. Z.; Riyaz-Ul-Hassan, S. Identification and Bioactive Potential of Endophytic Fungi Isolated from Selected Plants of the Western Himalayas. *Springer Plus* **2013,** *2*, 8.

Qadri, M.; Rajput, R.; Abdin, M. Z.; Vishwakarma, R. A.; Riyaz-Ul-Hassan, S. Diversity, Molecular Phylogeny and Bioactive Potential of Fungal Endophytes Associated with the Himalayan Blue Pine (*Pinus wallichiana*). *Microb. Ecol.* **2014,** *67*, 877–887.

Richardson, D. M. (Ed.) *Ecology and Biogeography of Pinus*; Cambridge University Press: Cambridge, 2000.

Richardson, D. M.; Rundel, P. W.; Jackson, S. T.; et al. Human Impacts in Pine Forests: Past. *Ann. Rev. Ecol. Syst.* **2007,** *38*, 275–297.

Rua, M. A.; Wilson, E. C.; Steele, S.; Munters, A. R.; Hoeksema, J. D.; Frank, A. C. Associations between Ectomycorrhizal Fungi and Bacterial Needle Endophytes in *Pinus radiata*: Implications for Biotic Selection of Microbial Communities. *Front. Microbiol.* **2016,** *7*, 399.

Ryberg, M.; Matheny, P. B. Asynchronous Origins of Ectomycorrhizal Clades of Agaricales. *Proc. R Soc. B: Biol. Sci.* **2012,** *279*, 2003–2011.

Sharma, A.; Sharma, L.; Goyal, R. A Review on Himalayan Pine Species: Ethnopharmacological, Phytochemical and Pharmacological Aspects. *Pharmacogn. J.* **2018,** *10* (4), 611–619.

Shukla, S. T.; Habbu, P. V.; Kulkarni, V. H.; Jagadish, K. S.; Pandey, A. R.; Sutariya, V. N. Endophytic Microbes: A Novel Source for Biologically/Pharmacologically Active Secondary Metabolites. *Asian J. Pharmacol. Toxicol.* **2014,** *2*, 1–16.

Singh, H.; Saklani, A.; Lal, B. Ethnobotanical Observations on Some Gymnosperms of Garhwal Himalaya, Uttar Pradesh, India. *Econ. Bot.* **1990,** *44* (3), 349–354.

Smith, S. E.; Read, D. *Mycorrhizal Symbiosis*, 3rd ed.; Academic Press: New York, 2008.

Staley, J. T.; Castenholz, R. W.; Colwell, R. R.; Holt, J. G.; Kane, M. D.; Pace, N. R.; Saylers, A. A.; Tiedje, J. M. *The Microbial World: Foundation of the Biosphere*; American Academy of Microbiology: Washington, DC, 1997; p 32.

Strobel, G. A. Endophytes as Sources of Bioactive Products. *Microbes Infect.* **2003,** *5*, 535–544.

Strobel, G. A.; Daisy, B. Bioprospecting for Microbial Endophytes and Their Natural Products. *Microbiol. Mol. Biol. Rev.* **2003,** *67*, 491–502.
Suske, J.; Acker, G. Internal Hyphae in Young, Symptomless Needles of *Picea abies*: Electron Microscopic and Cultural Investigation. *Can. J. Bot.* **1986,** *65*, 2098–2103.
Tedersoo, L.; May, T. W.; Smith, M. E. Ectomycorrhizal Lifestyle in Fungi: Global Diversity, Distribution, and Evolution of Phylogenetic Lineages. *Mycorrhiza* **2010,** *20*, 217–263.
Thakur, N. S.; Gupta, A.; Chauhan, V. K. Studies on Quality Characteristics of Chilgoza Nut and Oil. *Int. J. Farm. Sci.* **2015,** *5* (1), 78–82.
Troup, R. S. *The Silviculture of Indian Trees*, Vol. III; Clarendon: Oxford, 1921.
Turner, T. R.; James, E. K.; Poole, P. S. The Plant Microbiome. *Genome Biol.* **2013,** *14*, 209.
Wani, Z. A.; Ashraf, N.; Mohiuddin, T.; Riyaz-Ul-Hassan, S. Plant-Endophyte Symbiosis, an Ecological Perspective. *Appl. Microbiol. Biotechnol.* **2015,** *99*, 2955–2965.

CHAPTER 5

Exploring the Endophytic Microbiome of Saffron (*Crocus sativus*)

ZAHOOR AHMED WANI

Department of Botany, Govt. Degree College Kishtwar, Kishtwar, Jammu & Kashmir, India

E-mail: zawani1986@gmail.com

ABSTRACT

Crocus sativus L. is an important aromatic and medicinal plant belonging to Iridaceae family of Magnoliophyta class of monocots. *C. sativus* is a perennial plant comprising of a subterranean part known as corm/bulb, leafy vegetative shoot, and purple-colored flowers. *C. sativus* being a triploid genotype is a sexually sterile plant and hence propagated only by vegetative means only. The place of origin and evolution of *C. sativus* is not clear; however, its cultivation and utilization in Mediterranean area date back to 2500–1500 BC. The cultivation and production of *C. sativus* is constantly declining worldwide for the last few decades due to various biotic and abiotic factors. One of the most important factors that influence plant health is the endophytic community harbored by the host plant. Recently there are various reports of the application of microbes of endophytic origin for sustainable cultivation and crop management of saffron and also yield bioactive natural products for pharmacological and industrial applications.

5.1 INTRODUCTION

Crocus sativus L. belongs to the family Iridaceae of Magnoliophyta class of monocots. *C. sativus* has a triploid genotype ($2n = 3x = 24$) and is a sexually sterile plant as it produces abnormal gametes (Negbi et al. 1989). It is an autumn-flowering perennial plant propagated only by vegetative means via manual "divide-and-set" of the underground corms/bulbs (Negbi et al. 1989). The most important part of this plant is the dried red stigma known as saffron,[1] which has been used as a medicinal herb and spice since time immemorial (Javadi et al. 2013; Baba et al. 2015a). *Crocus* synthesizes a unique set of compounds known as apocarotenoids[2] that are synthesized in the stigma part of the plant (Ashraf et al. 2015). The apocarotenoids of saffron are crocin, picrocrocin, and safranal that are responsible for color, flavor, and aroma of saffron, respectively (Kumar et al. 2009). Furthermore, it is the only plant that produces these apocarotenoids in significant quantities (Ashraf et al. 2015). Owing to its high demand in dye, perfumery, and flavoring industries, it is one of the most expensive spices in the world and is recognized as *Red gold*. The apocarotenoids are derived from carotenoid—zeaxanthin—by enzymatic oxidative cleavage, and the enzymes responsible for this process are known as carotenoid cleavage dioxygenases (CCDs). These enzymes (CCDs) recognize and specifically cleave one or two double bonds in carotenoid molecule (Rubio-Moraga et al. 2008). In addition to the apocarotenoids, *C. sativus* also contains other secondary metabolites like carotenoids, phenolics, flavonoids, sugars, vitamins, and around 150 volatile organic compounds (VOCs). The diverse compositions of *C. sativus* metabolites contribute an important role in plant development and adaptation to various stress conditions. Apart from this, saffron metabolites are reported to have tremendous therapeutic properties, and their pharmacological importance has been appreciated both by the traditional Avicenna's Canon of Medicine (al-Qanun fi al-tib) as well as modern scientific reports (Bhargava, 2011; Hosseinzadeh and Nassiri-Asl, 2013; Baba et al. 2015a).

Saffron is primarily cultivated in Iran, Spain, India, Greece, Morocco, Italy, Turkey, and France (Fernandez et al. 2004). In India, commercial cultivation of saffron occurs mainly in Kashmir.[3] Saffron is being cultivated in Kashmir since 750 AD. However, the cultivation of *Crocus* is observing a constant decline worldwide, including J&K state due to various factors (Gresta et al. 2008). It is reported that there was a decrease of 83% in the area and 72% in the productivity of saffron in a single decade in Kashmir (Wani et al. 2016). This decline in cultivation of saffron worldwide, due to poor agronomic practices and disease management together with lack of breeding approaches, is a matter of concern to the saffron growers as well as agricultural scientists. Efforts are being made to understand the

[1] In Kashmir region of the Himalayas, saffron is locally known as "Kong" in Kashmiri and "Zafran" in Urdu.
[2] Apocarotenoids are the degradation products of the carotenoids.
[3] Kashmir is a Himalayan valley in the state of Jammu and Kashmir (J&K) in India.

biology of the crop and replace the traditional practices of saffron cultivation with modern technology–driven cultivation approach. The sterile nature of *Crocus* is hampering the conventional breeding approach, and modern biotechnological approach has also failed to deliver due to lack of established transformation protocols in *Crocus*, and now plant–microbe relationships are being explored as an alternative for sustainable cultivation of *Crocus* (Wani et al. 2016, 2017). The cultivation of *Crocus* is restricted to specific agroclimatic regions with a specific temperate and climate, and also the propagation of *Crocus* occurs by means of underground corms. It is expected that the microbiome associated with *Crocus* might have significant influence on the adaptation and functioning of the plant.

5.2 *CROCUS SATIVUS*: EVOLUTION, DISTRIBUTION, AND PRODUCTION

Crocus is derived from the Greek word *Corycus,* the name of an area in Cilcia in the eastern Mediterranean. The word "saffron" is derived from French term safran and the Latin word safranum. It is also related to the Italian *Zafferano* and Spanish *Azafran* and Arabic *Asfar or Zafran* (Kumar et al. 2009). *Crocus* genus consists of about 90 species of perennials, growing from corms, and many of them are considered economically valuable. *C. sativus* is a genetically monotypic clone (Busconi et al. 2015). *C. sativus* is believed to have evolved from *Crocus cartwrightianus* by autotriploidy or from *C. thomasii* and *C. pallasii* by allotriploidy (Caiola and Canini, 2010). Since *C. sativus* is a triploid species ($2n = 3X = 21$), there is not proper segregation of chromosomes in meiosis resulting in abnormal gamete formation. Due to aberrant meiotic behavior, *C. sativus* is self-incompatible and male sterile and, hence, incapable of independent sexual reproduction. Therefore, the propagation of *C. sativus* is by vegetative multiplication via manual "divide-and-set" of the underground corms/bulbs or by interspecific hybridization (Negbi et al. 1989).

The cultivation and use of saffron as a spice and medicinal plant in the Mediterranean area date back to 2500–1500 BC (Fernández, 2004). Although the place of origin of saffron plant is still not clear, it has probably originated in Greece, Iran, or Asia Minor, later spreading to India, China, the Mediterranean basin, and Eastern Europe. Presently, the main saffron-producing countries are Iran, Greece, Spain, Italy, and India (Kashmir) (Fernández, 2004). The historical accounts of saffron cultivation in Kashmir date back to 750 AD (Kumar et al. 2009). The cultivation of saffron in J&K is restricted to specific ecological niches with almost same edaphic and climatic conditions. The soils are heavily textured with silt clay loam, which are well drained, calcareous in nature. Also, these are slightly alkaline with pH ranging from 6.3 to 8.3 and with electrical conductivity between 0.09 and 0.30 dsm^{-1} (Nazir et al. 1996). The world's total annual saffron production is estimated to be 300 tons per year, of which Iran contributes about 80%. Spain is the second-largest producer and contributes about 10–12% of world's production followed by India (3.3%), Greece (2.0%), and Morocco (0.3%) (Kumar et al. 2009). The Kashmir region in India produces saffron mostly for domestic use.

5.3 MORPHOLOGY AND PHENOLOGY OF *CROCUS SATIVUS*

C. sativus is a small geophyte, comprising a subterranean part known as corm/bulb, leafy vegetative shoot, and purple-colored flowers (Fig. 1.1). The corm of *C. sativus* is an underground stem covered with fibrous sheaths known as tunics. The corm produces three types of roots, absorbing roots (fibrous), contractile roots, and adventitious roots. The leaves of *C. sativus* are radical, long, slender, grass-like, and channeled, with curved and fringed margins. Each plant produces one to three purple flowers having three violet sepals and three similar petals together. The plant is considered to be hysteranthous, as the flowers arise directly from corms (Kumar et al. 2009). Saffron flower bears three red-colored trilobed stigmas and three yellow-colored anthers. The commercially important parts of *Crocus* are stigma and corm. The mother/primary corm produces one or more daughter/secondary corm-lets that are dependent on the nourishment provided by the mother corm and eventually replace them in subsequent years. Vegetative cultivation offers advantage in maintaining the genetic characteristics of the plant, but it does not allow genetic improvement. Flower formation is directly related to corm size: small-sized corms take 3–4 years to flower, while graded corms (>8-g weight) produce flower in 1 year.

C. sativus is a perennial plant with its life cycle divided into three distinct phases. The life cycle of saffron is adapted to the climate of the Mediterranean region and is quite similar in all countries, but they differ in only the timing of events. The life cycle of saffron cultivated in J&K has the dormant phase extending from March to October, followed by the generative phase extending for less than a month, which is followed by the vegetative phase extending from November to February (Fig. 5.2). Corm development and bud sprouting occur during the dormant phase, flowering takes place during the generative phase, and leaf development is the main activity during the vegetative phase (Wani et al. 2016). Saffron shows summer dormancy, which is related to superior survival and high persistence under severe drought conditions. This strategy of saffron is of great ecological significance, particularly in view of the climate change resulting in increased temperature and drought conditions.

FIGURE 5.1 Morphology of *C. sativus* L.

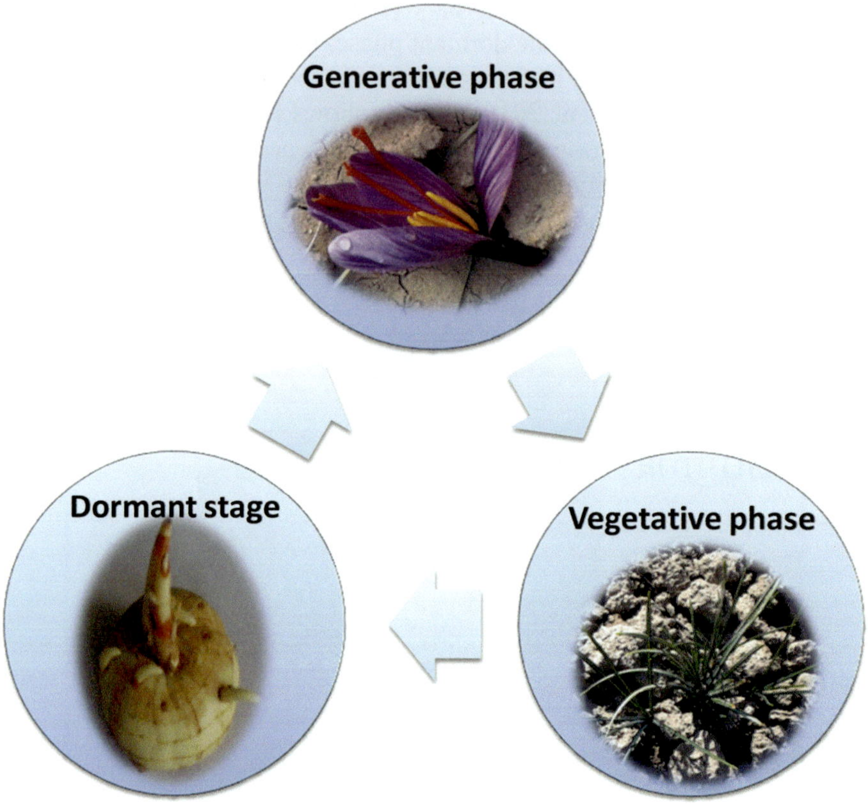

FIGURE 5.2 Phenology of *C. sativus* L.

5.4 DECLINE IN SAFFRON PRODUCTIVITY: CAUSES AND CONCERNS

C. sativus is an important high-value crop plant. The cultivation and production of *Crocus* is constantly declining worldwide for the last few decades. The cultivation of saffron is restricted to specific agroclimatic regions with temperate climate. The replacement of corms, breaking dormancy, transition from vegetative to reproductive stage, development of floral bud, floral emergence, etc. are all tightly regulated by various environmental factors like temperature, irrigation, sunlight, etc. It is reported that temperature regulates the growth and flowering of *Crocus* by affecting the enzyme activity in plant metabolism. Saffron production does not require much water, but it is reported that first irrigation is very important for flower emergence and length of flowering period of saffron. However, the declining trend in saffron production and quality is mainly attributed to poor agronomic practices and disease management together with lack of breeding approaches. *C. sativus* has remained outside the realm of genetic improvement because of its sterile nature. Moreover, biotechnological approaches have failed to deliver, because transformation protocol has not been established so far. There are a few reports where genes involved in the flowering and apocarotenoid biosynthetic pathway have been cloned and characterized (Rubio-Moraga et al. 2004; Frusciante et al. 2014; Baba et al. 2015b). Also a few transcription factors regulating the biosynthesis of these compounds have been identified and cloned (Ashraf et al. 2015). However, none of these genes have been taken forward for transforming *Crocus* for enhanced production of apocarotenoids.

An important aspect for sustainable cultivation of saffron is the adequate production of healthy corms, which is extremely important to guarantee flower production. However, the corms in their natural environment are constantly under siege from a multitude of disease-causing organisms, including viruses, bacteria, nematodes, and especially fungi. Several fungal species belonging to genera *Fusarium*, *Rhizoctonia*, *Penicillium*, *Macrophomina*, *Aspergillus*, *Sclerotium*, *Phoma*, *Stromatinia*, *Cochliobolus*, *Rhizopus*, *Porostereum*, *Talaromyces*, *Epicoccum*, etc. are reported to be associated with saffron diseases (Ahrazem et al. 2010; Wani et al. 2016). Considerable work has been done on pathogens causing diseases in *C. sativus* worldwide (Cappelli and Di Minco, 1999; Palmero et al. 2014; Gupta and Vakhlu, 2015; Wani et al. 2016, 2017). Corm rot caused by *Fusarium oxysporum* is the most destructive disease in saffron, causing severe performance losses in most saffron fields (Cappelli, 1994). The symptoms of corm rot include pigmentation, and in the later stages of the disease, tissue desiccation takes place. Infected plants die off early, which results in reduction of corm yield, flowering, and stigma production. The corm rot disease was first detected in Japan (Yamamoto et al. 1954). Corm rot disease is currently widespread throughout the saffron-producing countries, causing substantial yield losses. With a disease incidence of 100% and severity ranging from 6 to 46%, corm rot disease results in reduced plant growth and yield of saffron in Kashmir as well (Husaini et al. 2010).

To avoid pathogen attack, *Crocus* corm has developed several physical and chemical barriers, as well as a system of active defense reactions. Recently a new chitinase, SafchiA, isolated from corms of *C. sativus,* is reported to play an important role in saffron defense response induced by fungal (*F. oxysporum* f. sp. *tuberose*) infection and mediates inhibition of fungal growth under *in vitro* condition (López and Gomez-Gomez, 2009). *C. sativus* is characterized by the presence of saponins in stigma and corm tissues, where they seem to play an antifungal role. The ability of a plant to resist diseases is also dependent on soil conditions such as structure, compaction, drainage, temperature, and level of biological activity, along with cultural practices such as planting date and application of fertilizers or herbicides (Ahrazem et al. 2010). The corm rot caused by *Fusarium* and other fungi in saffron grown in Kashmir is being managed by using chemical fungicides such as carbendazim (broad-spectrum benzimidazole fungicide), myclobutanil (triazole chemical), mancozeb (subclass of carbamate), bavistin (50% WP carbendazim), and tecto (benzimidazole fungicide). However, the deleterious impact of these chemicals on the environment as well as human beings is well established. These chemicals also affect the beneficial microflora associated with the plant and put selection pressure for the evolution of resistant pathotypes. Therefore, biological control is gaining importance for integrated pest/disease management. There is a diverse community of microorganisms (endophytes) that interact positively with plants in agricultural systems in relation to their nutrition and ability to resist biotic and abiotic stress. The endophytes have the potential to be manipulated such that the benefits of their positive effects are harnessed.

5.5 PLANT–MICROBE ASSOCIATION IN *C. SATIVUS*

A lot of work is being done on *C. sativus* to understand the biology of the plant. However, work on plant–microbe interaction in *Crocus* is gaining momentum for the last few years. There are various reports of the application of microbes with established plant-growth-promoting properties on the production of saffron. The antagonistic potential of *Trichoderma viride* isolates collected from soil was investigated against *Crocus* corm rot pathogen, *F. oxysporum* (Mir et al. 2011). In Spain, the application of *Bacillus subtilis* FZB24 spore solution to saffron corms significantly increased leaf length, flower per corm, total stigma biomass and decreased the time required for corms to sprout. Moreover, significant increase in the quantity of picrocrocin, crocetin, and safranal compounds is reported, when the plants are soil drenched with *B. subtilis* FZB24 spore solution 14 weeks after the sowing (Sharaf-Eldin et al. 2008). Aytekin and Acikgoz (2008) reported that the production of saffron can be increased by treatment of corms with a synthetic hormone (polystimulin A6 and K) and microorganism-based materials like biohumus. Recently, a *Bacillus amyloliquefaciens* strain W2 collected from rhizospheric soil was found effective against corm rot caused by *F. oxysporum* (Gupta and Vakhlu, 2015).

Although the accessions of *C. sativus* cultivated in different regions show little genetic variability, the yield and productivity of saffron vary considerably. This could be attributed to variations in agricultural practices and various biotic and abiotic factors. One of the most important factors that influence plant health is the endophytic community harbored by the host plant. The cultivation of *Crocus* is restricted to specific agroclimatic regions with temperate climate, and also the propagation of *Crocus* is by means of underground corms. Therefore, the microbiome associated with *Crocus* might have significant influence on the adaptation and functioning of the plant. Thus, it is imperative to understand the patterns of distribution and community structure of endophytes of *C. sativus*, as well as their interactions with the host plant, for sustainable agriculture and crop management of this high-value medicinal and aromatic plant.

5.6 ENDOPHYTES OF *C. SATIVUS*

C. sativus harbors a huge diversity of fungal and bacterial endophytes. Wani et al. reported the fungal endophytic community of *C. sativus* cultivated in J&K, India. A total of 294 fungal endophytes were isolated from *Crocus* corms, which were grouped into 100 morphotypes based on phenotypic characters like growth pattern, colony texture, and colony color, as well as morphology of conidia and conidiophores. Molecular phylogenetic studies based on ITS1-5.8S-ITS2 ribosomal gene sequence analyses assigned these endophytes into 36 distinct species, spreading over 19 genera (Table 5.1). The diversity and composition of the endophytic community was almost similar across different sites in J&K state. It may be due to uniformity of both the genetic makeup of *Crocus* and edaphic factors across different sites. However, the diversity and composition of the endophytic community varied temporally at the two different phonological stages of *Crocus* lifecycle. It was higher at the dormant than at the vegetative stage, indicating influence of host/corm health status on the endophytic diversity (Fig. 5.3). This may be explained by the fact that during the vegetative stage, the corms get flaccid, nutrient deficient, and relatively inactive, thus supporting the growth of fewer endophytes inside the corm tissues. In addition, the corms remain in the vegetative stage during the winter season, which is marked by snowfall and low temperatures, thus creating conditions that are less favorable for the growth of the endophytes (Wani et al. 2016).

TABLE 5.1 The Table Presents the 36 Different Fungal Endophytes (their Genbank Accession Numbers and Isolation Frequency) Isolated from *C. sativus*.

ITS genotype	Molecular identification (GenBank accession no.)	Isolation frequency (%)
ITS1	*Aspergillus flavipes* (KR135119)	3.7
ITS2	*Trichoderma harzianum* (KR135120)	3.4
ITS3	*Cadophora malorum* (KR135121)	12.9
ITS4	*Fusarium oxysporum* (KR135122)	4.1
ITS5	*Alternaria alternata* (KR135123)	4.4
ITS6	*Penicillium pinophilum* (KR135124)	3.7
ITS7	*Paecilomyces tenuis* (KR135125)	1.7
ITS8	*Porostereum* sp. (KR135126)	1
ITS9	*Talaromyces pinophilus* (KR135127)	0.3
ITS10	*Aspergillus dimorphicus* (KR135128)	1
ITS11	*Aspergillus terreus* (KR135129)	1.4
ITS12	*Aspergillus iizukae* (KR135130)	1.4
ITS13	*Aspergillus pseudodeflectus* (KR135131)	3.7
ITS14	*Fusarium incarnatum* (KR135132)	0.3
ITS15	*Alternaria* sp. (KR135133)	0.3
ITS16	*Fusarium solani* (KR135134)	1.4
ITS17	*Talaromyces verruculosus* (KR135135)	2.4
ITS18	*Eucasphaeria* sp. (KR135136)	1.7
ITS19	*Penicillium canescens* (KR135137)	2
ITS20	*Talaromyces cellulolyticus* (KR135138)	9.5
ITS21	*Penicillium* sp. (KR135139)	0.7
ITS22	*Penicillium chrysogenum* (KR135140)	0.7
ITS23	*Epicoccum nigrum* (KR135141)	1
ITS24	*Phialophora mustea* (KR135142)	15
ITS25	*Penicillium griseofulvum* (KR135143)	9.2
ITS26	*Ilyonectria robusta* (KR135144)	0.3
ITS27	*Alternaria brassicae* (KR135145)	0.3
ITS28	*Mortierella alpina* (KR135146)	2
ITS29	*Penicillium* sp. (KR135147)	1
ITS30	*Acremonium* sp. (KR135148)	2.4
ITS31	*Cladosporium silenes* (KR135149)	1
ITS32	*Fusarium tricinctum* (KR135150)	1.7
ITS33	*Leptodontidium orchidicola* (KR135151)	2.4
ITS34	*Botrytis fabiopsis* (KR135152)	0.3
ITS35	*Paecilomyces marquandii* (KR135153)	0.3
ITS36	*Gloeosporium* sp. (KR135154)	1

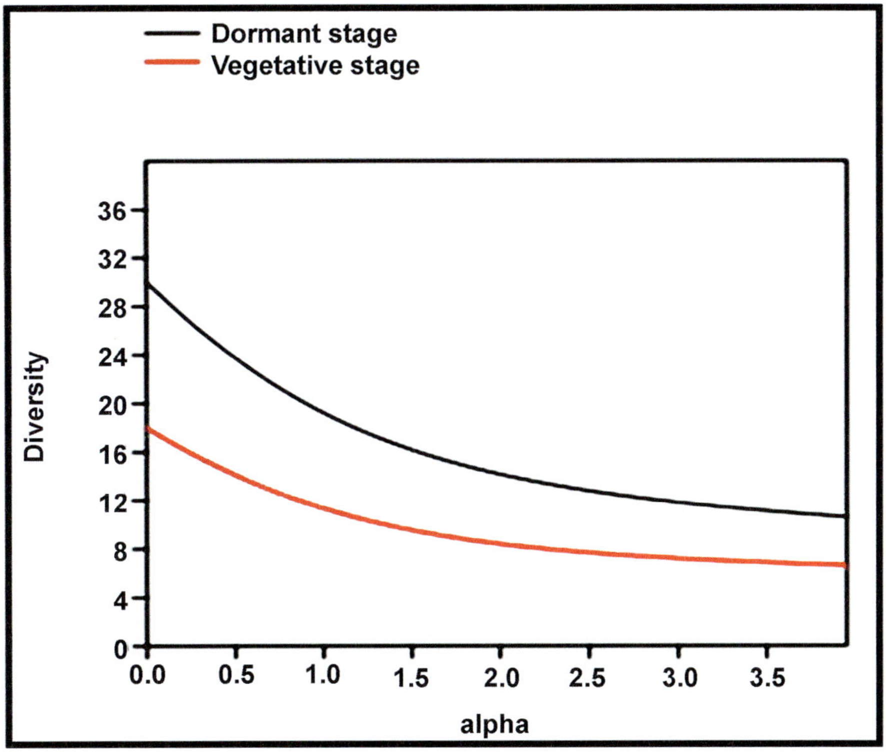

FIGURE 5.3 Diversity profile graph of fungal endophytes at two stages of *Crocus* life cycle. The black line indicates diversity profile at dormant stage, while the red line indicates the diversity profile at vegetative stage. The plot indicates clearly that the diversity of endophytes associated with *C. sativus* is higher during the dormant stage of its life cycle.

The saffron microbiome was dominated by dark septate endophytes (DSEs) with an isolation frequency of more than 30%, particularly *Phialophora mustea* and *Cadophora malorum* being the most dominant endophytes (Table. 5.1). Interestingly molecular phylogeny assigned these DSEs into a single clad, indicating a strong effect of the host genotype on the selective recruitment of endophytes (Fig. 5.4). This indicates host–endophyte specificity in the *Crocus* plant vis-à-vis *P. mustea* and *C. malorum*, and these species are the most preferred endophytes of the host. These associations might have developed over centuries of cultivation of saffron and transmitted vertically as the host is propagated only by vegetative means using corms (Wani et al. 2016).

Plants growing in different geographical regions are confronted with different environmental challenges. These environmental cues in combinatorial

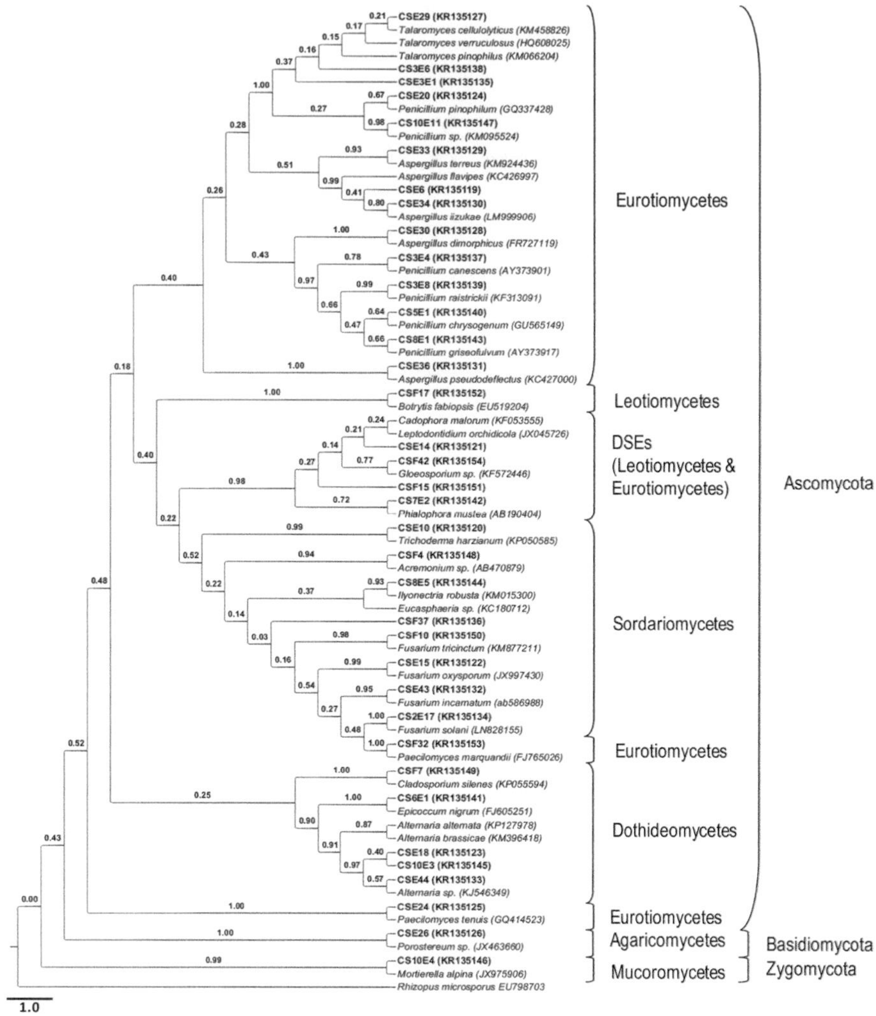

FIGURE 5.4 Phylogeny of endophytes of *Crocus sativus* using maximum parsimony analysis based on ITS1-5.8S-ITS2 sequence. Only strain names with accession numbers are provided in the phylogenetic tree for the endophytes isolated in this study. The tree is rooted with *Rhizopus microsporus* (a zygomycete, EU798703).

effect with host genotype shape the endophytic diversity harbored by the host plants (Arnold, 2007; Wani et al. 2015). As saffron cultivation in Kashmir is completely rainfed, with no scientific irrigation system in place, the plants usually suffer summer drought conditions, which may inevitably lead to production of reactive oxygen species (ROS). The higher colonization of

DSEs in the corms of *Crocus* indicates an ecological significance, as it is reported that the melanized hyphae[4] are considered to be of importance for the host to survive stress conditions. The cell-wall melanin can trap and eliminate oxygen radicals generated during abiotic stress. Also the DSEs associated with *Crocus* produce significant amount of indoleacetic acid (IAA), and it is reported that IAA increases colonization efficiency of the endophytes, possibly via interference with the host defense system (Navarro et al. 2006). The production of IAA or related compounds may be an important property for plant colonization by endophytes. Therefore, the endophytes, *P. mustea*, and *C. malorum* are efficient colonizers in *C. sativus* and may confer tolerance to the host against a variety of environmental stress factors. Also, *P. mustea* and *C. malorum* isolates showed intraspecific strain variations, indicating that these symbiotic associations are species specific rather than strain specific (Wani et al. 2016).

Some endophytic strains recovered from the *Crocus* corm were identified as being members of commonly observed genera of soil fungi, for example, *Fusarium*, *Penicillium*, *Talaromyces*, *Trichoderma*, and *Paecilomyces*. These fungi are characteristically free-living saprophytes that can also be opportunistic root endophytes or latent pathogens.[5] Pathogenicity assay indicated some of the endophytes of *Crocus* as latent pathogens, as they displayed virulence with varying levels of severity under both *in vitro* and *in vivo* conditions (Fig. 5.5). For instance, endophytes like *Alternaria alternata*, *Epicoccum nigrum*, *F. oxysporum*, *Acremonium* sp., *Penicillium pinophilum, Talaromyces cellulolyticus* displayed moderate-to-high virulence under both *in vitro* and *in vivo* conditions. However, *Aspergillus pseudodeflectus*, *Botrytis fabiopsis*, *Penicillium canescens, Porostereum* sp., *Paecilomyces marquandii, Talaromyces pinophilus*, and *Talaromyces verruculosus* displayed low virulence under *in vivo* condition and therefore considered low-risk pathogens (Wani et al. 2016).

Recently some studies on bacterial endophytes and rhizospheric bacterial associates of *C. sativus* are reported by culture-dependent and culture-independent approaches (Ambardar and Vakhlu, 2013; Ambardar et al. 2014; Sharma et al. 2015). Sharma and colleagues isolated cultivable bacterial endophytes from saffron plant and assessed for plant-growth-promoting activities. Molecular and phylogenetic analysis grouped the

[4] Melanized hyphae are a characteristic feature of dark septate endophytes, as they have melanin pigment present in their hyphae.

[5] They live as normal endophytes in the host plant but can turn pathogenic under stress condition or produce disease symptom in the host plant upon reinfection.

54 bacterial isolates into 11 different taxa, namely, *Bacillus licheniformis*, *B. subtilis*, *B. cereus*, *B. humi*, *B. pumilus*, *Paenibacillus elgii*, *B. safensis*, *Brevibacillus* sp., *Pseudomonas putida*, *Staphylococcus hominis*, and *Enterobacter cloacae*. *B. licheniformis* was the dominant endophyte in both leaves and corms of saffron. Ambardar and colleagues reported the bacteria associated with rhizosphere, cormosphere, and bulk soil of saffron, using cultivation-independent 16S rRNA gene-targeted metagenomic approach. Saffron during flowering stage revealed the presence of 22 genera, but none of the genus was common in all the three samples. Bulk soil bacterial community was represented by 13 genera with *Acidobacteria* being dominant genus, while as rhizospheric bacterial community was represented by eight different genera with *Pseudomonas* being the dominant genus, and cormospheric bacterial community comprised six different genera, dominated by the genus *Pantoea* (Ambardar et al. 2014).

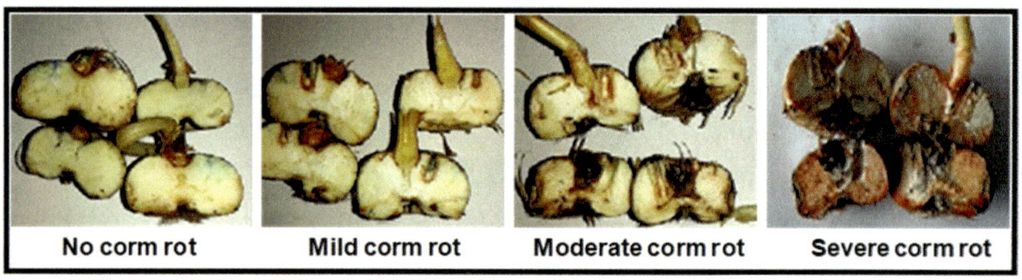

FIGURE 5.5 Corms of *C. sativus* reinfected with endophytes produced rotting symptoms with different levels of severity.

5.7 BIOACTIVE POTENTIAL OF *CROCUS* ENDOPHYTES

Endophytes are proficient producers of bioactive metabolites and drug-like molecules. Thus, they represent a huge bio-resource for the isolation of novel bioactive molecules for applications in medicine, agriculture, and industry (Porras-Alfaro and Bayman, 2011). This is not surprising in the light of their evolution over millions of years in diverse ecological niches and natural habitats. Extracts from several endophytes of *Crocus* showed promising antimicrobial activities. Four new metabolites, Phialomustin A–D isolated and characterized from an endophyte (*P. mustea* CS7E2) of *C. sativus*, are reported to have potential antimicrobial and anticancer activities (Nalli et al. 2015). Furthermore, a unique quinazoline alkaloid with cytotoxic and antifungal activities is isolated from *Penicillium vinaceum,* an endophyte of *C. sativus* (Zheng et al. 2012). Several endophytes are reported to inhibit the growth of plant pathogenic fungi, thereby indicating a strong biocontrol potential, which can be harnessed to control corm rot and other microbial diseases after carrying out further studies particularly under field conditions. By virtue of the antimicrobial activities, the endophytes may be imparting resistance to the host plant against microbial diseases. Also, these properties may be helping them to dominate the microbial populations in the corresponding ecological niches leading to their efficient colonization in the plants. Another benefit, which these endophytes provide to the host plant, is the production of the plant growth hormones. Phytohormone production by endophytes is probably the best studied mechanism of plant growth promotion, leading to morphological and architectural changes in plant hosts, thus contributing to the overall growth and development of the plant.

An oleaginous fungal endophyte, *M. alpina* CS10E4, isolated from *C. sativus* produces polyunsaturated fatty acids (PUFAs), including arachidonic acid (AA). *M. alpina* CS10E4 shifts the metabolic flux of *Crocus* toward enhanced production of apocarotenoids by modulating the expression of key genes of apocarotenoid pathway. Furthermore, *M. alpina* CS10E4 enhanced tolerance to corm rot disease by releasing arachidonic acid, which acts as conserved defense signal and induces jasmonic acid production in endophyte treated *Crocus* corms (Wani et al. 2017). A basidiomycete, *Porostereum* sp. CSE26 produces chlorinated aromatic compounds (CAMs), that is, 3-chloro-4-methoxybenzaldehyde and 2, 3-dichlorophenyl isothiocyanate, having phytotoxic activity against *Arabidopsis* plants. It is presumed that these compounds may be acting as pathogenic determinants of *Porostereum* sp. CSE26 (Wani et al. 2018).

The bacterial endophytes associated with *C. sativus* also display significant bioactivity. Sharma and colleagues reported that out of the 54 bacterial endophytes investigated for enzyme production, 81% isolates showed lipase activity, 57% isolates showed cellulase, 48% isolates showed protease, 38% isolates showed amylase, 33% isolates showed chitinase, and 29% isolates showed pectinase activity. These bacterial endophytes were investigated for plant-growth-promoting potential, and it is reported that 24% of the isolates were phosphate solubilizers, 86% showed siderophore production, and 80% showed phytohormone production (Sharma et al. 2015). The endophytic communities associated with *C. sativus* produce a diverse array of biomolecules like phytohormone, enzymes, anticancer, antimicrobial, and phytotoxic compounds, etc. Therefore, the endophytes associated with *C. sativus* can be harnessed to develop agro-technologies for sustainable cultivation of saffron and also yield bioactive natural products for pharmacological and industrial applications.

5.8 CONCLUSION

C. sativus is an important medicinal and aromatic plant. It is the only plant species that produces apocarotenoids like crocin, picrocrocin, and safranal in significant amounts. These compounds impart organoleptic properties to saffron, making it world's costliest spice. This plant has remained outside the realm of genetic improvement because of its sterile nature. Poor agronomic practices and disease management together with lack of breeding approaches have led to declining trend in saffron production and quality. This advocates the need to explore other possibilities for enhancing the production of *Crocus* apocarotenoids. The plant–endophyte interface provides an important ecological marketplace for harnessing the potential of endophytes to produce compounds of therapeutic potential or exert their positive influence on plants to enhance the production of specialized metabolites of plant origin. *C. sativus* harbors a great diversity of fungal and bacterial endophytes. These endophytes produce a diverse array of bioactive molecules that can be harnessed for pharmacological and industrial applications.

KEYWORDS

- **saffron**
- **Kashmir**
- **bioactive**
- **phytopathogens**
- **endophytes**

REFERENCES

Ahrazem, O.; Rubio-Moraga, A.; Castillo-Lopez, R.; Trapero-Mozos, A.; Gómez-Gómez. L. *Crocus sativus* Pathogens and Defence Responses. In *Functional Plant Science and*

Biotechnology; Husaini, A. M. Eds., Saffron Global Science Books, Ltd, London, 2010; pp 81–90.

Ambardar, S.; Vakhlu, J. Plant Growth Promoting Bacteria from *Crocus sativus* Rhizosphere. *World J. Microbiol. Biotechnol.* **2013,** 1–9.

Ambardar, S.; Sangwan, N.; Manjula, A.; Rajendhran, J.; Gunasekaran, P.; Lal, R,; Vakhlu, J. Identification of Bacteria Associated with Underground Parts of *Crocus sativus* by 16S rRNA Gene Targeted Metagenomic Approach. *World J. Microbiol. Biotechnol.* **2014,** *30*, 2701–2709.

Arnold, A. E. Understanding the Diversity of Foliar Fungal Endophytes: Progress, Challenges, and Frontiers. *Fungal Biol. Rev.* **2007,** *21*, 51–66.

Ashraf, N.; Jain, D.; Vishwakarma, R. A. Identification, Cloning and Characterization of an Ultrapetala Transcription Factor *CsULT1* from *Crocus*, A Novel Regulator of Apocarotenoid Biosynthesis. *BMC Plant Biol.* **2015,** *15*, 25.

Aytekin, A.; Acikgoz, A.O. Hormone and Microorganism Treatment in the Cultivation of Saffron (*Crocus sativus* L.) Plants. *Molecules* **2008,** *13*, 1135–1146.

Baba, S. A.; Malik, A. H.; Wani, Z. A.; Sumji, T.; Shah, Z.; Ashraf, N. Phytochemical Analysis and Antioxidant Activity of Different Tissue Types of *Crocus sativus* and Oxidative Stress Alleviating Potential of Saffron Extract in Plants, Bacteria, and Yeast. *S. Afr. J. Bot.* **2015a,** *99*, 80–87.

Baba, S. A.; Mohiuddin, T.; Basu, S.; Swarnkar, M. K.; Malik, A. H.; Wani, Z. A.; Abbas, N.; Singh, A. K.; Ashraf, N. Comprehensive Transcriptome Analysis of *Crocus sativus* for Discovery and Expression of Genes Involved in Apocarotenoid Biosynthesis. *BMC Genomics* 2015b, *16*, 698.

Bhargava, V. K. Medicinal Uses and Pharmacological Properties of *Crocus* sativus Linn (saffron). *Int. J. Pharm. Pharm. Sci.* **2011,** *3*, 322–326.

Busconi, M.; Colli, L.; Sánchez, R. A.; Santaella, M.; Pascual, M. D.; Santana, O.; Roldán, M.; Fernández, J. A. AFLP and MS-AFLP Analysis of the Variation Within Saffron Crocus (*Crocus sativus* L.) Germplasm. *PLoS One* **2015,** *10*, e0123434.

Caiola, M. G.; Canini, A. Looking for Saffron's (*Crocus sativus* L.) Parents. *Funct. Plant Sci. Biotechnol.* **2010,** *4*, 1–14.

Cappelli, C. Occurrence of *Fusarium oxysporum* f sp. Gladioli on Saffron in Italy. *Phytopathol. Mediterr.* **1994,** *33*, 93–94.

Cappelli, C.; Di Minco, G. Three-Years of Trials on Saffron Diseases in Abruzzo (Central Italy) *Crocus sativus* L. *Inf. Fitopatol.* **1999,** *49*, 27–32.

Fernández, J. A. Biology, Biotechnology and Biomedicine of Saffron. *Recent Res. Dev. Pl. Sci.* **2004,** *2*, 127–159.

Frusciante, S.; Diretto, G.; Bruno, M.; et al. Novel Carotenoid Cleavage Dioxygenase Catalyzes the First Dedicated Step in Saffron Crocin Biosynthesis. *Proc. Natl. Acad. Sci. U.S.A.* **2014,** *111*, 12246–122451.

Gresta, F.; Lombardo, G. M.; Siracusa, L.; Ruberto, G. Saffron, An Alternative Crop for sustainable agricultural systems: A Review. *Agron. Sustain. Dev.* **2008,** *28*, 95–112.

Gupta, R.; Vakhlu, J. Native *Bacillus amyloliquefaciens* W2 as a Potential Biocontrol for *Fusarium oxysporum* R1 Causing Corm Rot of *Crocus sativus. Eur. J. Plant. Pathol.* **2015,** *143*, 123–131.

Hosseinzadeh, H.; Nassiri-Asl, M. Avicenna's (Ibn Sina) the Canon of Medicine and Saffron (*Crocus sativus*): A Review. *Phytother. Res.* **2013,** *27*, 475–483.

Husaini, A. M.; Hassan, B.; Ghani, M. Y.; Teixeira da Silva, J. A.; Kirmani, N. A. Saffron (*Crocus sativus* L. Kashmirianus) Cultivation in Kashmir: Practices and Problems. *Funct. Plant Sci. Biotechnol.* **2010,** *4*, 108–115.

Javadi, B.; Sahebkar, A.; Emami, S. A. A Survey on Saffron in Major Islamic Traditional Medicine Books. *Iran J. Basic Med. Sci.* **2013,** *16*, 1–11.

Kumar, R.; Singh, V.; Devi, K.; Sharma, M.; Singh, M. K.; Ahuja, P. S. State of Art of Saffron (*Crocus sativus* L.) Agronomy, A Comprehensive Review. *Food Rev. Int.* **2009,** *25*, 44–85.

López, R. C.; Gómez-Gómez, L. Isolation of a New Fungi and Wound-Induced Chitinase Class in Corms of *Crocus sativus. Plant Physiol. Biochem.* **2009,** *47*, 426–434.

Mir, G. H.; Devi, L. S.; Ahmad, S.; Kumar, V. M.; Williams, P. Antagonistic Potential of Native Isolates of *Trichoderma viride* on Corm Rot Pathogen Complex of Saffron (*Crocus sativus*) in Kashmir. *Plant Pathol. J.* **2011,** 10.

Nalli, Y.; Mirza, D. N.; Wani, Z. A.; Wadhwa, B.; Mallik, F. A.; Raina, C.; Chaubey, A.; Riyaz-Ul-Hassan, S.; Ali, A. Phialomustin A-D, New Antimicrobial and Cytotoxic Metabolites from an Endophytic Fungus, *Phialophora mustea. RSC Adv.* **2015,** *5* (115), 95307–95312.

Navarro, L.; Dunoyer, P.; Jay, F.; Arnold, B.; Dharmasiri, N.; Estelle, M.; Voinnet, O.; Jones, J. D. G. A Plant miRNA Contributes to Antibacterial Resistance by Repressing Auxin Signaling. *Science* **2006,** *312*, 436–439.

Nazir, N. A.; Khitrov, N. B.; Chizkhikova, N. P. Statistical Evaluation of Soil Properties Which Influence Saffron Growth in Kashmir. *Eurasian Soil Sci.* **1996,** *28*, 120–138.

Negbi, M.; Dagan, D.; Dror, A.; Basker, D. Growth, Flowering, Vegetative Reproduction and Dormancy in the Saffron Crocus (*Crocus sativus* L). *Israel J. Bot.* **1989,** *38*, 95–113.

Palmero, D.; Rubio-Moraga, A.; Galvez-Patón, L.; Nogueras, J.; Abato, C.; Gómez-Gómez, L.; Ahrazem, O. Pathogenicity and Genetic Diversity of *Fusarium oxysporum* Isolates from Corms of *Crocus sativus. Ind. Crops Prod.* **2014,** *61*, 186–192.

Porras-Alfaro, A.; Bayman, P. Hidden Fungi, Emergent Properties: Endophytes and Microbiomes. *Annu. Rev. Phytopathol.* **2011,** *49*, 291–315.

Rubio-Moraga, A.; et al. Cytosolic and Plastoglobule-Targeted Carotenoids Dioxygenases from *Crocus sativus* are Both Involved in Beta-Ionone Release. *J. Biol. Chem.* **2008,** *283*, 24816–24825.

Rubio-Moraga, A.; Nohales, P. F.; Pérez, J. A.; Gómez-Gómez, L. Glucosylation of the Saffron Apocarotenoid Crocetin by a Glucosyltransferase Isolated from *Crocus sativus* Stigmas. *Planta* **2004,** *219*, 955–966.

Sharaf-Eldin, M.; Elkholy, S.; Fernandez, J.; Junge, H.; Cheetham, R.; Guardiola, J.; Weathers, P. *Bacillus subtilis* FZB24 Affects Quantity and Quality of Saffron (*Crocus sativus* L.). *Planta Med.* **2008,** *74*, 1316–1320.

Sharma, T.; Kaul, S.; Dhar, M. K. Diversity of Culturable Bacterial Endophytes of Saffron in Kashmir, India. *Springerplus* **2015,** *4*: 661.

Wani, Z. A.; Ashraf, N.; Mohiuddin, T.; Riyaz-Ul-Hassan, S. Plant-Endophyte Symbiosis, An Ecological Perspective. *Appl. Microbiol. Biotechnol.* **2015,** *99*, 2955–2965.

Wani, Z. A.; Mirza, D. N.; Arora, P.; Riyaz Ul Hassan, S. Molecular Phylogeny, Diversity, Community Structure, and Plant Growth Promoting Properties of Fungal Endophytes Associated with the Corms of Saffron Plant: An Insight into the Microbiome of *Crocus sativus* Linn. *Fungal Biol.* **2016,** *120*, 1509–1524.

Wani, Z. A.; Kumar, A.; Sultan, P.; Bindu, K.; Riyaz-Ul-Hassan, S.; Ashraf, N. *Mortierella alpina* (CS10E4), An Oleaginous Fungal Endophyte of *Crocus sativus* L. Enhances Apocarotenoid Biosynthesis and Stress Tolerance in Host Plant. *Sci. Rep.* **2017,** *7*(1), 8598.

Wani, Z. A.; Ahmad, T.; Nalli, Y.; Ali, A.; Singh, A. P.; Vishwakarma, R. A.; Ashraf, N.; Riyaz-Ul-Hassan, S. *Porostereum* sp., Associated with Saffron (*Crocus sativus* L.), is a Latent Pathogen Capable of Producing Phytotoxic Chlorinated Aromatic Compounds. *Curr. Microbiol.* **2018.** doi: 10.1007/s00284-018-1461-9.

Yamamoto, W.; Omatsu, T.; Takami, K. Studies on the Corm Rots of *Crocus sativus* L. On Saprophytic Propagation of *Sclerotinia gladioli* and *Fusarium oxysporum* f. sp. Gladioli on Various Plants and Soils. *Sci. Rep. Hyogo Univ. Agric.* **1954,** *1*, 64–70.

Zheng, C. J.; Li, L.; Zou, J. P.; Han, T.; Qin, L. P. Identification of a Quinazoline Alkaloid Produced by *Penicillium vinaceum*, An Endophytic Fungus from *Crocus sativus. Pharm. Biol.* **2012,** *50*, 129–133.

CHAPTER 6

Exploring Endophytic Microbiome of *Glycyrrhiza* sp.

PALAK ARORA[1] and ZAHOOR AHMED WANI[2*]

[1]*302 Sec 1, Channi Himmat, Jammu Tawi, India-180015*

[2]*Department of Botany, Govt. Degree College Kishtwar, Kishtwar, Jammu & Kashmir, India*

[*]*Corresponding author. E-mail: zawani1986@gmail.com*

ABSTRACT

Licorice is an important medicinal and aromatic plant and due to its ethnopharmacological value, it is also known as "the king of Chinese medicines". The endophytic microbiome of *G. glabra* comprises diverse group fungi mainly belonging to Ascomycota. *Phoma* and *Fusarium* species dominate the endophytic community of the host plant. The endophytes displayed differential tissue specificity and the diversity of endophytes varies with geographical location. Several endophytes possess plant growth–promoting traits, whereas none was pathogenic to the plant. All the endophytic taxa enhanced plant secondary metabolites under in vitro conditions, indicating these endophytes may have an important role to play in growth and development of the host plant. The interaction between the endophytes with the host plant needs to be explored further, which might lead to microbial formulations for the sustainable cultivation and productivity of plant.

6.1 INTRODUCTION

Glycyrrhiza meaning "sweet root" has its origin from the Greek words "glykos" (sweet) and "rhiza" (root). Licorice is the common name of *Glycyrrhiza glabra, and* locally it is known as Mulethi. *G. glabra* is a perennial herb belonging to the legume family (Fabaceae). It has a subcosmopolitan distribution in Asia, Australia, Europe, and the Americas. The leading producers of licorice include India, Iran, Italy, Afghanistan, China, Pakistan, Iraq, Azerbaijan, Uzbekistan, Turkmenistan, and Turkey. *Glycyrrhiza* comprises approximately 20 species, but only two species that is, *Glycyrrhiza uralensis* and *G. glabra* are reported to possess potential pharmacological applications (Barghi and Siljak-Yakovlev, 1990). Licorice is an important medicinal and aromatic plant and due to its ethanopharmacological value, it is used in traditional/folk medicine to cure various ailments. Therefore, it is also known as "the king of Chinese medicine" (Sharma and Agrawal, 2013). Due to its medicinal value and sweet taste, licorice finds its application in pharmaceutical and food industry. The root part of *G. glabra* is used to extract a sweet flavor that is used in herbalism and traditional medicine. *G. glabra* is a perennial herbaceous plant that grows to a height of 0.5–1.5 m (Fig. 6.1). They have deep stoloniferous root system, which in arid areas might be of several meters long (Kushiev et al. 2005). The flowers are long, purple to whitish blue in color, and are produced in a loose inflorescence. The fruit is an oblong pod containing several seeds. The propagation of *G. glabra* occurs mostly by vegetative means as the germination rate of seeds is very low (Thirugnanam et al. 2008). *Glycyrrhiza* grows best in well-drained soils with full sunny days and is harvested in autumn after two to three years of planting.

Apart from the medicinal and aromatic importance, *Glycyrrhiza* is also used as sweeteners and flavoring agent in food and confectionery industry. Other applications of *Glycyrrhiza* include use in skin care, personal care, and cosmetics. *Glycyrrhiza* is also known for promising antitumor, antimicrobial, antiviral, anti-inflammatory, antidiabetic, immunoregulatory, and hepatoprotective and neuroprotective activities (Park et al. 2004; Yang et al. 2017). *Glycyrrhiza* plant tolerates harsh environmental conditions and has been used in remediation of abandoned saline soils in hungry steppes of central Asia (Habibjon et al. 2005). *Glycyrrhiza* plant is nodulated by rhizobial bacteria that fix atmospheric nitrogen and help in promoting the growth of host plant (Li et al. 2012).

6.2 MEDICINAL VALUE OF *GLYCYRRHIZA*

From the ancient medical history of Ayurveda, *G. glabra* is known as a medicine and a flavoring herb to overcome the unpleasant flavors of other medications. Greek botanists reported the medicinal importance of the plant

FIGURE 6.1 Morphology of *Glycyrrhiza glabra* plant.

and recommended it for gastric and peptic ulcers. In Asia and Europe, the extract of licorice is also used to treat psoriasis. Licorice is used to relieve "Vata" and "Kapha" inflammations, eye and throat infections, arthritic conditions, and hepatic problems in Indian Ayurveda system.[1] Licorice is also used in the treatment of various other ailments like acidity, leucorrhea, jaundice, bronchitis, diarrhea, and fever (Sheth, 2005; Kaur et al. 2013). Volatile compound of licorice root is an important ingredient in medicinal oils used for the treatment of rheumatism, hemorrhagic diseases, epilepsy, and paralysis. *G. glabra* is a good pain reliever, a remedy for discomfort caused by acrid matter in the stomach due to its alkalizing effect (Chopra

[1] The Ayurvedic Pharmacopoeia of India, 2001.

and Chopra, 1958). It is an excellent tonic and is also used as demulcent in catarrh of the genitourinary passages (Nadkarni, 1976). In Asia, *Glycyrrhiza* is also commonly used in folk medicine as an anti-inflammatory agent on neutrophil functions, including reactive oxygen species (ROS) generation.

The pharmaceutical value of *Glycyrrhiza* lies in its capacity to produce a variety of secondary metabolites comprising terpenes, flavonoids, isoflavonoids, chalcones, coumarins, and polysaccharides (Seki et al. 2011). *Glycyrrhiza* also produces an array of volatile compounds, out of which about 35% are terpenoids with octanoic acid, paeonol, octadecane, benzaldehyde, α-terpineol, and 4-terpineol as its main constituents (Miyazawa and Kameoka, 1990). Rhizome of the plant is the main site for biosynthesis of glycyrrhizin and glycyrrhetinic acid (Hayashi and Sudo, 2009). The sweetness in licorice comes from glycyrrhizin, which is 30–35 times sweeter than sugar. The isoflavene glabrene and the isoflavene glabridin found in the root of licorice are phytoestrogens. The importance of this plant in modern medicine has increased in the recent past due to the activity of glycyrrhizin against hepatocellular carcinogenesis and prostate cancers (Thirugnanam et al. 2008). The medicinal value of *Glycyrrhiza* and renewed interest in glycyrrhizin have led to its increased demand. The primary active ingredient of liquorice root extract is glycyrrhizin (glycyrrhizic acid; glycyrrhizinate), which constitutes 10–25% of the total metabolic contents. Glycyrrhizin is a saponin compound comprising a triterpenoid aglycone, glycyrrhetic acid (glycyrrhetinici acid) conjugated to a disaccharide of glucuronic acid. Glycyrrhizin and glycyrrhetic acid can exist in the 18α and 18β stereoisomer forms.[2] It is reported that glycyrrhizin acts as a quenching agent of free radicals and also as blocking agent of lipid peroxidation (Rahman and Sultana, 2007).

The methanol extract of aerial parts of *G. glabra* exhibited antibacterial activity against various species of bacteria, including *Helicobacter pylori* (Fukai et al. 2002). Glabridin, glabrene, and licochalcone A exhibited antimicrobial activity against *Helicobacter pylori* under in vitro condition (Motsei et al. 2003). The alcohol extract of the root of *G. glabra* is found to possess antifungal activity against *Candida albicans* (Hojoa and Satob, 2002) and against other fungi *Arthrinium sacchari* and *Chaetomium funicola* (Fatima et al. 2009). Various constituents with anti-oxidant capacity were isolated from *G. glabra* that includes isoflavene, hispaglabridin A, hispaglabridin B, glabridin, and 4¢-*O*-methylglabridin; the two chalcones, isoprenylchalcone derivative and isoliquiritigenin; and the isoflavone, formononetin. Among these compounds, glabridin was found as the most abundant and potent

[2] Food Chemicals Codex, 2003.

anti-oxidant. Glycyrrhizin and glabridin inhibit the generation of ROS by neutrophil at the site of inflammation (Wang and Nixon, 2001). Glycyrrhizic acid also displays antiviral activity; it inactivates herpes simplex virus particles irreversibly (Pompei et al. 1979).

The ethanol extract obtained from the root of *G. glabra* exhibited anticonvulsant activities (Ambawade et al. 2002). The aqueous extract of liquorice had a significant effect in alleviating liver functions as well as restoring hepatic tissue in acute liver diseases. The extract of *G. glabra* is used as an alternative to bismuth, which is commonly known to play protective role against acid and pepsin secretions (Asl et al. 2008). Liquorice accelerates the metabolism of cells in the bone marrow erythroid stem and increases resistance to stress in animals. It is also reported to exhibit antiplatelet aggregation effect (Yu et al. 2005).

6.3 ENDOPHYTES OF *GLYCYRRHIZA*

Glycyrrhiza plant has symbiotic rhizobial association that helps in fixation of atmospheric nitrogen. Isolation of rhizobacteria from the roots of different species of *Glycyrrhiza* is reported, for example, *Mesorhizobium tianshanense* is isolated from *Glycyrrhiza pallidiflora.* Li and colleagues reported various symbiotic and other endophytic bacteria associated with *Glycyrrhiza* sp. (Li et al. 2012). A total of 159 endophytic bacteria are isolated from the root nodules of wild perennial *Glycyrrhiza* plant growing in China. The isolated bacterial symbionts are grouped into nodulating and nonnodulating species. The nonnodulating endophytic bacteria include *Agrobacterium* sp., *Enterobacter cloacae*, *Paenibacillus* sp., *Phyllobacterium bourgognense*, *Phyllobacterium* sp., and *Rhizobacterium daejeonense*. The nodulating rhizobacteria include *Mesorhizobium mediterraneum*, *M. tianshanense*, *Mesorhizobium* group A, *Mesorhizobium* group B, *Sinorhizobium meliloti*, *Rhizobium gallicum*, *R. galegae*, *R. leguminosarum*, *R. cellulosilyticum*, *R. giardini*, *Phyllobacterium* sp. Five distinct *Mesorhizobium* groups represented true symbionts of the host plant, and the majority of strains have a role in inducing N_2-fixing nodules (Li et al. 2012). Zhao et al. (2016) isolated 126 and 92 actinobacteria strains from wild perennial liquorice plants *Glycyrrhiza inflata* and *G. glabra*, respectively. Most of the strains belong to the genus *Streptomyces*. The endophytic strains belonging to genus *Micromonospora* and *Rhodococcus* were isolated from both the *Glycyrrhiza* species, while as the endophytic strains belonging to genus *Tsukamurella* were isolated only from *G. glabra* (Zhao et al. 2016).

Zhao and colleagues investigated the actinobacterial diversity associated with *G. inflata.* They isolated actinobacteria from 1-year old and 3-year old *G. inflata* plants and recovered 36 and 52 endophytic actinobacterial strains, respectively, showing distinct morphological characteristics. The strains isolated from 1-year old plants belong to the orders *Streptomycetales*, *Corynebacteriales*, *Micromonosporales*, and *Micrococcales*. Most of them belong to the genus *Streptomyces.* The strains isolated from 3-year old plants were more diverse than those from 1-year old plants and belong to the orders, *Streptomycetales, Micromonosporales, Micrococcales, Propionibacteriales, and Streptosporangiales.* These strains were represented by 10 genera: *Streptomyces, Micromonospora, Actinokineospora, Arthrobacter, Actinomadura, Oerskovia, Cellulomonas, Nocardioides, Promicromonospora, and Rhodococcus* (Zhao et al. 2018). In another study, Li et al. (2018) isolated 116 endophytic bacteria from wild populations of *G. uralensis* in China. Molecular phylogeny based on 16S ribosomal gene sequence acquisition assigned these endophytic bacteria into 35 species belonging to 20 distinct genera. All the endophytic strains belong to *Firmicutes*, *Actinobacteria*, and *Proteobacteria.* Most of the isolates belong to the genus *Bacillus* in *Firmicutes*, but the highest genus-level diversity was in *Actinobacteria* (Li et al. 2018). The endophytic bacteria exhibited a number of plant growth-promoting activities, including auxin synthesis, diazotrophy, siderophore production, phosphate and potassium solubilization, and production of hydrolytic enzymes (Li et al. 2018).

Arora et al. (2019) made a comprehensive effort to understand the diversity and community structure of fungal endophytes associated with *G. glabra*. A total of 266 fungal endophytic isolates were isolated from *G. glabra,* which are grouped into 100 morphotypes based on phenotypic characters like growth pattern, colony texture, colony color as well as morphology of conidia and conidiophores. Molecular phylogenetic studies based on ITS1-5.8S-ITS2 ribosomal gene sequence analyses assigned these endophytes into 38 distinct species, spreading over 21 genera (Table 6.1) (Arora et al. 2019). The diversity of endophytes in *G. glabra* was relatively high in aboveground part with a colonization frequency of 58.6% as compared to the underground part with a colonization frequency of 53.2%. Out of the 38 endophytic fungal species recovered from *G. glabra*, 16 endophytic species are specific to the aboveground part of the plant only, which include *Fusarium oxysporum, Talaromyces verruculosus*, *Alternaria alternata*, *Curvularia aeria*, *Didymella bryoniae*, *Botrytis cinerea*, *Alternaria* sp., *Alternaria brassicae*, *Cladosporium tenuissimum*, *Fusarium equiseti*, *Aspergillus flavus*, *Cladosporium cladosporioides*, *Stagonosporopsis cucurbitacearum*, *Aspergillus terreus*, *Xylaria* sp., and

TABLE 6.1 The 36 Different Fungal Endophytes (Their GenBank Accession Numbers and Isolation Frequency) Isolated from *Glycyrrhiza glabra.*

ITS genotypes	Molecular identification (GenBank accession no.)	Isolation frequency (%)
ITS1	*Phoma* sp. (KY419531)	12.03
ITS2	*Phoma macrostoma* (KY419532)	7.89
ITS3	*Diaporthe* sp. (KY419533)	7.14
ITS4	*Phomopsis* sp. (KY419534)	4.89
ITS5	*Fusarium oxysporum* (KY419535)	4.51
ITS6	*Phoma exigua* (KY419536)	4.51
ITS7	*Fusarium incarnatum* (KY419537)	3.38
ITS8	*Colletotrichum* sp. (KY419538)	3.38
ITS9	*Talaromyces verruculosus* (KY419539)	3.38
ITS10	*Rhizoctonia* sp. (KY419540)	3.38
ITS11	*Alternaria alternata* (KY419541)	3.01
ITS12	*Diaporthe terebinthifolii* (KU168142)	2.63
ITS13	*Botryosphaeria dothidea* (KY419542)	2.63
ITS14	*Curvularia aeria* (KY419543)	2.63
ITS15	*Mucor circinelloides* (KY419544)	2.63
ITS16	*Fusarium solani* (KY419545)	2.26
ITS17	*Lasiodiplodia theobromae* (KY419546)	2.26
ITS18	*Didymella bryoniae* (KY419547)	2.26
ITS19	*Macrophomina phaseolina* (KY419548)	2.26
ITS20	*Rhizopus oryzae* (KY419549)	2.26
ITS21	*Fusarium avenaceum* (KY419550)	2.26
ITS22	*Botrytis cinerea* (KY419551)	2.26
ITS23	*Fusarium brachygibbosum* (KY419552)	1.88
ITS24	*Lasiodiplodia pseudotheobromae* (KY419553)	1.5
ITS25	*Mucor hiemalis* (KY419554)	1.5
ITS26	*Alternaria* sp. (KY419555)	1.5
ITS27	*Alternaria brassicae* (KY419556)	1.5
ITS28	*Cladosporium tenuissimum* (KY419557)	1.13
ITS29	*Alternaria tenuissima* (KY419558)	1.13
ITS30	*Fusarium equiseti* (KY419559)	1.13
ITS31	*Alternaria porri* (KY419560)	1.13
ITS32	*Aspergillus flavus* (KY419561)	0.75
ITS33	*Cladosporium cladosporioides* (KY419562)	0.75
ITS34	*Alternaria burnsii* (KY419563)	0.75
ITS35	*Stagonosporopsis cucurbitacearum* (KU168143)	0.38
ITS36	*Aspergillus terreus* (KY419564)	0.38
ITS37	*Xylaria* sp. (KY419565)	0.38
ITS38	*Bionectria* sp. (KY419566)	0.38

Bionectria sp. A total of 10 endophytic fungal species are specific to the belowground part only, which include *Fusarium incarnatum*, *Rhizoctonia* sp., *Mucor circinelloides*, *Lasiodiplodia theobromae*, *Macrophomina phaseolina*, *Lasiodiplodia pseudotheobromae*, *Mucor hiemalis*, *Alternaria tenuissima*, *Alternaria porri,* and *Alternaria burnsii*, while as the remaining 12 endophytic fungal strains are common to both aboveground and belowground parts of the plant (Fig. 6.2). This indicates that the endophytic colonization in *G. glabra* displays tissue specificity with respect to most of the endophytes. It is presumed that the aboveground and belowground tissues of the plant represent two distinct ecological niches with different metabolic cues and micro-environments, consequently shaping their microbiota differently (Arora et al. 2019). Further, multivariate analysis of the endophytes of *G. glabra* samples collected from different locations showed that the geographical location plays an important role in the recruitment of fungal communities in *G. glabra*. It is also supported by cluster analysis of endophytes isolated from *G. glabra* using different media; it showed higher similarity between the fungal communities isolated from only one location irrespective of the media used. Therefore, it is suggested that the recruitment of endophytes in *G. glabra* is influenced by many factors, including the host plant status and geographical location (Arora et al. 2019).

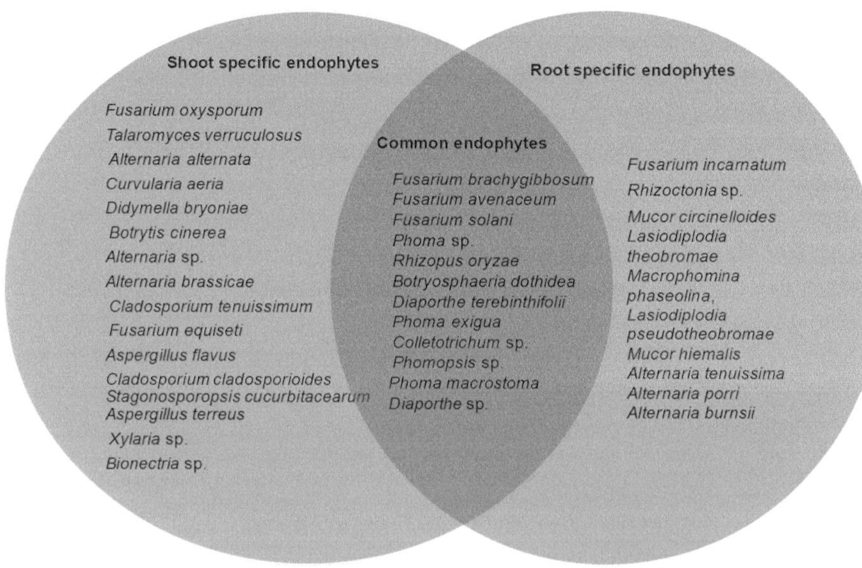

FIGURE 6.2 Venn diagram showing differential recruitment/colonization of fungal endophytes in shoots and root tissues of *Glycyrrhiza glabra*.

The endophytic community structure in *G. glabra* indicates most of the endophytes belong to the fungal class Dothideomycetes, followed by Sordariomycetes, Mucoromycetes, Eurotiomycetes, Agaricomycetes, Euascomycetes, and Leotiomycetes. The most abundant class is Ascomycota with 18 genera, followed by Zygomycota with 2 genera (*Mucor* and *Rhizopus*) and Basidiomycota with 1 genus (*Rhizoctonia*) (Fig. 6.3). The fungal genera *Phoma* and *Fusarium* constitute the two most dominant groups with an isolation frequency of 24.4% and 15.4%, respectively (Arora et al. 2019). It is pertinent to mention that these two fungal genera are common to both aboveground and belowground parts of the host plant. Therefore, it is presumed that *G. glabra* has strong affinity toward establishing symbiotic association with fungi belonging to the genera *Phoma* and *Fusarium.* These endophytes can be regarded as the preferred/true endophytes of *G. glabra.* To substantiate this hypothesis, there is dire need to study the diversity and community structure of the endophytes associated with *G. glabra* from other places, so as to have a better understanding of the true endophytes of *G. glabra.* Also, the mechanism of establishment of symbiotic associations by the host plant with the endophytes needs to be studied in order to understand the preferential plant–microbe interactions and their implications on the growth and metabolism of the host plant.

6.4 BIOACTIVE POTENTIAL OF FUNGAL ENDOPHYTES OF *G. GLABRA*

Endophytes carry out specific functions in nature thus helping the host plant to survive in specific environment, particularly under stress conditions. Therefore, endophytes are expected to produce antimicrobial compounds to inhibit the growth of plant pathogens and also compete with other organisms for survival in their specific niches (Wani et al. 2016). In order to understand, if the endophyte provides the host plant any resistance to phytopathogens and subsequently exploited for the production of antimicrobial compounds, the endophyte is assessed in dual/coculture assay against a panel of plant pathogens. The endophyte displaying potent activity against plant pathogens in coculture assay infers that the endophyte may be helpful to the host plant in providing resistance against pathogens. It is reported that the extract of some fungal endophytes, particularly *S. cucurbitacearum* GG1F1, showed potential antimicrobial activity. The chemo profiling of *S. cucurbitacearum* GG1F1 resulted in the isolation of two thiodiketopiperazine molecules with potential antimicrobial and biofilm inhibition activities against several human pathogens, particularly *Staphylococcus aureus* ATCC 29213 and

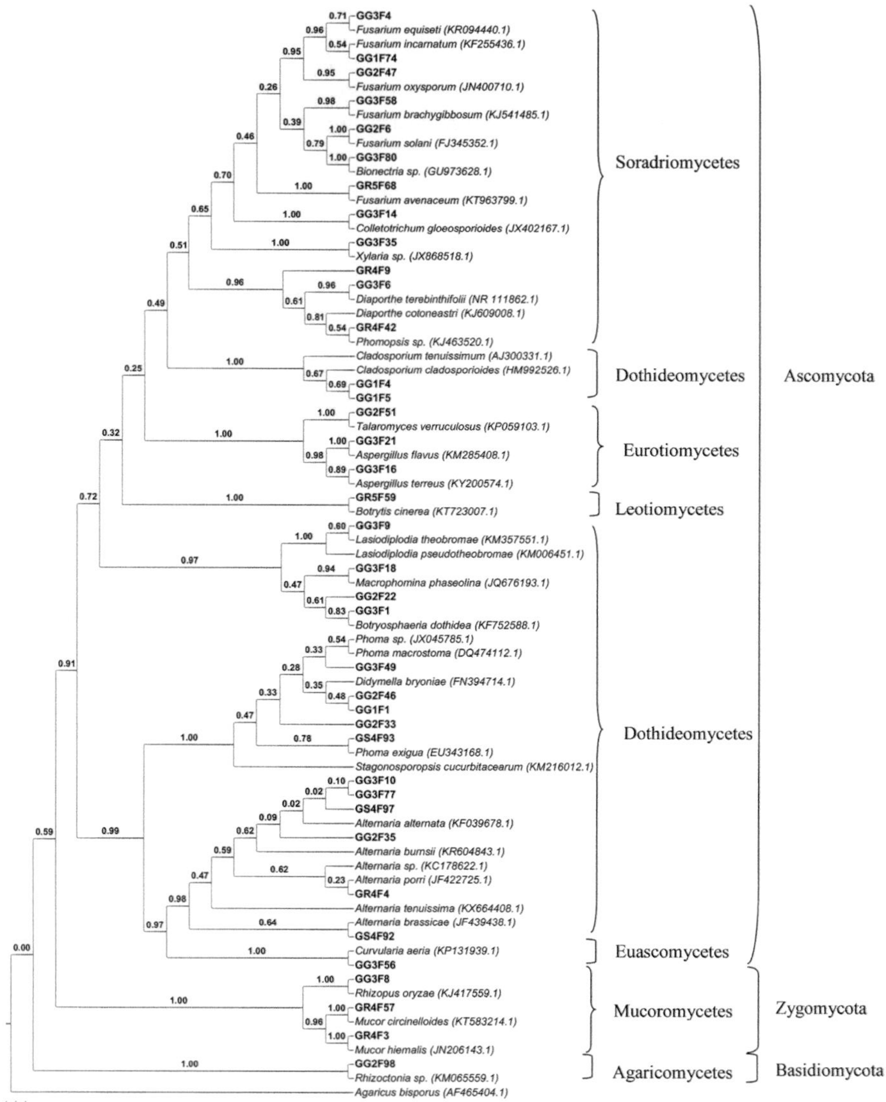

FIGURE 6.3 Phylogenetic analysis of fungal endophytes associated with *Glycyrrhiza glabra*. The strain numbers of the isolates obtained in this study are presented in bold font. The tree is rooted with *Agaricus bisporus* (an agaricomycetes, AF465404.1).

Staphylococcus pyogenes MTCC 442 (Arora et al. 2016). Apart from this, two new hydroxylated unsaturated fatty acids designated as diapolic acid A–B along with known compounds xylarolide and phomolide G were recovered from the endophytic fungus, *Diaporthe terebinthifolii* GG3F6. Xylarolide is found to exhibit potential cytotoxic activity against the breast cancer cell line T47D and moderate antifungal activity against *C. albicans* (Nalli et al. 2016). An antitubercular molecule, Fusarubin is recovered from an endophyte (*Fusarium solani*) of *G. glabra*. Fusarubin showed good activity against *Mycobacterium tuberculosis* strain H37Rv with MIC value of 8 $\mu g\ ml^{-1}$ and is suggested to be a potential drug for tuberculosis (Shah et al. 2017).

Endophytes are reported to enhance growth and development of the host plant by virtue of other biochemical properties (Rodriguez et al. 2009). An important aspect of the endophytes is the production of plant growth hormones, like Indole acetic acid (IAA), gibberellins, etc. All the endophytic strains isolated from *G. glabra* produces IAA under in vitro condition in varying concentrations (Arora et al. 2019). Phytohormone production by endophytes is probably the best-studied mechanism of plant growth promotion, contributing to overall growth and development of the plant. IAA is found to be involved in biotic and abiotic stress tolerance in the host plant, while it also plays an important role in colonization of the endophytes in plant tissues (Wani et al. 2015). This may be the reason for the endophytes being efficient producers of IAA. More than 50% of the endophytes isolated from *G. glabra* are able to produce siderophore, indicating their plant growth–promoting potential. The endophytes of *G. glabra* are also reported to produce hydrolytic enzymes that are believed to play some role in colonization of endophytes in the host tissue. It is also reported that no pathogenesis symptom was detected in the host plant upon reinfection with the isolated endophytes. This is in confirmation that *G. glabra* is rarely infected by any pathogens (Janke, 2004). Some endophytes of *G. glabra* are reported to enhance the production of secondary metabolites, particularly the total phenolic and flavonoid content, in the host plant under in vitro condition. Phenolics, flavonoids, and carotenoids are responsible for the anti-oxidant potential of the plant (Shahidi and Ambigaipalan, 2015). Therefore, it is hypothesized that the endophytes associated with *G. glabra* are involved in enhancing the anti-oxidant potential of the host plant, thereby increasing its pharmaceutical value. However, this hypothesis needs to be investigated further in future research work.

6.5 CONCLUSION

The endophytic microbiome of *G. glabra* comprises diverse group fungi mainly belonging to Ascomycota. *Phoma* and *Fusarium* species dominate the endophytic community of the host plant. The endophytes displayed differential tissue specificity and the diversity of endophytes varies with geographical location. Several endophytes possess plant growth–promoting traits, whereas none was pathogenic to the plant. All the endophytic taxa enhanced plant secondary metabolites under in vitro conditions, indicating these endophytes may have an important role to play in growth and development of the host plant. The interaction between the endophytes with the host plant needs to be explored further, which might lead to microbial formulations for the sustainable cultivation and productivity of plant. Also, the application of culture-independent metagenomics approach for the characterization of whole microbiome may add to the knowledge on endophytes of *G. glabra* and give a better understanding of the microbes associated with the plant.

KEYWORDS

- **licorice**
- **endophytes**
- **anti-microbial**
- **plant-microbe interaction**
- **ethnopharmacological**

REFERENCES

Ambawade, S. D.; Kasture, V. S.; Kasture, S. B. Anticonvulsant Activity of Roots and Rhizomes of *Glycyrrhiza glabra*. *Indian J. Pharmacol.* **2002,** *34*, 251–255.

Arora, P.; Wani, Z. A.; Nalli, Y.; Ali, A.; Riyaz-Ul-Hassan, S. Antimicrobial Potential of Thiodiketopiperazine Derivatives Produced by *Phoma* sp., An Endophyte of *Glycyrrhiza glabra* Linn. *Microb. Ecol.* **2016.** DOI:10.1007/s00248-016-0805-x.

Arora, P.; Wani, Z. A.; Ahmad, T.; Sultan, P.; Gupta, S.; Riyaz-Ul-Hassan, S. Community Structure, Spatial Distribution, Diversity and Functional Characterization of Culturable Endophytic Fungi Associated with *Glycyrrhiza glabra* L. *Fungal Biol.* **2019.** https://doi.org/10.1016/j.funbio.2019.02.003

Asl, M. N.; Hosseinzadeh, H. Review of Pharmacological Effects of *Glycyrrhiza* sp. and Its Bioactive Compounds. *Phytother. Res.* **2008,** *22*, 709–724.

Barghi, N.; Siljak-Yakovlev, S. Karyological Study in Three Species of *Glycyrrhiza* Genus (*G. glabra*, *G. lepidota* and *G. echinata*). *Caryologia* **1990,** *43*, 223–234.

Fatima, A.; Gupta, V. K.; Luqman, S.; Negi, A. S.; Kumar, J. K.; Shanker, K.; Saikia, D.; Srivastava, S.; Darokar, M. P.; Khanuja, S. P. Antifungal Activity of *Glycyrrhiza glabra* Extracts and Its Active Constituent Glabridin. *Phytother. Res.* **2009,** *23*, 1190–1193.

Fukai, T.; Marumo, A.; Kaitou, K.; Kanda, T.; Terada, S.; Nomura, T. Anti-Helicobacter Pylori Flavonoids from Licorice Extract. *Life Sci.* **2002,** *71*: 1449–1463.

Habibjon, K.; Noble, A.; Iskandar, A.; Uktam, T. Remediation of Abandoned Saline Soils Using *Glycyrrhiza glabra*: A Study from the Hungry Steppes of Central Asia. *Int. J. Agric. Sustain.* **2005,** *3*, 102–113.

Hayashi, H.; Sudo, H. Economic Importance of Licorice. *Plant Biotechnol. J.* **2009,** *26*, 101–104.

Hojoa, H.; Satob, J. Antifungal Activity of Licorice (*Glycyrrhiza glabra*) and Potential Applications in Beverage Foods. *Foods Ingredients J. Jpn.* **2002,** 203.

Janke, R. Farming a Few Acres of Herbs: Licorice, Kansas State University. 2004. https://www.bookstore.ksre.ksu.edu/pubs/mf2616.

Kushiev, H.; Noble, A. D.; Abdullaev, I.; Toshbekov, U. Remediation of Abandoned Saline Soils using *Glycyrrhiza glabra*: A Study from the Hungry Steppes of Central Asia. *Int. J. Agric. Sustain.* **2005,** *3*, 102–113.

Li, L.; et al. Biogeography of Symbiotic and Other Endophytic Bacteria Isolated from Medicinal *Glycyrrhiza* Species in China. *FEMS Microbiol. Ecol.* **2012,** *79*, 46–68.

Li, L.; Mohamad, O. A. A.; Ma, J.; et al. Synergistic Plant–Microbe Interactions between Endophytic Bacterial Communities and the Medicinal Plant *Glycyrrhiza uralensis F.* Antonie van Leeuwenhoek **2018.** https://doi.org/10.1007/s10482-018-1062-4

Miyazawa, M.; Kameoka, H. Volatile Flavour Components of *Glycyrrhizae radix* (*Glycyrrhiza glabra* L. var. glandulifera Regel et Herder) from China. *Flavour Fragr. J.* **1990,** *5*, 157–160.

Motsei, M.; Lindsey, K.; Van Staden, J.; Jäger, A. Screening of Traditionally Used South African Plants for Antifungal Activity Against *Candida albicans*. *J. Ethnopharmacol.* **2003,** *86*, 235–241.

Nalli, Y.; Arora, P.; Wadhwa, B.; Malik, F. A.; Vishwakarma, R. A.; Riyaz-Ul-Hassan, S.; Ali, A. Diapolic acid A-B from an Endophytic Fungus, *Diaporthe terebinthifolii* Depicting Anti-Microbial and Cytotoxic Activity. *J. Antbiot.* **2016.** doi:10.1038/ja.2016.109.

Park, H. Y.; Park, S. H.; Yoon, H. K.; Han, M. J.; Kim, D. H. Anti-Allergic Activity of 18β-Glycyrrhetinic Acid-3-*O*-β-D-Glucuronide. *Arch. Pharmacal. Res.* **2004,** *27*, 57.

Pompei, R.; Flore, O.; Marccialis, M. A.; Pani, A.; Loddo, B. Glycyrrhizic Acid Inhibits Virus Growth and Inactivates Virus Particles. *Nature* **1979,** *281*, 689–690.

Rodriguez, R. J.; White, J. F. Jr.; Arnold, A. E.; Redman, R. S. Fungal Endophytes, Diversity and Functional Roles. *New Phytol.* **2009,** *182*, 314–330.

Seki, H.; Sawai, S.; Ohyama, K.; Mizutani, M.; Ohnishi, T.; Sudo, H.; Fukushima, E. O.; Akashi, T.; Aoki, T.; Saito, K.; Muranaka, T. Triterpene Functional Genomics in Licorice for Identification of CYP72A154 Involved in the Biosynthesis of Glycyrrhizin. *Plant Cell* **2011,** *23*: 4112–4123.

Shah, A.; Rather, M. A.; Hassan, Q. P.; et al. Discovery of Anti-Microbial and Anti-Tubercular Molecules from *Fusarium solani*: An Endophyte of *Glycyrrhiza glabra*. *J. Appl. Microbiol.* **2017,** *122*, 1168–1176.

Shahidi, F.; Ambigaipalan, P. Phenolics and Polyphenolics in Foods, Beverages and Spices: Antioxidant Activity and Health Effects–A Review. *J. Funct. Foods* **2015,** *18*, 820–897.

Sharma, V.; Agrawal, R. C. *Glycyrrhiza glabra*—A Plant for the Future. *Mintage J. Pharm. Med. Sci.* **2013,** *2*(3), 15–20.

Thirugnanam, S.; Xu, L.; Ramaswamy, K.; Gnanasekar, M. Glycyrrhizin Induces Apoptosis in Prostate Cancer Cell Lines DU-145 and LNCaP. *Oncol. Rep.* **2008,** *20*, 1387–1392.

Wang, Z. Y.; Nixon, D. W. Licorice and Cancer. *Nutr. Cancer* **2001,** *39*: 1–11.

Wani, Z. A.; Ashraf, N.; Mohiuddin, T.; Riyaz-Ul-Hassan, S. Plant Endophyte Symbiosis, An Ecological Perspective. *Appl. Microbiol. Biotechnol.* **2015,** *99*, 2955–2965.

Wani, Z. A.; Mirza, D. N.; Arora, P.; Riyaz Ul Hassan, S. Molecular Phylogeny, Diversity, Community Structure, and Plant Growth Promoting Properties of Fungal Endophytes Associated with the Corms of Saffron Plant: An Insight into the Microbiome of *Crocus sativus* Linn. *Fungal Biol.* **2016,** *120*, 1509–1524.

Yang, R.; Yuan, B. C.; Ma, Y. S.; Zhou, S.; Liu, Y. The Anti-Inflammatory Activity of Licorice, A Widely used Chinese Herb. *Pharm. Biol.* **2017,** *55*, 5–18.

Yu, Z.; Ohtaki, Y.; Kai, K.; Sasano, T.; Shimauchi, H.; Yokochi, T.; Takada, H.; Sugawara, S.; Kumagai, K.; Endo, Y. Critical Roles of Platelets in Lipopolysaccharide-Induced Lethality: Effects of Glycyrrhizin and Possible Strategy for Acute Respiratory Distress Syndrome. *Int. Immunopharmacol.* **2005,** *5*, 571–580.

Zhao, K.; et al. Isolation and Antimicrobial Activities of Actinobacteria Closely Associated with Liquorice Plants *Glycyrrhiza glabra* L. and *Glycyrrhiza inflata* BAT. in Xinjiang, China. *Microbiol.* **2016,** *162*, 1135–1146.

Zhao, K.; Li, J.; Zhang, X.; et al. Actinobacteria Associated with *Glycyrrhiza inflata* Bat. Are Diverse and Have Plant Growth Promoting and Antimicrobial Activity. *Sci. Rep.* **2018,** *8,* 13661.

CHAPTER 7

Endohyphal Bacteria: Endosymbiotic Partner of Fungal Endophytes

PALAK ARORA[1] and ZAHOOR AHMED WANI[2*]

[1]*302 Sec 1, Channi Himmat, Jammu Tawi, India-180015*

[2]*Department of Botany, Govt. Degree College Kishtwar, Jammu & Kashmir 182204, India*

[*]*Corresponding author. E-mail: zawani1986@gmail.com*

ABSTRACT

The endohyphal bacteria (EHB) occur in living hyphae of phylogenetically diverse fungal endophytes isolated from various plant lineages and in multiple biogeographic provinces. EHB may be playing an important role in modulating the secondary metabolism of the fungus and loss of the endohyphal partner may result in attenuation of cultures for the production of key metabolites. This interaction represents the third component of plant–endophyte association, thus making it a tripartite association between the plant, fungus and bacteria. This tripartite relationship has a great impact on the host plant's diversity, metabolism, and ecology. It is reported that the endohyphal bacteria play a complementary protective role for the host fungus and the host plant under various stress conditions. This remarkable symbiotic association has been poorly studied and the implications of the presence of endohyphal symbiotic partner on the metabolism of fungus and host plant are inadequately understood.

7.1 INTRODUCTION

Fungi are present in various ecological niches as free-living organisms as well as in association with other organisms and play diverse ecological functions.

Fungi interact with bacteria in specific ecological niches and their association has positive effects on agriculture and environment. Apart from the most frequently observed microbial cell-cell interactions, endosymbiotic associations are also reported where bacteria reside within the fungal hyphae. This phenomenon was first discovered in mycorrhizal fungi wherein "bacterium-like organelles" (BLOs) were detected inside the fungal mycelia (MacDonald and Chandler, 1981). Recent studies have shown that foliar endophytes frequently harbor highly diverse bacteria, now called endohyphal bacteria (EHB). The endohyphal bacteria occur in living hyphae of phylogenetically diverse fungal endophytes isolated from various plant lineages and in multiple biogeograpic provinces (Arora and Riyaz-Ul-Hassan, 2018).

Fungi are regarded as incredible chemical factories producing numerous bioactive molecules. It is estimated that less than 5% of fungi are characterized so far, while a majority of the members belonging to the fungal kingdom remain to be discovered and/or characterized (Staley et al. 1997). There is an unending demand for novel antimicrobial agents due to a rapid increase in drug-resistant microbes and recurring infectious diseases. Hence, microbiologists are screening various ecological niches, including extreme environments like ocean beds, geothermal vents, and cold desserts, in search of novel fungal strains with promising bioactive potential (Staley et al. 1997; Wang et al. 2015). The source of isolation has a profound effect on the secondary metabolism of the microbe (Riyaz-Ul-Hassan et al. 2012). In practice, microbiologists face a challenging task in maintaining the fungal cultures in the active form for a longer time to produce key metabolites consistently, as the fungal cultures tend to show decline in the production of secondary metabolites in laboratory conditions after sometime. One of the possible factors for this decline in the production of secondary metabolites might be the presence and maintenance of EHB within the hyphae of fungal cultures. Thus, EHB may be playing an important role in modulating the secondary metabolism of the fungus and loss of the endohyphal partner may result in attenuation of cultures for the production of key metabolites. This remarkable symbiotic association has been poorly studied and the implications of the presence of endohyphal symbiotic partner on the metabolism of fungus and host plant are inadequately understood.

7.2 DETECTION OF EHB

The presence and viability of the EHB is detected by the Live/Dead stain and fluorescence in situ hybridization (FISH). Live/Dead strain is a mixture of SYTO9 green-fluorescent nucleic acid stain and the red-fluorescent nucleic acid stain, propidium iodide. Examination of fungal mycelia with molecular probe labeled Live/Dead fluorescent stain provides evidence that bacteria within fungal hyphae are viable and shows that such bacteria occurs within living fungal tissues (Stewart et al. 1995). FISH is a molecular technique employing fluorescent probe that binds to specific position of the chromosome with a high degree of sequence complementarity. In FISH, a bacteria-specific probe (EUB338), which is a universal 16S rRNA gene oligonucleotide, is used to confirm the presence of EHB in the fungal hyphae (Hoffman and Arnold, 2010). In this technique, fresh mycelium is harvested and dehydrated using 70% of ethanol. Dehydrated mycelium is kept at 46°C for 90 min with hybridization buffer using 40% of formamide hybridization stringency with EUB338 probe. After incubation, mycelium is mounted on glass slides using 4,6-diamidino-2-phenylindole (DAPI) as counter strain and examined under microscope equipped with a confocal system and a laser.

7.3 ISOLATION OF EHB

To isolate an EHB from fungal mycelia, the fungus is cultured in potato dextrose broth (PDB) medium and incubated at 37°C with 100 rpm in a shaker incubator. Mycelia from 7-day old culture are mechanically sheared used autoclaved mortar-pestle followed by centrifugation. The supernatant is plated onto different media, like Nutrient Agar (NA), Potato Dextrose Agar (PDA), and Water Agar (WA), and incubated at 37°C. The emerged bacterial colonies are picked up, purified, and preserved (Partida-Martinez et al. 2007). Fungus can be cured of EHB by cultivation on 2% of malt extract agar (MEA) supplemented with four antibiotics: ampicillin (100 μg/mL), kanamycin (50 μg/mL), tetracycline (10 μg/mL), ciprofloxacin (40 μg/mL), etc. It can be confirmed that the fungal culture is free of EHB by using EHB detection techniques.

7.4 HOW EHB INVADE THE FUNGAL MYCELIA?

Despite a growing number of described endohyphal symbionts, there is a lack of information on the mechanism that allows integration of EHB within the fungus or entry of EHB into fungal hyphae. Moebius and colleagues in 2014 reported new bacterial fusion processes that include the secretion of chitinolytic enzymes. Chitin is one of the main components of the fungal cell wall; it is presumed that the chitinolytic enzyme might be playing a significant role in the active intrusion of bacteria into fungal cells. To establish an intimate association, physical contact between the endohyphal partner and host fungus should take place at the right time and right place besides many other factors (Bright and Bulgheresi, 2010). It is reported that bacteria attach themselves to the fungus even before the chitinolytic enzymes are produced and secreted. The fungal cell wall penetration is a more melting-like, mild process without damaging the hyphae (Moebius et al. 2014). The two possible molecular mechanisms involved in the interaction and infection process of EHB into fungal hyphae are reported to be a type two secretion system (T2SS) and secretion of chitinase. It is suggested that T2SS is involved in the secretion of extracellular enzyme chitinase and the translocation of the associated proteins by the bacteria, which in turn softens the cell wall of fungi helping the bacteria to invade into the host fungus.

7.5 TYPES OF EHB

Initially termed "bacteria-like organisms", EHB was discovered in a vesicular-arbuscular mycorrhizal (VAM) fungus, *Acaulospora laevis* (Mosse, 1970). Later, the presence of EHB was found in several other fungi, including nonmycorrhizal fungi. The endohyphal symbionts are widespread in animals as well, particularly in insects. This is an emerging field of science, although several insights have been obtained on the interactions between Glomeromycota and their endosymbionts. Investigations on endophytic fungi revealed widespread presence of EHB in fungi. The EHB are obtained from several classes of fungi and are found to live as facultative symbionts. Recent studies have indicated that EHB are widespread in rhizospheric fungi belonging to diverse fungal phyla, for example, mycorrhizal and pathogenic fungi from Glomeromycota, Basidiomycota, Mucoromycota, and in highly diverse Ascomycota that infect roots, stem, and leaves as endophytes.

7.5.1 EHB ASSOCIATED WITH GLOMEROMYCOTA

The first EHB was discovered in a mycorrhizal fungus, *A. laevis*—belonging to phylum Glomeromycota. This EHB, also known as mycorrhiza helper bacteria, plays an important role in the stimulation of the presymbiotic growth of the mycelium. The EHB increase the survival and/or germination of fungal spores and also increase the receptivity of fungal signals by the root by stimulating the root mycelium recognition, and an alteration in the physicochemical properties that would aid in the mycorrhizal formation. There are many reports of isolation of bacterial strains from the surface-sterilized ectomycorrhiza, of which 80% are found to possess a positive effect on mycorrhiza establishment and 20% are neutral. There are reports of association of EHB with *Glomus caledonius*, *A. laevis*, *Glomus mosseae*, *Gigaspora margarita*, *Gigaspora heterogama,* and an unidentified white reticulate vesicular-arbuscular mycorrhizal spore (Macdonald et al. 1982).

In Glomeromycota, two types of EHB have widespread distribution—a rod-shaped Gram-negative bacterium, *Candidatus Glomeribacter gigasporarum* (CaGg) and a coccoid *Mollecultes* related endobacterium (Mre). The genome sequencing of CaGg revealed that it has a reduced genome and lacks metabolic pathways for important amino acids. It depends upon its fungal host for carbon, phosphorus, and nitrogen supply and most of the amino acids are imported from the host fungus. Though the fungus feeds the endobacteria, while the fungus is itself an obligate biotroph dependent on the host plant. However, the benefit of the endobacterium to the fungal partner is not well understood (Desirò et al. 2018). It is reported that the bacterial endosymbionts of arbuscular mycorrhizal (AM) fungi are involved in the synthesis of vitamin B12, antibiotics, and toxin-resistant molecules providing the ecological fitness to the fungal host (Salvioli et al. 2012). The presence of endohyphal bacteria-CaGg is reported in *G. margarita, G. gigantean, G. decipiens, Scutellospora persica,* and *S. castanea* (Bianciotto et al. 2003).

Another important endohyphal bacterial taxon clustering with the *Mollicutes* and encompassing the *Mycoplasmatales* and *Entomoplasmatales* is reported from various members of the Glomeromycota (Naumann et al. 2010). AMF can harbor both the types of endobacteria, namely, CaGg and Mre. However, the Mre are more abundant, variable, and prone to recombination as compared to CaGg. Molecular evolution patterns of Mre residing within AMF revealed that the diversity of Mre is divergent within the individual AMF, but there is not much differentiation between Mre associated with AMF from different continents. Mre can also be present as parasites of AMF Glomeromycota and this stable association is the outcome of a combination of both the transmissions modes, vertical and horizontal.

A very rare and unique association is reported between a fungus (*Geosiphon pyriformis*) and a cyanobacterium (*Nostoc punctiforme*) (Kluge, 2002). *G. pyriformis* is a monotypic species belonging Glomeromycota. This species is characterized by the formation of arbuscular mycorrhiza and it forms unicellular, multinucleated bladders containing the cyanobacterium.

N. punctiforme cells are not enclosed by fungal cell wall but they live freely in the cytoplasm of the host fungus (Schüssler, 2002). The stable association between the interacting partners is a result of the exchange of sugars by cyanobacterium and essential products required for the photosynthesis of cyanobacterium by the fungal host (Schüßler, 2012). However, the mechanism behind the exchange of metabolites between the two partners is yet to be explored.

7.5.2 EHB ASSOCIATED WITH ZYGOMYCOTA

One of the well-studied endohyphal association is between a fungal phytopathogen (*Rhizopus microsporus*) and its bacterial endosymbiont (*Burkholderia endofungorum*). The fungus is the causative agent of rice seedling blight and it produces a mycotoxin known as rhizoxin. Interestingly, the rhizoxin is not biosynthesized by the fungus itself, but by the endosymbiotic bacteria of the genus *Burkholderia* residing inside the fungal hyphae. However, the enzymes required to synthesize 2, 3-oxirane ring of the precursor in the rhizoxin biosynthetic pathway are produced by the host fungus. Thus, both the interacting partners play an important role in the biosynthesis of this phytotoxin. This interesting breakthrough unveiled a remarkably complex symbiotic-pathogenic relationship involving a tripartite association between plant–fungus–bacteria (Partida-Martinez and Hertweck, 2005). Similarly, the endosymbiont *B. endofungorum* was found to produce rhizonins, which are cyclopeptides and were wrongly regarded as mycotoxins (Lackner et al. 2009). Later, it was reported that the absence of the bacterial endosymbiont affects vegetative reproduction of the fungal host, *R. microsporus*. Formation of sporangia and spores is restored only upon reintroduction of the endobacteria to the fungal culture (Partida-Martinez et al. 2007). This showed a complex interaction, which broadens the scope of plant–fungus associations leading to its third component.

The endosymbiont resides within the fungal cytosol, as shown by transmission electron microscopy (TEM), confocal laser scanning microscopy, and freeze–fracture EM (Partida-Martinez et al. 2007). The endosymbiont bacterium produces potential antimitotic macrolides, which are then converted to phytotoxic rhizoxin by the host fungus. This phytotoxin is the causative agent of rice seedling blight that weakens or kills the rice plants (Scherlach et al. 2012). As rhizoxin possesses potential antimycotic activity against a number of eukaryotic fungi, the host *R. microsporus* must be resistant to the toxin produced by its endosymbiont. The antimycotic activity of rhizoxin is by attacking the β-tubulin and the presence of four amino acids, namely, isoleucine, valine, serine, and alanine, in the β-tubulin sequences is pivotal for conferring resistance against the toxin. The evolution of rhizoxin resistance is an important key to maintain the stable association between the fungal host and its bacterial endosymbiont (Schmitt et al. 2008). Another important novel EHB, *Mycoavidus cysteinexigens* gen. nov., is found to be associated with *Mortierella elongate* (Sato et al. 2010). The endobacterium (*M. cysteinexigens*) lack key genes involved in cysteine biosynthesis and glycolytic pathway, therefore the endobacterium was isolated from the host using cysteine supplemented media.

7.5.3 EHB ASSOCIATED WITH BASIDIOMYCETE

The first report of association of EHB with a basidiomycete is that of the fungal host—*Laccaria bicolor* and bacterial endosymbiont—*Paenibacillus* sp. (Bertaux et al. 2003). The endosymbiont inhabited both live and dead cells of the fungal host. However, the role of endobacterium in the ectomycorrhizal symbiosis is not clear. Another important association of EHB with a Basidiomycete is a Gram-positive nitrogen-fixing bacterium *Bacillus pumilus,* present inside the hyphae of a pathogenic fungal strain *Ustilago maydis.* The endobacterium helps the fungal host in nitrogen fixation. The ability of the fungus *U. maydis* to fix nitrogen was confirmed by nitrogenase activity and incorporation of 15N into the cells (Ruiz-Herrera et al. 2015).

EHB association with the fungal species belonging to the order *Sebacinales* is quite diverse with three distinct bacterial genera, namely, *Paenibacillus*, *Acinetobacter*, and *Rhodococcus*, associated with *Sebacina vermifera,* whereas *Rhizobium radiobacter* was found to be associated with the fungus—*Piriformospora indica* (Guo et al. 2017). The endohyphal bacterium is protected inside the host fungus and in turn the endobacterium increases the fitness of the fungus.

7.5.4 EHB ASSOCIATED WITH ASCOMYCOTA

Hoffman and Arnold (2010) reported the presence of phylogenetically diverse endobacterial symbionts within the hyphae of ascomycetous fungi. There are growing evidences of the presence of EHB in diverse lineages of Ascomycota, including Eurotiomycetes, Dothideomycetes, Sordariomycetes, Pezizomycetes, and multiple functional groups. They are widespread in foliar endophytes and soil-borne ascomycetous fungi. A fungal endophyte *Pestalotiopsis* sp. harbors an endohyphal bacteria *Luteibacter* sp. that influences its ability to produce auxin and certain hydrolytic enzymes. Another fungal endophyte *Microdiplodia* sp. was also found to be in association with *Luteibacter* sp. as endosymbiont partner. The EHB–fungus association was resynthesized under *in vitro* conditions and it was observed that the EHB infections were initiated and maintained more often under low-nutrient culture conditions (Arendt et al. 2016). The endohyphal bacteria *Chitinophaga* sp. associated with the fungus *Fusarium keratoplasticum* alters the substrate (particularly carbon) use efficiency of the fungal host (Shaffer et al. 2017).

Pakvaz and Soltani (2016) explored the occurrence, diversity, and bioactive potential of diverse endosymbiotic bacteria associated with fungal endophytes of *Cupressus sempervirens*. Almost, 31% of the fungal endophytes isolated from *C. sempervirens* harboured endohyphal bacteria. More generally, studies of EHB in Ascomycota and other fungi have been focused primarily on Proteobacteria, Firmicutes, and Mollicutes, leaving gaps with regard to the potential for symbiotic modulation of the fungal phenotypes by members of other bacterial lineages.

7.6 FUNCTIONAL DIVERSITY OF EHB

The EHB symbioses represent the third component of plant–endophyte association, thus making it a tripartite association between the plant, fungus, and bacteria. This tripartite relationship has a great impact on the host plant's diversity, metabolism, and ecology (Fig. 7.1). It is reported that the endohyphal bacteria play a complementary protective role for the host fungus and the host plant under various stress conditions (Hoffman and Arnold, 2010, Pakvaz and Soltani, 2016). It is also reported that EHB influence the phytohormone production in the host fungus; for example, the indole acetic acid (IAA) production is enhanced significantly by an endohyphal bacterium identified as *Luteibacter* sp. (Hoffman et al. 2013). They also help regulate key components of host reproductive machinery and are responsible for the production of characteristic phytotoxins by the host fungus. In rice seedling blight, the phytotoxin (rhizoxin) was produced by the *Burkholderia*—an EHB residing inside the hyphae of *Rhizopus* sp. (Partida-Martinez and Hertweck, 2005). Similarly, an endosymbiont *B. endofungorum* was found to produce rhizonins, which are cyclopeptides and were wrongly regarded as mycotoxins (Lackner et al. 2009). It is also reported that in the absence of the endosymbionts, the host fungus is not capable of vegetative reproduction. Formation of sporangia and spores is restored only upon reintroduction of the EHB (Partida-Martinez et al. 2007). The EHB of *G. margarita* enhances the fungal sporulation, raises the fungal bioenergetic capacity by increasing ATP production, and elicits mechanisms to detoxify reactive oxygen species. These indicate the importance of EHB in host metabolism, reproduction, and stress tolerance (Salvioli et al. 2016). An EHB identified as *B. pumilus* is reported to confer nitrogen-fixing ability to the host fungus *U. maydis* (Ruiz-Herrera et al. 2015). In case of some AM fungi, EHB is found to facilitate phosphate acquisition and transport. The EHB in association with mycorrhiza[1] are considered to be an important aid toward controlled mycorrhization technique in forestry, by extending their use in poor nursery soils, saving fungal inoculum, improving the mycorrhizal quality, or suppressing the use of soil disinfectants (Garbaye, 1994). Recently, an interesting study on the symbiotic association between *R. microsporus* and its bacterial endosymbiont *Burkholderia* inferred that this symbiosis shifts from mutualism to antagonism in response to change in the status of lipid metabolism in the host fungus. The endosymbiotic relationship is mutualistic, when the lipid metabolism and phosphatidic acid–producing enzymes are activated, while it turns antagonistic when the phosphatidic acid–producing enzymes are inhibited (Lastovetsky et al. 2016).

These findings illustrate that this tripartite association between plant–fungus–bacteria is a complex interaction, which broadens the scope of plant–endophyte associations. The beneficial endohyphal associations are widespread but the underlying mechanisms are still unexplored due to problems in isolation, cultivation, and maintenance of all types of endosymbionts. The formation and consequences of the fungal and bacterial associations are very complex involving intricate molecular cross talks. This complex tripartite association should be studied in detail and the functional diversity of the events involved in these associations could be explored using integrated "omics" technologies, like genomics, transcriptomics, metabolomics, and proteomics.

7.7 CONCLUSION

Recent studies on "fungal–bacterial" associations, wherein the bacteria reside within the fungal hyphae, unfolded a novel chapter in microbial ecology.

[1] The EHB in association with mycorrhiza were earlier known as mycorrhiza helper bacteria (MHB).

These associations are more common and important than previously thought. Moreover, many of these associations are central to agriculture, forestry, and bioremediation. Though few of these unique symbioses have been studied to a significant extent, but this discipline is still in its infancy. There is a lack of knowledge about the mechanism of association and the molecular cross talk involved remains unexplored due to the microscopic scale of the associating partners, the complexity of the communities involved, and the intricate nature of the association. Therefore, it is imperative to work in a systems biology approach allowing for discovery and characterization of the molecular mechanisms and novel links between endohyphal bacterial partners and the fungal host.

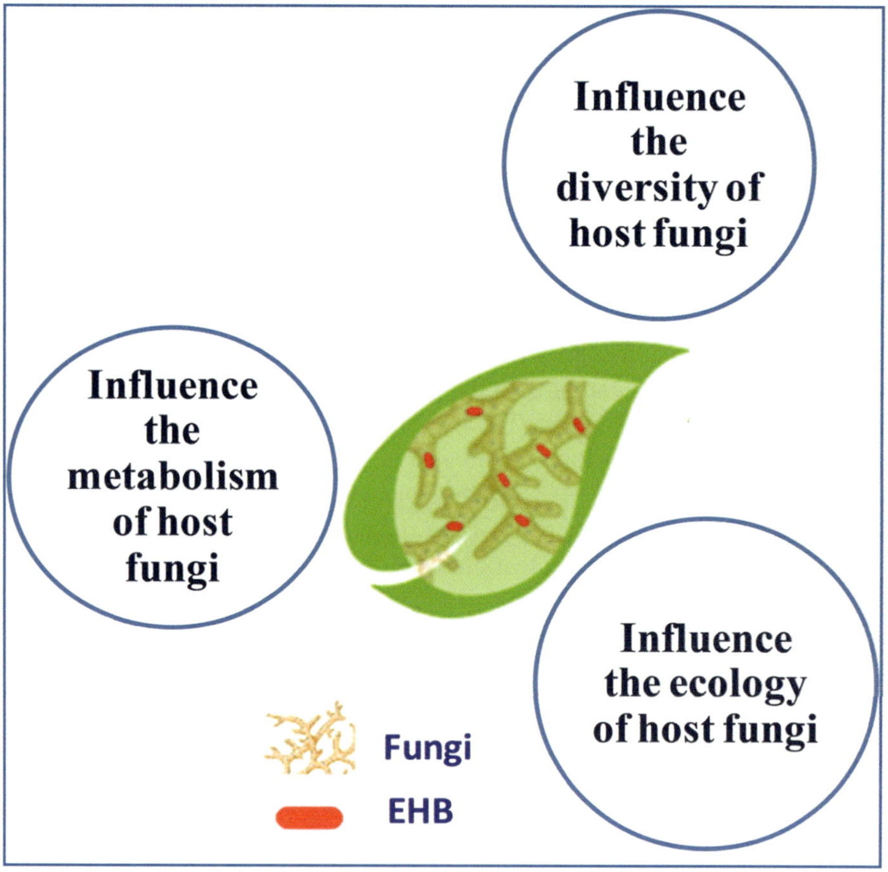

FIGURE 7.1 Diagrammatic representation of the role of endohyphal bacteria on the host fungus.

KEYWORDS

- **endohyphal bacteria**
- **endophytes**
- **symbiosis**
- **secondary metabolism**
- **tripartite association**

REFERENCES

Arendt, K. R.; Hockett, K. L.; Araldi-Brondolo, S. J.; Baltrzus, D. A.; Arnold, A. E. Isolation of Endohyphal Bacteria From Foliar Ascomycota and In Vitro Establishment of Their Symbiotic Associations. *Appl. Environ. Microbiol.* **2016,** *82*, 2943–2949.

Arora, P.; Riyaz-Ul-Hassan, S. Endohyphal Bacteria; The Prokaryotic Modulators of Host Fungal Biology. *Fungal Biol. Rev.* **2018.** https://doi.org/10.1016/j.fbr.2018.08.003.

Bertaux, J.; Schmid, M.; Prevost-Boure, N. C.; Churin, J. L.; Hartmann, A.; Garbaye, J.; Frey-Klett, P. In Situ Identification of Intracellular Bacteria Related to *Paenibacillus* spp. in the Mycelium of the Ectomycorrhizal Fungus Laccaria Bicolor S238N. *Appl. Environ. Microbiol.* **2003,** *69*, 4243–4248.

Bianciotto, V.; Lumini, E.; Bonfante, P.; Vandamme, P. *Candidatus Glomeribacter gigasporarum'gen.* nov., sp. nov., an Endosymbiont of Arbuscular Mycorrhizal Fungi. *Int. J. Syst. Evol. Microbiol.* **2003,** *53*, 121–124.

Bright, M.; Bulgheresi, S. A Complex Journey: Transmission of Microbial Symbionts. *Nat. Rev. Microbiol.* **2010,** *8*, 218.

Desirò, A.; Hao, Z.; Liber, J. A.; Benucci, G. M. N.; Lowry, D.; Roberson, R.; Bonito, G. Mycoplasma-Related Endobacteria within Mortierellomycotina Fungi: Diversity, Distribution and Functional Insights into Their Lifestyle. *ISME J.* **2018,** *12*(7), 1743–1757.

Garbaye, J. Tansley Review no. 76 Helper Bacteria: A New Dimension to the Mycorrhizal Symbiosis. *New Phytol.* **1994,** *128* (2), 197–210.

Guo, H.; Glaeser, S. P.; Alabid, I.; Imani, J.; Haghighi, H.; Kampfer, P.; Kogel, K. H. The Abundance of Endofungal Bacterium *Rhizobium radiobacter* (syn. *Agrobacterium tumefaciens*) Increases in Its Fungal Host *Piriformospora indica* during the Tripartite Sebacinalean Symbiosis with Higher Plants. *Front. Microbiol.* **2017,** *8*, 629.

Hoffman, M. T.; Arnold, A. E. Diverse Bacteria Inhabit Living Hyphae of Phylogenetically Diverse Fungal Endophytes. *Appl. Environ. Microbiol.* **2010,** *76*, 4063–4075.

Hoffman, M. T.; Gunatilaka, M. K.; Wijeratne, K.; Gunatilaka, L.; Arnold, A. E. Endohyphal Bacterium Enhances Production of Indole-3-Acetic Acid by a Foliar Fungal Endophyte. *PLoS One* **2013,** *8*, e73132.

Kluge, M. A Fungus Eats a Cyanobacterium: The Story of the *Geosiphon pyriformis* Endocyanosis. *Proc. Royal Ir. Acad.* **2002,** *102*, 11–14.

Lackner, G.; Partida-Martinez, L. P.; Hertweck, C. Endofungal Bacteria as Producers of Mycotoxins. *Trends Microbiol.* **2009,** *17*, 570–576.

Lastovetsky, O. A.; Gaspar, M. L.; Mondo, S. J.; LaButti, K. M.; Sandor, L.; Grigoriev, I. V.; Henry, S. A.; Pawlowska, T. E. Lipid Metabolic Changes in an Early Divergent Fungus Govern the Establishment of A Mutualistic Symbiosis with Endobacteria. *Proc. Natl. Acad. Sci.* **2016,** *113*, 15102–15107.

MacDonald, R. M.; Chandler, M. R. Bacterium-Like Organelles in the Vesicular-Arbuscular Mycorrhizal Fungus *Glomus caledonius*. *New Phytol.* **1981,** *89*, 241–246.

MacDonald, R.; Chandler, M. R.; Mosse, B. The Occurrence of Bacterium-Like Organelles in Vesicular-Arbuscular Mycorrhizal Fungi. *New Phytol.* **1982,** *90*(4), 659–663.

Moebius, N.; Uzum, Z.; Dijksterhuis, J.; Lackner, G.; Hertweck, C. Active Invasion of Bacteria Into Living Fungal Cells. *Elife* **2014,** *3*.

Mosse, B. Honey-Coloured, Sessile Endogone Spores: II. Changes in Fine Structure During Spore Development. *Archiv. Fur. Mikrobiol.* **1970,** *74*, 129–145.

Naumann, M.; Schußler, A.; Bonfante, P. The Obligate Endobacteria of Arbuscular Mycorrhizal Fungi Are Ancient Heritable Components Related to the Mollicutes. *ISME J.* **2010,** *4*, 862–871.

Pakvaz, S.; Soltani, J. Endohyphal Bacteria from Fungal Endophytes of the Mediterranean Cypress (*Cupressus sempervirens*) Exhibit In Vitro Bioactivity. *For. Pathol.* **2016,** *46*, 569–581.

Partida-Martinez, L. P.; Hertweck, C. Pathogenic Fungus Harbours Endosymbiotic Bacteria for Toxin Production. *Nature* **2005,** *437*, 884–888.

Partida-Martinez, L. P.; Monajembashi, S.; Greulich, K. O.; Hertweck, C. Endosymbiont Dependent Host Reproduction Maintains Bacterial-Fungal Mutualism. *Curr. Biol.* **2007,** *17*, 773–777.

Ruiz-Herrera, J.; Leon-Ramırez, C.; Vera-Nunez, A.; et al. A Novel Intracellular Nitrogen-Fixing Symbiosis Made by *Ustilago maydis* and *Bacillus* spp. *New Phytol.* **2015,** *207*, 769–777.

Salvioli, A.; Ghignone, S.; Novero, M.; Navazio, L.; Bagnaresi, P.; Bonfante, P. Symbiosis with an Endobacterium Increases the Fitness of a Mycorrhizal Fungus, Raising its Bioenergetic Potential. *ISME J.* **2016,** *10*, 130.

Sato, Y.; Narisawa, K.; Tsuruta, K.; Umezu, M.; Nishizawa, T.; Tanaka, K.; Yamaguchi, K.; Komatsuzaki, M.; Ohta, H. Detection of *Betaproteobacteria* inside the Mycelium of the Fungus *Mortierella elongata*. *Microbes Environ.* **2010,** *25* (4), 321–324.

Scherlach, K.; Busch, B.; Lackner, G.; Paszkowski, U.; Hertweck, C. Symbiotic Cooperation in the Biosynthesis of a Phytotoxin. *Angew. Chem.* **2012,** *124* (38), 9753–9756.

Schmitt, I.; Partida-Martinez, L. P.; Winkler, R.; Voigt, K.; Einax, E.; Dolz, F.; Sabine, T.; Johannes, W.; Hertweck, C. Evolution of Host Resistance in a Toxin-Producing Bacterial-Fungal Alliance. *ISME J.* **2008,** *2* (6), 632.

Schüssler, A. Molecular Phylogeny, Taxonomy, and Evolution of *Geosiphon pyriformis* and Arbuscular Mycorrhizal Fungi. *Plant Soil* **2002,** *244*, 75–83.

Schüßler, A. The *Geosiphon-Nostoc* Endosymbiosis and Its Role as a Model for Arbuscular Mycorrhiza Research. In: *Fungal Associations*. Springer, Berlin, Heidelberg, 2012; pp 77–91.

Shaffer, J. P.; U'Ren, J. M.; Gallery, R. E.; Baltrus, D. A.; Arnold, A. E. An Endohyphal Bacterium (*Chitinophaga Bacteroidetes*) Alters Carbon Source Use by *Fusarium Karatoplasticum* (*F. solani* Species Complex, Nectriaceae). *Front. Microbiol.* **2017,** *8*, 350.

Wang, Y. T.; Xue, Y. R.; Liu, C. H. A Brief Review of Bioactive Metabolites Derived from Deep-Sea Fungi. *Mar. Drugs* **2015,** *13* (8), 4594–4616.

CHAPTER 8

Polyketides: Bioactive Secondary Metabolites, Biosynthesis, and Their Modulation

MASROOR QADRI

Entomology and Nematology Department, University of Florida, 1881 Natural Area Drive, Steinmetz Hall, Gainesville, FL 32611-0620, United States

E-mail: qadrimasroor6@gmail.com

ABSTRACT

Endophytic microorganisms are an important source of bioactive secondary metabolites with enormous potential for the discovery of new molecules for drug discovery, industrial use and agricultural applications. Secondary metabolites display enormous chemical complexity and diversity produced by a few common biosynthetic pathways. Some of the classes of secondary metabolites mainly include polyketides, non-ribosomal peptides, alkaloids, flavonoids, steroids, terpenes, and indole terpenes. Polyketides (PKs) are a group of natural secondary metabolites which have been regarded as one of the richest "drug gold mine" groups since many of them are used as therapeutic drugs. PKs are extremely diverse in both structure and biological activity and are synthesized with the help of enzymes called polyketide synthase (PKS). In past few decades, investigations have been made to understand the biosynthesis of these polyketides and, consequently, to manipulate the genes encoding the polyketide synthases (PKSs) for the production of PK compounds with novel structures. Determining the genetic potential of endophytes for the synthesis of polyketides is a very important tool for estimating their biosynthetic proficiency. It also provides important information about the ecological role of fungal polyketide metabolites. In this chapter, we review the diverse classes of polyketides, their biosynthesis and various approaches for engineering the biosynthetic pathways for increasing the metabolite production.

8.1 INTRODUCTION

Polyketides belong to a diverse class of natural products with complex structures and various biological functions. They are assembled from units of small carboxylic acids such as acetate and propionate by successive decarboxylative condensations like fatty acids (Khosla et al., 1999). The complex structures with diverse patterns of functional groups arise due to partial processing of intermediates, while in fatty acid biosynthesis, unfunctionalized alky chains are generated by complete reduction of the intermediates. Additional complexity in polyketides occurs due to the use of different substrates for starter and chain extension, the generation of chiral centers, and further modification of functional groups, such as cyclization, reduction, and oxidation (Khosla et al., 1999). Polyketides have been studied from several aspects of chemistry and biology. This is largely due to a wide range of chemical structures, biological activities that impact human health (Simpson, 1995), and the mystery surrounding their presence in biological systems that produce them (Vining, 1992). Polyketides have diverse biological activities and are commonly used for their antibiotic and pharmacological properties (Castoe et al., 2007). Some of these polyketides include antibacterial (e.g., tetracycline, griseofulvin), antitumor agents (e.g., enediyne), immunosuppressants (e.g., rapamycin), and cholesterol-lowering agents (e.g., lovastatin, compactin) (Amnuaykanjanasin et al., 2005). The most striking characteristic of the polyketide class of natural products is their structural diversity. The range of structures includes macrolides and aliphatic and alicyclic compounds, as well as simple or complex polycyclic aromatic compounds. The structural diversity is further heightened by the existence of hybrids composed of polyketides condensed with amino acids or peptides, which add further to the range of pharmacological activities (Du and Shen, 2001). Polyketides have been isolated from numerous sources including fungi, bacteria, and plants. Filamentous fungi are prolific producers of diverse polyketides such as 6-methylsalicylic acid, lovastatin, rapamycin, and tetracycline. (Fig. 8.1), many of which exhibit interesting and important biological activities (Fujii, 1999). The antimicrobial properties of many polyketides implicate competition with other microbes, and they act as selective pressure required for maintenance of their biosynthetic genes in the population. However, the reason they are so fascinating to researchers is that the polyketides are so complex and diverse in structure, but their building blocks are some of the simplest molecules.

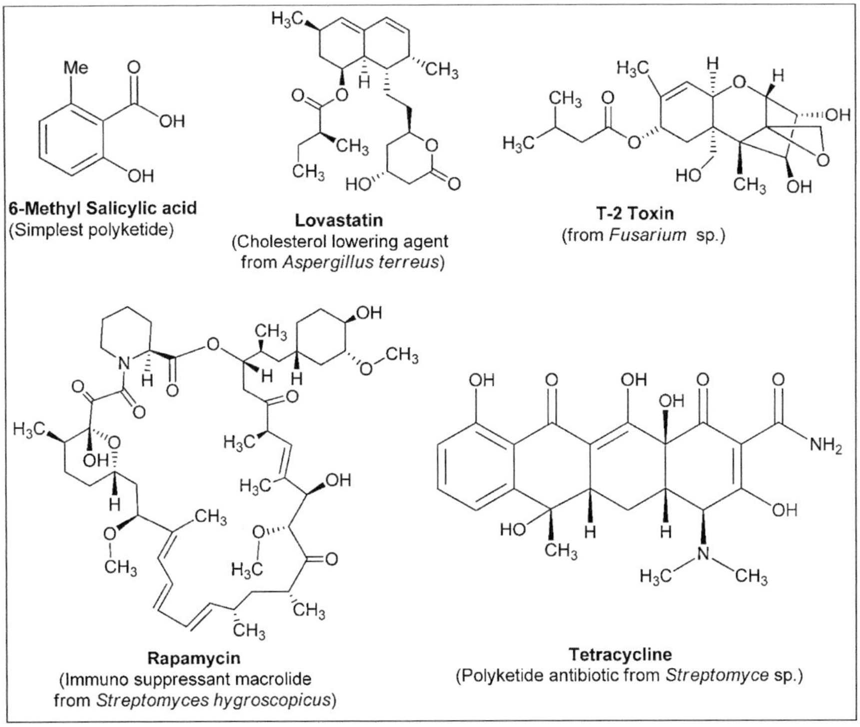

FIGURE 8.1 Structures of some polyketides isolated from microbial species.

Polyketide rifamycin, which was discovered in 1959, made huge impact on the pharmaceutical industry for patients suffering from mycobacterial infections. Rifamycin B has been given to patients suffering from leprosy, tuberculosis, and mycobacterial infections related to AIDS/HIV (Floss and Yu, 1999). Another polyketide, erythromycin, is an antibiotic isolated from *Saccharopolyspora erythraea*. This polyketide is a macrolide and has been used as a model for the production of new antibiotics (Ray et al., 2004). Rapamycin is another important polyketide, which is often administered to patients receiving organ transplants. This polyketide inhibits cell cycle progression, thereby preventing tumor cell growth (Decker et al., 2003). Another polyketide with anticancer activity is epothilone B, a 16-membered macrolide. This polyketide promotes tubulin polymerization in vitro and the stabilization of microtubules against Ca^{2+}-dependent depolymerization

(Altmann, 2003). Lovastatin is another therapeutic agent that is used to control the irregular cholesterol levels. It is often administered to patients to increase the levels of low-density lipoprotein activity, thereby leading to reduction in the synthesis of cholesterol (Xie et al., 2006). The enzymes that are involved in the assembly of polyketides by microorganisms are called polyketide synthase (PKS), and these enzymes are receiving much attention as access to them has improved through molecular methods.

8.2 POLYKETIDE BIOSYNTHESIS

PKS are the enzymes used for the synthesis of polyketides and their synthesis is much similar to that of the fatty acid biosynthesis. The major differences between fatty acid and polyketide biosynthesis are the greater variety of acyl precursors used to make polyketides, as well as the greater potential for differential ketoreduction, dehydration, and reduction after each new extender unit is added to the polyketide chain (Hopwood and Sherman, 1990). Polyketides are synthesized through repetitive Claisen condensation reactions, which involves combining together small organic acids, such as acetic acid and malonic acid, by an enzyme called ketosynthase. Before the assembly of polyketide chain, the building units, acetate, propionate, malonate, or methylmalonate are activated in the form of coenzyme A (CoA) esters, such as acetyl-CoA and malonyl-CoA. The starter unit, acetyl-CoA, is then condensed with malonyl-CoA, resulting in a chain of four carbon atoms with the loss of one carbon dioxide (CO_2) molecule. Only two carbon atoms are added into the chain in each round of condensation with malonyl-CoA. Selection of different number and type of starter and extender units diversifies the structure of the polyketide chain (Hopwood, 2004). After every Claisen condensation reaction, the resulting β-keto ester is successively reduced by enoylreduction, ketoreduction, and dehydration resulting in a saturated chain. The β-keto groups in the growing polyketide chain may also be left untreated, partially reduced, or fully reduced, thereby giving different levels of reduction (Hopwood, 2004).

After the carbon chain attains a specific length, it is released from the PKS by thiolysis or acyl transfer (Wakil, 1989) and cyclized to give different folding patterns. The "tailoring" enzymes including cyclase's, transferases (e.g., C-, O-, and *N*-methyltransferases, acyltransferases, and glycosyltransferases), cytochrome P450-type oxygenase's, and FAD(H)- or NADP(H)-dependent oxidoreductases can act on the PKS-derived intermediate to yield the final biologically active product (Pfeifer and Khosla, 2001).

8.2.1 STEPS INVOLVED IN POLYKETIDE BIOSYNTHESIS

a) After activation of the starter units, that is, acetic acid and malonic acid, as the corresponding CoA esters, they are attached to the PKS modules by specific acyl transferases. The enzyme ketosynthase, which has starter unit acetyl-CoA attached to its active site (cysteine thiol), catalyzes condensation and the acyl carrier protein (ACP) has extender unit malonyl-CoA attached to its thiol residue. During the condensation, one carbon from malonyl-CoA is lost as carbon dioxide resulting in a four-carbon chain attached to the ACP. The saturated chain is then transferred back from the ACP to the KS, this process is then iterated to produce a polyketide chain.

b) The keto group is first reduced to a hydroxyl, which is further reduced to a double bond, and finally resulting in a fully saturated carbon.

c) A complex polyketide contains keto group, hydroxyl group, double bond, and fully saturated carbon at different positions.

Initially, the experiments involved in determining the biosynthetic origin of secondary metabolites involved feeding labeled precursors to the producing organism in order to obtain labeled metabolites. A series of degradation experiments (for radiotracers) or analysis of the ^{13}C nuclear magnetic resonance (NMR) spectrum (for magnetic labels) of the labeled metabolite was then used to determine the location and distribution of incorporated label. The latter approach employing stable isotopes and NMR spectroscopy has been particularly productive in uncovering new biosynthetic pathways (Wright et al., 1977; Vederas, 1987). Such experiments remain essential to characterize the biosynthetic pathway of new metabolites. For example, biochemical experiments on cell-free extracts of the patulin-producer, *Penicillium patulum,* resulted in the isolation of the first PKS involved in patulin biosynthesis (Dimroth et al., 1970). However, in the last two decades, biochemical and genetic approaches have been employed to identify the enzymes and genes involved in secondary metabolism and specifically to unravel the process of polyketide biosynthesis. Such studies have resulted in the identification of a complete set of biosynthetic genes for a rapidly growing number of metabolites (Piel et al., 2000; Molnar et al., 2000).

8.3 CLASSIFICATION OF POLYKETIDE SYNTHASES (PKSs)

PKSs are multifunctional enzymes that possess several domains such as β-ketoacyl synthase (KS), an acyltransferase (AT), and an ACP having different catalytic activities. Some PKSs have additional functional domains, which include β-ketoacyl reductase (KR), enoylreductase (ER), dehydratase (DH), methyl esterase (ME), methyl transferase (MT), and thioesterase (TE) (Amnuaykanjanasin et al., 2005). PKSs have traditionally been defined into two classes: type I PKSs and type II PKSs. Type I PKSs are multifunctional modular systems responsible for the biosynthesis of macrolactones, polyethers, and polyenes (Bibb et al., 1989; Fernandez-Moreno et al., 1992; Yu et al., 1994). Type II PKSs are a cluster of small distinct monofunctional proteins that catalyze the biosynthesis of bacterial aromatic polyketides. However, cloning and sequencing of more PKS genes resulted in the discovery of fungal and plant PKSs during last one decade. Therefore, PKSs were reclassified with the addition of type III PKS and type I PKS were divided into two subclasses, modular type I PKS and iterative type I PKS (Hopwood, 1997).

A. *Type I PKS*

1. **Modular type I PKS:** They are large multifunctional polypeptides arranged in a modular fashion in which each module carries out one round of chain extension and subsequent β-keto processing. Specifically, each active site is only used once during the polyketide biosynthesis (Hertweck, 2009) (Fig. 8.2). They are mostly present in bacterial systems.
2. **Iterative type I PKS:** Iterative type I PKS possesses only one multidomain protein that is used repeatedly to catalyze multiple cycles of chain elongation and appropriate β-keto processing. In certain iterative type I PKS, one set of catalytic domains is able to vary the reduction level of β-keto groups during different extension cycles (Kennedy et al., 1999). These types of PKSs are involved in the biosynthesis of fungal metabolites such as 6-methylsalicyclic acid (Shoolingin-Jordan and Campuzano., 1999) and lovastatin (Staunton, 1998). Fungal iterative type I PKSs can be divided into two subclasses: nonreducing and reducing (Nicholson et al., 2001).

   2.1) The nonreducing (NR) PKSs (or called WA-type) do not have chemical reduction in their structure and are involved in the synthesis of fungal pigments such as melanin and aflatoxin. They have also been shown to be involved in the synthesis of a yellow conidial pigment intermediate—a heptaketide naphthopyrone, in *Aspergillus nidulans* (Mayorga and Timberlake 1992).

2.2) The reducing PKSs are involved in the synthesis of polyketides with different chemical reductions in their structure. They are of two types, partially reduced PKSs (PR-PKS) and highly reduced PKSs (HR-PKS). These PKSs have reducing domains such as ketoreductase (KR), which catalyze the β-keto reduction (a keto to a hydroxyl group), dehydratase (DH) that catalyzes dehydration (the hydroxyl to an enoyl group), and/or enoylreductase (ER) that catalyzes enoyl reduction (the enoyl to an alkyl group). PR-PKS only have KR and DH domains and they are also named as methylsalicylic acid synthase (MSAS)-type (Bingle et al., 1999), since most of them are involved in the synthesis of MSAs. The HR-PKS have all three reducing domains. In fungi, HR-PKSs have been further subdivided into four subclades (I, II, III, and IV), based on their characteristic domain structure (Kroken et al., 2003). The diversity of polyketides is increased due to the reduction/dehydration reactions that occur after each condensation step generating polyketides with alcohol, alkene, or ketone at specific positions along the chain.

B. Type II PKS

Type II PKSs are iterative in nature and the enzyme required for each biosynthetic step is encoded by a single gene. These PKSs have one set of a heterodimeric ketosynthase (KSα-KSβ) and an ACP, which act repeatedly to build a polyketide chain of correct length. Further cyclisation, reduction, and aromatization of the polyketide chain are performed by cyclase, ketoreductase, and aromatase, respectively. They are involved in catalyzing the biosynthesis of a wide range of polyfunctional aromatic compounds (Fig. 8.2).

C. Type III PKS

They are homodimeric synthases of the chalcone synthase (CHS) super-family known as CHS-like PKS. These PKSs do not have ACP and KS domains and use acyl CoA substrates directly. CHSs are abundantly present in higher plants and involved in the synthesis of a diverse set of biologically important phenylpropanoid metabolites (Schroder, 1999). Type III PKS were traditionally associated with plants but have also been recently discovered in several bacteria.

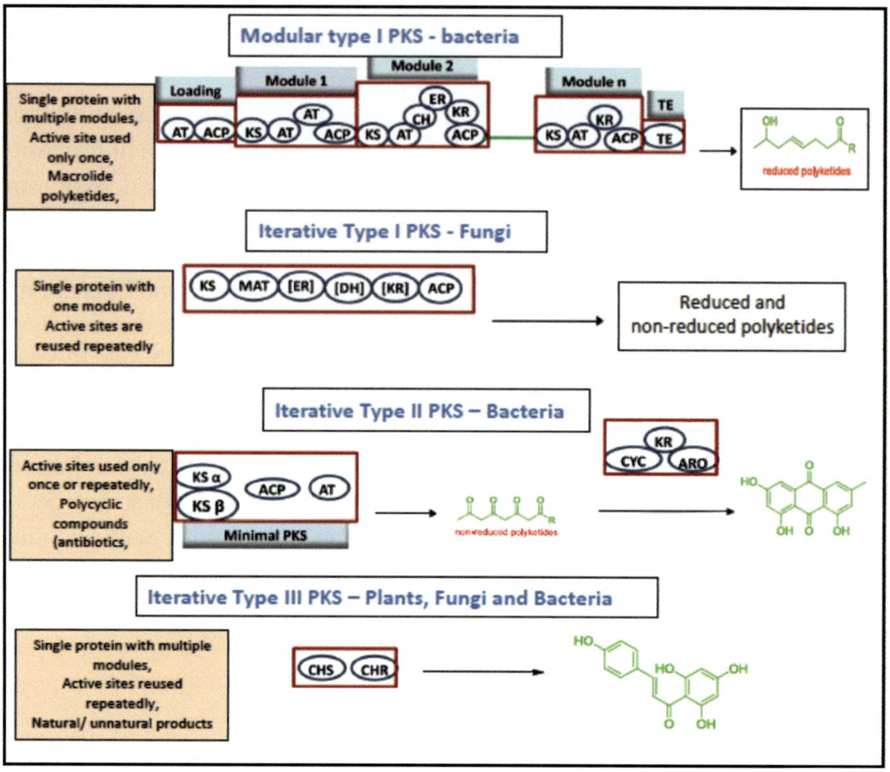

FIGURE 8.2 Different classes of polyketide synthases.

8.4 POLYKETIDE SYNTHASE GENE DIVERSITY IN FUNGI

The isolation and characterization of genes responsible for polyketide biosynthesis has attracted more interests in order to enrich the understanding of the relationship between chemical structures of compounds and functions of the related biosynthetic proteins. In the early genetic research, some gene clusters responsible for the biosynthesis of polyketide antibiotics were identified by comparing the differences in the genome of the wild-type strain and mutant strain blocked in a step in the biosynthetic pathway. The large number of available sequences from bacterial sources has facilitated comparative PKS gene analyses in this group, the results of which suggest their origin from duplicated fatty acid synthase (FAS) genes and subsequent distribution via horizontal gene transfer (HGT) (Vining, 1992; Hopwood, 1997). Since the first PKS gene was sequenced (Rudd and Hopwood, 1979), the subsequent explosion of PKS genetic information has mainly been obtained from actinomycetes, with relatively few PKS genes being reported from eukaryotes (Fujii, 1999). This is somewhat surprising considering the tremendous diversity of secondary metabolites that have been isolated from eukaryotes such as the filamentous fungi (Walton, 2000). Sequence analysis of different cloned biosynthetic gene clusters has revealed the conserved motifs in the same type of synthase enzymes. These conserved sequences have been used to synthesize primers, thereby enabling the isolation and identification of PKS genes from different species. The identification of these PKS genes exposes an opportunity to create novel chemical structures via engineering of PKS genes. A large number of fungal PKS genes have been detected using primers that are designed on the basis of conserved regions of the KS and also specific primers that identify different types of reducing and nonreducing PKS genes. For example, LC1/2c primers have been used to amplify KS of NR-type PKSs, LC3/5c recognize PR-type PKSs, and KS3/4c amplify HR-type PKSs (Bingle et al., 1999; Nicholson et al., 2001). Similarly, primers designed for the detection of CMT domain have been used for the identification of HR-PKS involved in biosynthesis of qualestatin from *Phoma* sp. (Hoffmeister and Keller, 2007). By the use of degenerate primers, it has also been possible to explore the diversity of polyketides in different ecological niches, for example, PKS genes from insect- and nematode-associated fungi (Lee et al., 2001), PKS genes from a nonsporulating endophytic fungi isolated from *Vaccinium macrocarpon* (Sauer et al., 2002), PKS genes from the lichenized genus *Lecanora* (Grube and Blaha, 2003), PKS genes from lichenized Ascomycetes *Pertussis* (Schmitt et al., 2008), PKS genes from *Aspergillus ochraceus*, and *Aspergillus carbonarius* producing ochratoxin A (Atoui et al., 2006), PKS genes from marine fungi (Mayer et al., 2007), and PKS genes from a group of tropical entomopathogenic fungi (Amnuaykanjanasin et al., 2005). An analysis of fungal genomes indicated that the PKS amino acid sequences are useful for phylogenetic analysis as they can form clades corresponding to NR, PR, and HR synthase architectures (Kroken et al., 2003).

8.5 MODULATION OF FUNGAL METABOLITE EXPRESSION

Microorganisms produce valuable secondary metabolites only under certain conditions and, therefore, cannot be detected upon culturing the organism on standardized laboratory media. Standard fungal fermentation methods including static or shake cultures on artificially defined media are unable to mimic the organism's natural habitat since they lack certain environmental signals required for inducing the expression of secondary metabolite genes (Chiang et al., 2009). The encoded secondary metabolites may also be produced at very low rates and, therefore, escape detection. The sequencing of fungal genomes and their bioinformatics analysis indicates that the biosynthetic potential of microorganisms is underinvestigated, as the number of gene clusters encoding secondary metabolites in many bacteria and fungi far exceeds the known secondary metabolites produced by these organisms (Cichewicz, 2010; Brakhage and Schroeckh, 2011). Therefore, a huge number of cryptic metabolites encoded by the microorganisms await discovery (Brakhage et al., 2009; Chiang et al., 2009). For instance, in *A. nidulans*, genome sequencing and mining enabled the identification of 53 putative secondary metabolite gene clusters (Von Dohren, 2009), which include 27 PKSs and 14 nonribosomal peptide synthetases (NRPSs), responsible for polyketide and nonribosomal peptide biosynthesis, respectively (Galagan et al., 2005). However, until today, the number of secondary metabolites identified from the fungus is much lower than the number of gene clusters encoded by them.

The study of the mechanisms that lead to the suppression of biosynthetic transcription, as well as the development of methodologies to induce the expression of these gene clusters, are of much significance in the search for new secondary metabolites. A number of approaches are being used for inducing the production of theses unknown metabolites. Some of these methods include the following sections.

8.5.1 OSMAC APPROACH

An indirect but successful tool to gain access to hidden natural products is the "one strain many compounds" (OSMAC) approach (Bode et al., 2002). This concept is based on the assumption that secondary metabolite production occurs as a specific response to a changed environment. The corresponding biosynthetic pathways are subsequently activated or upregulated as a consequence of these stimuli. Systematically varying fermentation parameters such as media composition, temperature, aeration, or shape of culturing flask has been successful in inducing the production of some of the unknown metabolites (Grond et al., 2002; Gross, 2007). For example, high concentrations of glucose, phosphate, or ammonium are generally regarded as repressors of secondary metabolism, while as low concentrations are regarded as inducers of secondary metabolism in microorganisms. However, in some cases, high phosphate concentrations might induce the production of certain metabolites (Gotoh et al., 1982). Even amino acids are described as potential inducers of secondary metabolites (Zahner and Kurth, 1982). In *A. ochraceus* DSM7428, aspinonene was designated as the main metabolite with the production of up to 8 mg/l (Fuchser et al., 1995). However, variation of the culture conditions and detailed analysis of the culture broth led to the isolation of 15 additional metabolites with production of up to 94 mg/l, and these metabolites result from five different biosynthetic pathways. This increase in the number and quantity of metabolites was due to the extensive use of different culture vessels (for example, shaking flasks, static liquid cultures in different fermenters) (Fuchser and Zeeck, 1996). Most of these new metabolites were based on different PKSs. Similarly, isoaspinonene, aspyrone, dihydroaspyrone, and dientriol represent putative variations of the aspyrone biosynthesis under varying culture conditions (Simpson and Holker, 1975; Staunton and Sutkowski, 1991a, b, c).

8.5.2 MANIPULATING EXPRESSION OF CLUSTER-SPECIFIC REGULATORY ACTIVATORS

Genes encoding the individual secondary metabolic pathways are normally clustered together and are under the control of specific regulatory genes within the genome (Walton, 2000). Therefore, novel secondary metabolites can be discovered by manipulating the expression of cluster-specific regulatory activators, for example, *tri6* involved in regulating trichothecene biosynthesis (Proctor et al., 1995), *aflR* involved in regulating aflatoxin biosynthesis (Yu et al., 1996), and *ctnA* for citrinin biosynthesis (Shimizu et al., 2007; Hoffmeister and Keller, 2007). It has been suggested that overexpressing the pathway-specific regulatory gene can be used to generate novel metabolites (Yu et al., 1996). The detection of a putative hybrid PKS/NRPS gene in the genome of *A. nidulans* with no corresponding natural product indicated that this gene locus is silent under standard fermentation conditions. A presumed activator gene was also identified within the cluster, and its homologous overexpression under the control of an inducible promoter resulted in the activation of the biosynthetic pathway and the production of two novel pyridine alkaloids, aspyridones A and B, isolated after scale-up fermentation (Bergmann et al., 2007).

8.5.3 MANIPULATION OF A GLOBAL REGULATOR OF SECONDARY METABOLISM

LaeA is a nuclear protein, globally regulating secondary metabolism in *Aspergillus* sp. It is required for sterigmatocystin (ST) biosynthesis, penicillin (PN) biosynthesis, as well as biosynthesis of gliotoxin in *A. nidulans* and mycelial pigments in *Aspergillus fumigatus* (Bok and Keller, 2004). In *A. nidulans,* it is also required for heterologous lovastatin (LOV) gene cluster expression and in *Aspergillus terreus* for the expression of native LOV gene cluster. The LaeA (loss of *aflR* expression) has been identified to function as a global regulator of secondary metabolic gene clusters in this genus (Bok and Keller, 2004). Interestingly, the protein has been found to be conserved in filamentous fungi, except in *Saccharomyces cerevisiae*, a fungus which does not produce any secondary metabolites. The loss of LaeA does not affect the morphological developmental processes unlike other genes that are involved in regulating secondary metabolism (Bok and Keller, 2004). However, loss of LaeA has been shown to influence the production of secondary metabolites such as decreased sterigmatocystin and penicillin production in *A. nidulans*, gliotoxin production in *A. fumigatus*, and sclerotial production in *Aspergillus flavus* (Kale et al., 2008).

The sequence analysis of LaeA gene has indicated that it encodes a methyltransferase and also shares some sequence similarity to arginine and histone methyltransferases. It has been suggested that this protein functions via chromatin remodeling since many of the targets of LaeA have subtelomeric locations (Keller et al., 2005). A comparison of the transcriptional profile of wild type, *ΔlaeA*, and complemented control strains of *A. fumigatus* revealed that 9.5% of the genome transcription is controlled by LaeA, and out of 22 secondary metabolite gene clusters, 13 gene clusters were positively regulated by LaeA. Importantly, seven of these gene clusters have subtelomeric location with a high degree of heterochromatin (Perrin et al., 2007). Therefore, genetically manipulating *laeA* in filamentous fungi may be used to increase the production of valuable metabolites or identify previously unknown metabolites in fungi or enable the improvement of strains by elimination of the fungal toxins.

8.5.4 COCULTURE OF MICROORGANISMS

Interaction between same or different species of microorganisms is assumed to induce the production of vast diversity of secondary metabolites. Therefore, applying the same approach in laboratory by performing mixed microbe's fermentation experiments can be used to produce diverse metabolites (Scherlach and Hertweck, 2009). For instance, coculturing of marine fungus *Pestalotia* sp. with an unidentified Gram-negative bacterium led to the isolation of pestalone—a new and potent benzophenone antibiotic (Cueto et al., 2001). In another study, coculturing of marine-derived *Emericella* sp. with the marine actinomycete *Salinispora arenicola* induced the production and isolation of two new cyclic depsipeptides, emericellamides A and B (Oh et al., 2007). Likewise, in *A. nidulans*, the production of orsellinic acid, its derivative lecanoric acid, and the cathepsin K inhibitors, F-9775A and F-9775B, encoded by previously unrecognized PKS gene cluster was found to be induced by the soil-dwelling bacterium *Streptomyces rapamycinicus* (Nutzmann et al., 2011).

8.5.5 GENE INACTIVATION

Diversity of secondary metabolites can also be increased by the genetic mapping of a fungal metabolic expression, followed by gene inactivation and subsequent examination of the metabolic profile produced by the mutant, as compared to the wild type. Using this approach in *A. nidulans* strain, six NRPSs were randomly selected and inactivated followed by analyzing the fungal metabolic expression to compare the difference in metabolic profile of the mutants and the control species (Chiang et al., 2008). RNA interference (RNAi) technology can also be utilized for inhibition of a specific gene involved in biosynthesis of metabolites enabling a broader control of their production. In *Penicillium chrysogenum*, gene suppression was shown to induce the production of a methyltransferase, which inhibited the production of meleagrin and enhanced the production of glandicolin B. The suppression of two other genes located within the same gene cluster was able to inhibit meleagrin and roquefortine C production, showing that this cluster is responsible for the biosynthesis of both metabolites (García-Estrada et al., 2011).

8.5.6 EPIGENETIC MANIPULATION

The fungal genome sequences reported so far have demonstrated that most of the putative metabolic biosynthetic gene clusters are located in the distal regions of chromosomes (Shwab et al., 2007). Importantly, these regions of fungal genomes are known to exist in a heterochromatin state and the transcription of constitutive genes is controlled by epigenetic regulation, which includes DNA methylation and histone modification. Histone modification and DNA methylation modify the chromatin structure, thereby regulating the gene expression. In *A. nidulans*, the disruption of histone deacetylase activity (Dhda) led to the transcriptional activation of gene clusters involved in the production of sterigmatocystin and penicillin (Shwab et al., 2007). The histone deacetylases (HDAC) and DNA methyltransferase (DMNT) inhibitors are the most commonly used chemicals to track biosynthetic silent pathways, as they are capable of activating silent gene clusters (Asai et al., 2012). Small molecule epigenetic modulators are applied to change the transcription rate of some genes or induce the expression of genes or gene clusters involved in the production of novel metabolites (Henrikson et al., 2009). This approach can also increase the quantity of metabolites already produced by the certain fungal species (Williams et al., 2008). Commonly used DNMT inhibitors include 5-azacytidine (5-AZA), and 5-aza-20-deoxycytidine, and the HDAC inhibitors include hydroxamic-acid containing compounds or cyclic peptides such as trichostatin A and trapoxin B, respectively (Cichewicz, 2010). Among the HDAC inhibitors, suberoylanilide hydroxamic acid (SAHA) is most commonly used. These substances are considered useful because of the fact that among the chemical changes that a DNA Histone may undergo, its acetylation is generally associated with gene activation. Histone acetyltransferases are responsible for this step, while as HDAC are involved in inactivating gene clusters. HDAC inhibitors such as SAHA prevent the inactivation by interacting with the catalytic site of histone-deacetylases. The most commonly used MT inhibitor, 5-AZA, interacts with methyltransferase, in turn resulting in DNA hypomethylation, which subsequently leads to chromatin restructuring and activation of gene clusters (Fisch et al., 2009).

In *Aspergillus niger,* epigenetic modulation by treatment with SAHA for about 2 weeks in a vermiculite-based semisolid medium led to the isolation of nygerone A, thereby emphasizing the importance of epigenetic modulation to induce the production of unknown metabolites (Henrikson et al., 2009). The same fungi when treated with a combination of SAHA and 5-AZA led to the production of secondary metabolites encoded by other silent genes (Fisch et al., 2009). Additionally, transcriptional rates of PKS, NRPS, and hybrid PKS-NRPS (HPN) biosynthetic gene clusters were increased after treatment with epigenetic modifiers to *A. niger* culture, whereas in absence of the modifiers less than 30% of these gene clusters were transcribed (Fisch et al., 2009). In *A. fumigatus*, a class II histone deacetylase (*hdaA*) was utilized to regulate the production of secondary metabolites. The suppression of the *hdaA* gene reduced the production of gliotoxin, a toxin produced by this fungus. On the other hand, overexpression of the *hdaA* gene promoted an increase in the production of gliotoxin (Lee et al., 2009).

Similarly, in *Neurospora crassa*, an increase in carotenoid production was achieved by addition of low doses of 5-AZA (≤30 μM) whereas higher doses (100 and 300 μM) decreased carotenoid levels and altered its reproductive structures (Kritsky et al., 2001). The same compound triggered the biosynthesis of two new galactose-conjugated polyunsaturated polyketides in *Diatrype* sp. (Cichewicz, 2010). Similarly, treatment of *Cladosporium cladosporioides* with 5-AZA stimulated the production of several oxylipins including (9Z, 12Z)-11-hydroxyoctadeca-9, 12-dienoic acid, its methyl ester, and a glycerol conjugate. In contrast, administration of SAHA yielded a complex series of perylenequinones including cladochromes, and calphostin B (Williams et al., 2008).

Epigenetic modulation using SAHA and AZA has also been shown to induce the production of volatile organic compounds (VOCs) by the endophytic fungus *Hypoxylon* sp. An 8-day-old *Hypoxylon* sp. culture revealed significant variations in the VOCs profiles, with the production of several new compounds when treated with epigenetic modulators (Riyaz-Ul-Hassan et al., 2012).

Small-molecule epigenetic modifiers have been very successful in inducing the production of novel natural products from fungi, which indicates that this is a very promising and rational approach for inducing the expression of cryptic biosynthetic pathways. Medium modification, changes in culture parameters and epigenetic modifiers result in the modulation of secondary metabolite production by regulating different gene expression patterns. This method has significant benefits compared to the currently available molecular or culture-dependent techniques. First, it allows for rapidly accessing potential reserves of cryptic fungal natural products in their native hosts. Second, this methodology can be readily implemented in most labs without extensive retooling. Thirdly, this approach significantly reduces the cost and effort of obtaining the natural products from silent biosynthetic pathways, since it does not require pre-screening of fungi using a variety of culture conditions (Cichewicz, 2010).

8.6 CONCLUSION

Endophytic fungi hold the promise of obtaining bioactive molecules from cryptic biosynthetic gene clusters, opening up an exciting area of research toward using them as a feasible alternative source of important phytochemicals. Due to the diverse biological activities of polyketides, it is important to unravel the polyketide biosynthetic machinery and based on the information examine the possible manipulation of this machinery to produce unnatural bioactive polyketides. With the development of gene sequencing technology and biosynthetic engineering or activation of the cryptic gene clusters through epigenetic manipulation, gene inactivation, regulation of global regulators and pathway-specific transcription factors (TFs), new bioactive molecules can be developing with improved activities. Elucidation of the molecular mechanisms behind the complex regulatory network will not only provide a deeper insight into how microbes translate environmental signals into secondary metabolite biosynthesis but will also warrant the identification of novel secondary metabolites and a deeper understanding of their ecological role.

KEYWORDS

- **endophytes**
- **secondar metabolites**
- **polyketides (PKs)**
- **polyketide synthases (PKSs)**
- **biosynthetic modulation**

REFERENCES

Altmann, K. H. Epothilone B and Its Analogs—A New Family of Anticancer Agents. *Mini Rev. Med. Chem.* **2003,** *3* (2), 149–158.

Amnuaykanjanasin, A.; Punya, J.; Paungmoung, P.; Rungrod, A.; Tachaleat, A.; Pongpattanakitshote, S.; Cheevadhanarak, S.; Tanticharoen, M. Diversity of Type I Polyketide Synthase Genes in the Wooddecay Fungus *Xylaria* sp. BCC 1067. *FEMS Microbiol. Lett.* **2005,** *251*, 125–136.

Asai, T.; Chung, Y. M.; Sakurai, H.; Ozeki, T.; Chang, F. R.; Yamashita, K.; Oshima, Y. Tenuipyrone, a Novel Skeletal Polyketide from the Entomopathogenic Fungus, *Isaria tenuipes*, Cultivated in the Presence of Epigenetic Modifiers. *Org. Lett.* **2012,** *14*, 513–515.

Atoui, A.; Dao, H. P.; Mathieu, F.; Lebrihi, A. Amplification and Diversity Analysis of Ketosynthase Domains of Putative Polyketide Synthase Genes in *Aspergillus ochraceus* and *Aspergillus carbonarius* Producers of Ochratoxin A. *Mol. Nutr. Food. Res.* **2006,** *50*, 488–493.

Bergmann, S.; Schumann, J.; Scherlach, K.; Lange, C.; Brakhage, A. A.; Hertweck, C. Genomics-Driven Discovery of PKS-NRPS Hybrid Metabolites from *Aspergillus nidulans*. *Nat. Chem. Biol.* **2007,** *3*, 213–217.

Bibb, M. J.; Biro, S.; Motamedi, H.; Collins, J. F.; Hutchinson, C. R. Analysis of the Nucleotide Sequence of the *Streptomyces glaucescens* tcmI Genes Provides Key Information about the Enzymology of Polyketide Antibiotic Biosynthesis. *EMBO J.* **1989,** *8*, 2727–2736.

Bingle, L. E.; Simpson, T. J.; Lazarus, C. M. Ketosynthase Domain Probes Identify Two Subclasses of Fungal Polyketide Synthase Genes. *Fungal Genet. Biol.* **1999,** *26*, 209–223.

Bode, H. B.; Bethe, B.; Hofs, R.; Zeeck, A. Big Effects from Small Changes: Possible Ways to Explore Nature's Chemical Diversity. *Chem. Bio. Chem.* **2002,** *3*, 619–627.

Bok, J. W.; Keller, N. P. LaeA, A Regulator of Secondary Metabolism in *Aspergillus* spp. *Eukaryot. Cell* **2004,** *3,* 527–535.

Brakhage, A. A.; Schroeckh, V. Fungal Secondary Metabolites-Strategies to Activate Silent Gene Clusters. *Fungal Genet. Biol.* **2011,** *48*, 15–22.

Brakhage, A. A.; Bergmann, S.; Schuemann, J.; Scherlach, K.; Schroeckh, V.; Hertweck, C. Fungal Genome Mining and Activation of Silent Gene Clusters. In *The Mycota XV*; Esser, K., Ed.; Springer-Verlag: Heidelberg, Germany, 2009; pp 297–303.

Castoe, T. A.; Stephens, T.; Noonan, B. P.; Calestani, C. A Novel Group of Type I Polyketide Synthases (PKS) in Animals and the Complex Phylogenomics of PKSs. *Gene* **2007,** *392*, 47–58.

Chiang, Y.; Szewczyk, E.; Nayak, T.; et al. Molecular Genetic Mining of the *Aspergillus* Secondary Metabolome: Discovery of the Emericellamide Biosynthetic Pathway. *Chem. Biol.* **2008,** *15*, 527–532.

Chiang, Y. M.; Lee, K. H.; Sanchez, J. F.; Keller, N. P.; Wang, C. C. Unlocking Fungal Cryptic Natural Products. *Nat. Prod. Commun.* **2009,** *11*, 1505–1510.

Cichewicz, R. H. Epigenome Manipulation as a Pathway to New Natural Product Scaffolds and Their Congeners. *Nat. Prod. Rep.* **2010,** *27*, 11–22.

Cueto, M.; Jensen, P. R.; Kauffman, C.; Fenical, W.; Lobkovsky, E.; Clardy, J. Pestalone, A New Antibiotic Produced by a Marine Fungus in Response to Bacterial Challenge. *J. Nat. Prod.* **2001,** *64*, 1444–1446.

Decker, T. H. S.; Ringshausen, I.; Bogner, C.; Oelsner, M.; Schneller, F.; Peschel, C. Rapamycin-Induced G1 Arrest in Cycling B-CLL Cells is Associated with Reduced Expression of Cyclin D3, Cyclin E, Cyclin A, and Survivin. *Blood* **2003,** *101*, 278–285.

Dimroth, P.; Walter, H.; Lynen, F. Biosynthese Von 6-Methylsalicylsaure. *Eur. J. Biochem.* **1970,** *13*, 98–110.

Du, L.; Shen, B. Biosynthesis of Hybrid Peptide-Polyketide Natural Products. *Curr. Opin. Drug. Discov. Devel.* **2001,** *4*, 215–228.

Fernandez-Moreno, M. A.; Martıinez, E.; Boto, L.; Hopwood, D. A.; Malpartida, F. Nucleotide Sequence and Deduced Function of a Set of Co-Transcribed Genes of *Streptomyces coelicolor* A3 (2) Including the Polyketide Synthase for the Antibiotic Actinorhodin. *J. Biol. Chem.* **1992,** *267*, 19278–19290.

Fisch, K. M.; Gillaspy, A. F.; Gipson, M.; Henrikson, J. C.; Hoover, A. R.; Jackson, L.; Najar, F. Z.; Wagele, H.; Cichewicz, R. H. Chemical Induction of Silent Pathway Transcription in *Aspergillus niger*. *J. Ind. Microbiol. Biotechnol.* **2009,** *36*, 1199–1213.

Floss, H. G.; Yu, T. Lessons from the Rifamycin Biosynthetic Gene Cluster. *Curr. Opin. Chem. Biol.* **1999,** *3*, 592–597.

Fujii, I. Polyketide Biosynthesis in Filamentous Fungi. In *Comprehensive Natural Products Chemistry*; Sankawa, U., Ed.; Elsevier: Oxford, 1999; Vol. 1; pp 409–441.

Fuchser, J.; Thiericke, R.; Zeeck, A. Biosynthesis of Aspinonene, A Branched Pentaketide Produced by *Aspergillus ochraceus*, Related to Aspyrone. *J. Chem. Soc. Perkin. Trans.* **1995,** *1*, 1663–1666.

Fuchser, J.; Zeeck, A. Aspinolides and Aspinonene/Aspyrone Co-Metabolites, New Pentaketides Produced by *Aspergillus ochraceus*. *Liebigs Ann./Recl.* **1996,** *1997*, 87–95.

Galagan, J. E.; Calvo, S. E.; Cuomo, C.; et al. Sequencing of *Aspergillus nidulans* and Comparative Analysis with *A. fumigatus* and *A. oryzae*. *Nature* **2005,** *438*, 1105–1115.

García-Estrada, C.; Ullán, R. V.; Albillos, S. M.; Fernández-Bodega, M. A.; Durek, P.; von Döhren, H.; Martín, J. F. A Single Cluster of Coregulated Genes Encodes the Biosynthesis of the Mycotoxins Roquefortine C and Meleagrin in *Penicillium chrysogenum*. *Chem. Biol.* **2011,** *18*, 1499–1512.

Gotoh, T.; Nakahara, K.; Iwami, M.; Aoki, H.; Imanaka, H. Studies on a New Immunoactive Peptide, FK-156 I, Taxonomy of the Producing Strains. *J. Antibiot.* **1982,** *35*, 1280–1285.

Grond, S.; Papastavrou, I.; Zeeck, A. Novel α-L-Rhamnopyranosides from a Single Strain of Streptomyces by Supplement-Induced Biosynthetic Steps. *Eur. J. Org. Chem.* **2002,** *2002*, 3237–3242.

Gross, H. Strategies to Unravel the Function of Orphan Biosynthesis Pathways: Recent Examples and Future Prospects. *Appl. Microbiol. Biotechnol.* **2007,** *75*, 267–277.

Grube, M.; Blaha, J. On the Phylogeny of Some Polyketide Synthase Genes in the Lichenized Genus *Lecanora*. *Mycol. Res.* **2003,** *107*, 1419–1426.

Henrikson, J. C.; Hoover, A. R.; Joyner, P. M.; Cichewicz, R. H. A Chemical Epigenetics Approach for Engineering the *In-Situ* Biosynthesis of a Cryptic Natural Product from *Aspergillus niger*. *Org. Biomol. Chem.* **2009,** *7*, 435–438.

Hertweck, C. The Biosynthetic Logic of Polyketide Diversity. *Angew. Chem. Int. Ed. Engl.* **2009,** *48*, 4688–4716.

Hoffmeister, D.; Keller, N. P. Natural Products of Filamentous Fungi: Enzymes, Genes, and their Regulation. *Nat. Prod. Rep.* **2007,** *24*, 393–416.

Hopwood, D. A. Genetic Contributions to Understanding Polyketide Synthases. *Chem Rev.* **1997,** *97*, 2465–2497.

Hopwood, D. A. Cracking the Polyketide Code. *PLoS Biol.* **2004,** *2*, 35.

Hopwood, D. A.; Sherman, O. H. Molecular Genetics of Polyketides and Its Comparison to Fatty Acid Biosynthesis. *Annu. Rev. Genet.* **1990,** *24*, 37–66.

Kale, S. P.; Milde, L.; Trapp, M. K.; Frisvad, J. C.; Keller, N. P.; Bok, J. W. Requirement of LaeA for Secondary Metabolism and Sclerotial Production in *Aspergillus flavus*. *Fungal Genet. Biol.* **2008,** *45*, 1422–1429.

Keller, N. P.; Turner, G.; Bennett, J. W. Fungal Secondary Metabolism—From Biochemistry to Genomics. *Nature Rev. Microbiol.* **2005,** *3*, 937–947.

Kennedy, J.; Auclair, K.; Kendrew, S. G.; Park, C.; Vederas, J. C.; Hutchinson, C. R. Modulation of Polyketide Synthase Activity by Accessory Proteins During Lovastatin Biosynthesis. *Science* **1999,** *284*, 1368–1372.

Khosla, C.; Gokhale, R. S.; Jacobsen, J. R.; Cane, D. E. Tolerance and Specificity of Polyketide Synthases. *Annu. Rev. Biochem.* **1999,** *68*, 219–253.

Kritsky, M. S.; Filippovich, S. Y.; Afanasieva, T. P.; Bachurina, G. P.; Russo. V. E. A. Effect of Inhibitors of Enzymatic DNA Methylation on the Formation of Reproductive Structures and Carotenoid Production in *Neurospora crassa*. *Appl. Biochem. Microbiol.* **2001,** *37*, 243–247.

Kroken, S.; Glass, N. L.; Taylor, J. W.; Yoder, O. C.; Turgeon, B. G. Phylogenomic Analysis of Type I Polyketide Synthase Genes in Pathogenic and Saprobic Ascomycetes. *PNAS, USA* **2003,** *100*, 15670–15675.

Lee, I.; Oh, J. H.; Shwab, E. K.; Dagenais, T. R. T.; Andes, D.; Keller, N. P. HdaA, A Class 2 Histone Deacetylase of *Aspergillus fumigatus*, Affects Germination and Secondary Metabolite Production. *Fungal Genet. Biol.* **2009,** *46*, 782–790.

Lee, T.; Yun, S. H.; Hodge, K. T.; Humber, R. A.; Krasnoff, S. B.; Turgeon, G. B.; Yoder, O. C.; Gibson, D. M. Polyketide Synthase Genes in Insect- and Nematode-Associated Fungi. *Appl. Microbiol. Biotechnol.* **2001,** *56*, 181–187.

Mayer, K. M.; Ford, J.; Macpherson, G. R.; Padgett, D.; et al. Exploring the Diversity of Marine-Derived Fungal Polyketide Synthases. *Can. J. Microbiol.* **2007,** *53*, 291–302.

Mayorga, M. E.; Timberlake, W. E. The Developmentally Regulated *Aspergillus nidulans* wA Gene Encodes a Polypeptide Homologous to Polyketide and Fatty Acid Synthases. *Mol. Gen. Genet.* **1992,** *235*, 205–212.

Molnar, I.; Schupp, T.; Ono, M.; Zirkle, R.; Milnamow, M.; et al. The Biosynthetic Gene Cluster for the Microtubule-Stabilizing Agents Epothilones A and B from *Sorangium cellulosum* So ce90. *Chem. Biol.* **2000,** *7*, 97–109.

Nicholson, T. P.; Rudd, B. A.; Dawson, M.; Lazarus, C. M.; Simpson, T. J.; Cox, R. J. Design and Utility of Oligonucleotide Gene Probes for Fungal Polyketide Synthases. *Chem. Biol.* **2001,** *8*, 157–178.

Nutzmann, H. W.; Reyes-Dominguez, Y.; Scherlach, K.; Schroeckh, V.; Horn, F.; Gacek, A.; Schumann, J.; Hertweck, C.; Strauss, J.; Brakhage, A. A. Bacteria-Induced Natural Product Formation in the Fungus *Aspergillus nidulans* Requires Saga/Ada-Mediated Histone Acetylation. *PNAS USA* **2011,** *108*, 14282–14287.

Oh, D. C.; Kauffman, C. A.; Jensen, P. R.; Fenical, W. Induced Production of Emericellamides A and B from the Marine-Derived Fungus *Emericella* sp. in Competing Co-Culture. *J. Nat. Prod.* **2007,** *70*, 515–520.

Perrin, R. M.; Fedorova, N. D.; Bok, J. W.; Cramer, R. A.; Wortman, J. R.; Kim, H. S.; Nierman, W. C.; Keller, N. P. Transcriptional Regulation of Chemical Diversity in *Aspergillus fumigatus* by LaeA. *PLoS Pathog.* **2007,** *3*, e50.

Pfeifer, B. A.; Khosla, C. Biosynthesis of Polyketides in Heterologous Hosts. *Microbiol. Mol. Biol. Rev.* **2001,** *65*, 106–118.

Piel, J.; Hertweck, C.; Shipley, P. R.; Hunt, D. M.; Newman, M. S.; Moore, B. S. Cloning, Sequencing and Analysis of the Enterocin Biosynthesis Gene Cluster from the Marine Isolate 'Streptomyces maritimus' Evidence for the Derailment of an Aromatic Polyketide Synthase. *Chem. Biol.* **2000,** *7*, 943–955.

Proctor, R. H.; Hohn, T. M.; McCormick, S. P.; Desjardins, A. E. Tri6 Encodes an Unusual Zinc Finger Protein Involved in Regulation of Trichothecene Biosynthesis in *Fusarium sporotrichioides* Produces Volatile Antimicrobials. *Mycology* **1995,** *1*, 179–186.

Ray, W. A.; Murray, K. T.; Meredith, S.; Narasimhulu, S. S.; Hall, K.; Stein, C. M. Oral Erythromycin and the Risk of Sudden Death from Cardiac Causes. *N. Engl. J. Med.* **2004,** *351*, 1089–1096.

Riyaz-Ul-Hassan, S.; Strobel, G. A.; Booth, E.; Knighton, B.; Floerchinger, C.; Sears, J. Modulation of Volatile Organic Compound Formation in the Mycodiesel Producing Endophyte-*Hypoxylon* sp. C1-4. *Microbiology* **2012,** *158*, 464–473.

Rudd, B. A. M.; Hopwood, D. A. Genetics of Actinorhodin Biosynthesis by *Streptomyces coelicolor* A3(2). *J. Gen. Microbiol.* **1979,** *114*, 35–43.

Sauer, M.; Lu, P.; Sangari, R.; Kennedy, S.; Polishook, J.; Bills, G.; An, Z. Estimating Polyketide Metabolic Potential Among Nonsporulating Fungal Endophytes of *Vaccinium macrocarpon*. *Mycol. Res.* **2002,** *1006*, 460–470.
Scherlach, K.; Hertweck, C. Triggering Cryptic Natural Product Biosynthesis in Microorganisms. *Org. Biomol. Chem.* **2009,** *7*, 1753–1760.
Schmitt, I.; Kautz, S.; Lumbsch, H. T. 6-MSAS-Like Polyketide Synthase Genes Occur in Lichenized Ascomycetes. *Mycol. Res.* **2008,** *112*, 289–296.
Shimizu, T.; Kinoshita, H.; Nihira, T. Identification and In Vivo Functional Analysis by Gene Disruption of ctnA, An Activator Gene Involved in Citrinin Biosynthesis in *Monascus purpureus*. *Appl. Environ. Microbiol.* **2007,** *73*, 5097–5103.
Shwab, E. K.; Bok, J. W.; Tribus, M.; Galehr, J.; Graessle, S.; Keller, N. P. Histone Deacetylase Activity Regulates Chemical Diversity in *Aspergillus*. *Eukaryot. Cell* **2007,** *6*, 1656–1664.
Shoolingin-Jordan, P. M.; Campuzano, I. D. G. In Sankawa, U., Ed.; *Comprehensive Natural Products Chemistry*; Elsevier: Oxford, 1999; pp 345–365.
Schroder, J. Probing Plant Polyketide Biosynthesis. *Nat. Struct. Biol.* **1999,** *6*, 714–716.
Simpson, T. J. Polyketide Biosynthesis. *Chem. Ind.* **1995,** *1995*, 407–411.
Simpson, T. J.; Holker, J. S. E. Biosynthesis of a Pyrone Metabolite of *Aspergillus melleus*. Application of Long-Range 13C–13C Coupling Constants. *Tetrahedron. Lett.* **1975,** *16*, 4693–4696.
Staunton, J.; Sutkowski, A. C. 17O NMR in Biosynthetic Studies: Aspyrone, Asperlactone and Isoasperlactone, Metabolites of *Aspergillus melleus*. *J. Chem. Soc. Chem. Commun.* **1991a,** *16*, 1106–1108.
Staunton, J.; Sutkowski, A. C. Biosynthesis of Aspyrone, A Metabolite of *Aspergillus melleus*: Advanced Precursor Studies to Identify the Product of the Polyketide Synthase. *J. Chem. Soc. Chem. Commun.* **1991b,** *16*, 1108–1110.
Staunton, J.; Sutkowski, A. C. The Polyketide Synthase (PKS) of Aspyrone Biosynthesis: Evidence for the Enzyme Bound Intermediates from Incorporation Studies with *N*-acetyl-cysteamine Thioesters in Intact Cells of *Aspergillus melleus*. *J. Chem. Soc. Chem. Commun.* **1991c,** *16*, 1110–1112.
Staunton, J. Combinatorial Biosynthesis of Erythromycin and Complex Polyketide. *Curr. Opin. Chem. Biol.* **1998,** *2*, 339–345.
Varga, J.; Rigo, K.; Kocsube, S.; Farkas, B.; Pal, K. Diversity of Polyketide Synthase Gene Sequences in *Aspergillus* species. *Res. Microbiol.* **2003,** *154*, 593–600.
Vederas, J. C. The Use of Stable Isotopes in Biosynthetic Studies. *Nat. Prod. Rep.* **1987,** *4*, 277–337.
Vining, L. C. Secondary Metabolism, Inventive Evolution and Biochemical Diversity—A Review. *Gene* **1992,** *115*, 135–140.
Von Dohren, H. A Survey of Nonribosomal Peptide Synthetase (NRPS) Genes in *Aspergillus nidulans*. *Fungal Genet. Biol.* **2009,** 46, 45–52.
Wakil, S. J. Fatty Acid Synthase, A Proficient Multifunctional Enzyme. *Biochemistry* **1989,** *28*, 4523–4530.
Walton, J. D. Horizontal Gene Transfer and the Evolution of Secondary Metabolite Gene Clusters In Fungi: A Hypothesis. *Fungal Genet. Biol.* **2000,** *30*, 167–171.
Williams, R. B.; Henrikson, J. C.; Hoover, A. R.; Lee, A. E.; Cichewicz, R. H. Epigenetic Remodeling of the Fungal Secondary Metabolome. *Org. Biomol. Chem.* **2008,** *6*, 1895–1897.
Wright, J. L. C.; Vining, L. C.; McInnes, A. G.; Smith, D. G.; Walter, J. A. Use of 13C in Biosynthetic Studies, The Labelling Pattern in Tenellin Enriched from Isotope-Labelled Acetate, Methionine, and Phenylalanine. *Can. J. Biochem.* **1977,** *55*, 678–685.

Xie, X.; Watanabe, K.; Wojcicki, W. A.; Wang, C.; Tang, Y. Biosynthesis of Lovastatin Analogs with a Broadly Specific Acyltransferase. *Chem. Biol.* **2006,** *13*, 1161–1169.

Yu, T. W.; Bibb, M. J.; Revill, W. P.; Hopwood, D. A. Cloning, Sequencing, and Analysis of the Griseusin Polyketide Synthase Gene Cluster from Streptomyces griseus. *J. Bacteriol.* **1994,** *176*, 2627–2634.

Yu, J. H.; Butchko, R. A.; Fernandes, M.; Keller, N. P.; Leonard, T. J.; Adams, T. H. Conservation of Structure and Function of the Aflatoxin Regulatory Gene aflR from *Aspergillus nidulans* and *A. flavus*. *Curr. Genet.* **1996,** *29*, 549–555.

Zahner, H.; Kurth, R. Overproduction of Microbial Metabolites—The Supply of Precursors from the Intermediary Metabolism. In *Overproduction of Microbial Products*; Krumphanzl, V.; Sikyta, B.; Vanek, Z., Eds.; Academic Press: London, 1982; pp 167–179.

CHAPTER 9

Endophytes: An Asset for Extracellular Hydrolases Production

HIMANSHU KACHROO

Department of Biotechnology, Indian Institute of Technology Roorkee, Roorkee, Uttarakhand, India

E-mail: h_kachroo@bt.iitr.ac.in; himanshukachroo357@gmail.com

ABSTRACT

Abundant endophytes are discovered, isolated from various tissues of plants. With the advancement of biotechnology, research indicating the value-added scope of endophytes has increased globally. The intense diversity and unique host interaction capability of endophytes is an untapped source of generating novel compounds. Endophytes seem beneficial to the host plants in providing additional resources at the time of any biotic and abiotic stresses. They help in relieving stress through the production of valuable secondary metabolites, extracellular enzymes, and nutrients. Endophytes are proficient enough to resist the toxic substances released by host plants to limit their growth by producing extracellular enzymes. Extracellular hydrolases consist of cellulase, proteases, amylases, lipases, pectinases, xylanases, chitinases, which function outside the endophyte microbial cell in many biological and environmental stresses. The prospecting of endophytes for the production of extracellular hydrolases is encouraging. The likelihood of isolation and quantification of novel extracellular hydrolases will be satisfactory as substitutes for specialized industries. The current chapter emphasizes disclosure of the sources, types of hydrolase enzymes, and the outlook for promoting adequate research for enhancing the applications of endophyte-derived extracellular hydrolases.

9.1 INTRODUCTION

Endophytes are known for their ability to spend their life cycle inside host plant tissues without generating any obvious harm or symptoms (Bezerra et al., 2012; Kaul et al., 2013; Tan and Zou, 2001; Yadav et al., 2016). It is suggested that endophytic microbes are recognized source of genes, extracellular hydrolases, and other secondary metabolites (Leo et al., 2016). Endophytes help in promoting the resistance of plants toward drought, variation in temperature, pH and salinity, and heavy metals concentration in the soil (Brem and Leuchtmann, 2001; Schulz et al., 2002; Jalgaonwala et al., 2017). In addition to this, endophytes also give disease resistance to host plants against pathogen infection. This disease resistance is accomplished through production and release of antimicrobial compounds, extracellular enzymes, and stimulation of host defense (Hall et al., 1986; Pleban et al., 1995; Larkin et al., 1996; Benhamou et al., 2000). Endophytic fungal association with mangrove plants provides the host plant protection from adverse environmental conditions (Kumaresan and Suryanarayanan, 2002). This association further allows them to compete with saprobic fungi capable of decomposing senescent parts. The failure to exploit endophytes depends on our restrict understanding on the interaction of endophytes with their host plant. Microbial endophytes not only safeguard their host plant against invasion from other microorganisms, insects, and herbivores but also seem furthermore serviceable, for example, production of growth regulators, enzymes, and numerous other beneficial chemical compounds (Azevedo et al., 2000; Bezerra et al., 2012). Endophytes are reservoir for discovery of novel compounds (Correa et al., 2014). Various secondary metabolites such as alkaloids, cyclohexanes, flavonoids, hydrocarbons, quinines, and terpenes of excellent biological properties are produced by endophytes (Naik and Krishnamurthy, 2010; Ruma et al., 2013; Fernandes et al., 2015). They can lead to the discovery of compounds of extreme significance like antibiotics, antioxidants, immunomodulators, anticancer, and antiparasitic drugs. Production of extracellular enzymes by endophytes is faster than intracellular enzymes (Hankin and Anagnostakis, 1975). It is easier to extract extracellular enzymes in comparison to intracellular ones. *Trichoderma reesei* and its mutants can be employed for hemicellulases and cellulases commercial production (Nieves et al., 1998). Hydrolase enzymes are those enzymes that are proficient enough to catalyze hydrolysis of a particular substrate. Endophytes are well known for producing various extracellular hydrolase enzymes such as cellulases, esterases, lipases, pectinases, proteases, and xylanases (Suto et al., 2002; Bezerra et al., 2012).

Cellulases basically catalyze cellulolysis, which basically implements cellulose and certain polysaccharides degradation (Rana et al., 2019). Xylanases catalyze the endohydrolysis of 1,4-b-D-xylosidic linkages in xylan present in hemicelluloses of plants (Thomas et al., 2017). Lipases are involved in hydrolysis of fats and oils (Gopinath et al., 2013). Pectinases are responsible of hydrolysis of pectin polysaccharide found in plant cells (Garg et al., 2016). Phytases are capable of catalyzing hydrolysis of phytic acid to inositol phosphates, myoinositol, and inorganic phosphate (Kumar et al., 2016; Kaur et al., 2017). Extracellular hydrolases such as lipases, amylases, and proteases produced by endophytes are considered most significant and important due to their industrial applications (Traving et al., 2015). They are regarded more valuable in food industry, fermentation dye synthesis, and other biotechnological applications.

Endophytes are potentially useful for the production of extracellular hydrolases due to their ability to degrade the complex structures of lignocelluloses (Naik et al., 2019). Endophytes associated with crops are gaining attention for production of hydrolase enzymes for biomass conversion (Ferreira et al., 2008; Xiong et al., 2014; Castro et al., 2014). The demand for enzyme utilization mainly hydrolases is continuously growing specifically in food and beverage sectors (Bonugli-Santos et al., 2015). Enzymes are used from ancient time for the production of cheese, beer, wine, and vinegar (Kirk et al., 2002). To fulfill this tremendous demand, there lays an intense requirement for more and enhanced production of hydrolase enzymes on an industrial scale. In this chapter, we aim to broaden our understanding related to extracellular hydrolases enzymes from endophytic origin and their immense values in industrial applications.

9.2 EXTRACELLULAR HYDROLASES FROM ENDOPHYTES

Endophytes from various plant sources are able to produce various kinds of extracellular hydrolases (Esteves et al., 2014). Cellulase and hemicellulase producing endophytic fungi *Acremonium* is isolated from corn plant (Almeida et al., 2011). Phytase producing endophytes *Fusarium verticillioides* and *Rhizoctonia* species isolated from host plant *Glycine max* (Marlida et al., 2010). Amylase producing *Discosia* species isolated from *Calophyllum inophyllum* host plant (Hegde et al., 2011). Different hydrolases such as lipase, protease, amylase, cellulase, and xylanase produced by endophytes like *Talaromyces flavus, Mortierella hyalina, Paecilomyces variabilis*, and *Penicillin* isolated from *Potentilla fulgens, Osbeckia stellata, Osbeckia chinensis, Camellia caduca*, and *Schima khasiana* plant species (Bhagobaty and Joshi, 2012). Various hydrolases such as amylase, cellulase, lipase, and protease produced from different endophytes are categorized in Table 10.2.

Some strains of bacterial endophytes include *Bacillus, Azotobacter, Arthrobacter, Agrobacterium*, and *Enterobacter* producing extracellular hydrolases (Gray and Smith, 2005). A lot of endophytic fungal strains such as *Alternaria alternata, Aspergillus terreus, Hymenoscyphus ericae, Periconia, Acremonium*, and many more are producer of several extracellular hydrolases like xylanase, cellulase, hemicellulase, etc. (Burke and Cairney 1997; Harnpicharnchai et al., 2009; Wipusaree et al., 2011; Sorgatto et al., 2012; De Almeida et al., 2012). Around 155 endophytic strains are capable to produce extracellular hydrolase enzymes (Suto et al., 2002). Bacterial endophytes are reported for production of cellulase, protease, amylase, pectinase, etc. (Sturz et al., 2000; Carrim et al., 2006). Various bacterial endophytes are isolated from different medicinal plants (Vijayalakshmi et al., 2016). Two hydrolases producing fungal strains, that is, *Xylariaceae* and *Annulohypoxylon* species are reported from a medicinal plant *Hevea brasiliensis* (Gazis and Chaverri, 2010). Extracellular hydrolase producing endophytic strains from *Fusarium, Colletotrichum, Phoma*, and *Penicillium* species are derived from various anticancer medicinal plants (Chow and Ting, 2015). Some fungal endophytes are screened for extracellular cellulose activity isolated from Cameroonian medicinal plants (Toghueo et al., 2017).

9.3 NECESSITY FOR EXTRACELLULAR HYDROLASES PRODUCTION

Endophytes are proficient enough to produce many extracellular enzymes, including hydrolases. Extracellular enzymes are derived as a product of microorganism's cell growth. These enzymes are efficient for endophytes to continue their survival in extreme stressful habitat. Endophytes gain adaptability, survival efficiency, and utilization of their ecological niche conditions with the release of extracellular enzymes (Gopinath et al., 2005; Naik et al., 2019). Extracellular hydrolases are capable of performing their function outside the cell in various key processes (Khan et al., 2017). Endophytes produce enzymes such as proteases, lipases, and cellulases for their development as hydrolytic enzymes can help in obtaining nutrition (Torres et al., 2003; Sunitha et al., 2012). Endophytes are well known for their extraordinary ability to interact with peripheral and internal tissues of the plants through symbiotic mutualism (Sudheep et al., 2017). Extracellular enzymes like hydrolases produced by endophytic microbes can be useful for initiating symbiotic process (Hallmann et al., 1997). In most cases, endophytes are benefitted through host plants in terms of energy, nutrients, and shelter. However, endophytes indirectly prove advantageous to their respective hosts by synthesizing valuable substances like hormones, secondary metabolites, and extracellular enzymes (Barz et al., 1988; Sudheep et al., 2017). The production of extracellular hydrolases by endophytes could play a crucial role in the growth and development of host plants (Khan et al., 2016). Besides establishment of host symbiosis process, they initiate extracellular hydrolases production to neutralize plant pathogenic infection (Hallmann et al., 1997; Leo et al., 2016).

Various researchers are uncovering the fact that endophytes help plant to develop resistance against pathogenic agents like insects and microorganisms (Saikkonen et al., 1996; Sudheep et al., 2017). Implantation of endophytes from disease free host plant to disease prone plant enhances the fitness of disease prone plant (Sudheep et al., 2017). Extracellular hydrolase enzymes like protease, chitinase, and glucanase produced by endophytes are able to inhibit the growth of pathogens attacking the host plant. Endophytes isolated from internal tomato crown can produce extracellular hydrolases in vitro against plant pathogen *Verticillium dahliae* (Dhouib et al., 2019). Endophytes like any other microorganisms are invasive to plant tissues. The disproportionate condition of host plant and endophyte interaction would either result in disease in the host plant or the plant defense mechanism will be capable to kill the endophytes (Schulz and Boyle, 2005).The defense mechanism of the host plant determines whether the endophyte interaction is healthy or unhealthy. In order to overcome attack initiated by the host plant, they release extracellular hydrolases (Petrini et al., 1993; Reddy et al., 1996; Tan and Zou, 2001). This resistance mechanism against host plant invasion alternatively proves completely beneficial for the host plant (Tan and Zou, 2001; Leo et al., 2016). Studies show that both endophytic bacteria and fungi are capable of producing extracellular hydrolases that can target carbohydrates, lignin, proteins, and other macromolecules (Boer et al., 2005; Strong and Claus, 2011; Traving et al., 2015).

Endophytes release xylanases and cellulases in order to achieve degradation of lignocellulosic fibers. Similarly, ligninases and peroxidases are released to degrade lignin (Yadav et al., 2012; Shukla et al., 2016). Lignocellulose is the major structural component of plants (Howard et al., 2003). Hemicellulose, cellulose, and lignin are the cell wall components of plant (Naik et al., 2019). Release of exoenzymes by endophytic microbes facilitates degradation of organics such as cellulose, glucose, keratin, lignin, lipids, pectin, proteins present or produced by plants (Kudanga and Mwenje, 2005; Tomita, 2003; Rana et al., 2019).

9.4 ISOLATION AND QUANTIFICATION OF EXTRACELLULAR HYDROLASES

Endophytes capable of producing extracellular hydrolases can be isolated from various plants like *Pisum sativum* (pea), *Lycopersicum esculentum* (tomato), *Zea mays* (corn), *Triticum aestivum* (wheat), *Solanum tuberosum* (potato), and many more (Khan et al., 2017). The endophytes are isolated from stems, leaves, and roots of different plants. Plant tissues are thoroughly washed and surface sterilized by tween weak acids, sodium hypochlorite, ethanol, hydrogen peroxide, and distilled water (Sturz et al., 1998; Taechowisan et al., 2003; Bezerra et al., 2012). After rigorous surface sterilization, plant tissues are then inoculated in different sterilized nutrient agar to isolate different types of endophytes present in them. The endophytic microorganisms can be seen growing around the inoculated plant tissue in the microbial media; the different colonies according to morphology (size, shape, and color) are collected and grown individually on different microbial media plates (Rashid et al., 2012). These individual colonies are further subcultured to get pure culture for further screening. Both fungal and bacterial strains are identified after continuous maintenance of the culture till pure culture is obtained. Endophytic strains can be identified through 16s rRNA sequencing and internal transcribed spacer (ITS) region of bacteria and fungi (Khan et al., 2017). PCR amplification, Sanger sequencing, BLASTn, and a full detailed phylogenetic analysis are employed for achieving strain identification. Pure cultures are inoculated and incubated in nutrient medium containing all essentials required for enhanced growth of endophytes. Culture-containing flasks are grown with continuous shaking at different parameters (according to the endophytic microbial strain) (Bischoff et al., 2009; Sunitha et al., 2012). The culture broth will be further filtered, centrifuged, and supernatant collected will be analyzed to detect enzyme activity (Sunitha et al., 2012). The extracellular hydrolases produced by isolated endophytes can be detected in specific growth medium. It is reported that fluorogenic substrates like 4-methylumbelliferone (MUB) can be used for enzyme analysis purpose (Wallenstein and Weintraub, 2008; Khan et al., 2016). The quantification of the isolated enzymes can be done by implementing highly sensitive techniques. Advanced chromatographic techniques are employed for quantification of extracellular hydrolases (Khan et al., 2017). It is suggested that affinity chromatography can be implemented for quantification of amylase derived from *Bacillus licheniformis* (Mendu et al., 2005). DEAE-cellulose ion-exchange chromatography along with gel filtration chromatography is employed for quantification of endophytic fungi *Fusarium oxysporum* derived extracellular lipase (Panuthai et al., 2012). Step-by-step representation of isolation and quantification of extracellular hydrolases produced by endophytes is given in Fig. 9.1.

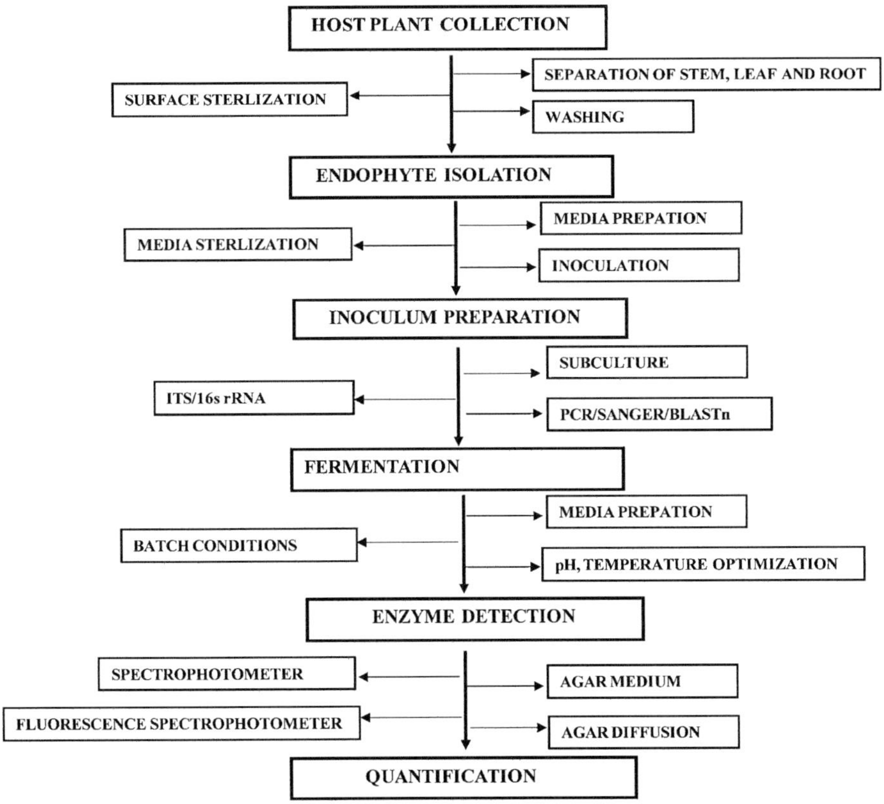

FIGURE 9.1 Step-by-step representation of isolation and quantification of extracellular hydrolases produced by endophytes.

9.5 EXTRACELLULAR HYDROLASES DETECTION

The extracellular hydrolases produced from different strains of endophytes (both fungal and bacterial) can be detected through several detection methods. It is reported that most common and preferred methods are agar medium and spectrophotometer detection (Ayob and Simarani 2016; Escudero et al., 2016; Kalyanasundaram et al., 2015). Detection method totally relies upon the type of endophytic strain (Sunitha et al., 2013). Particular type of endophytic strain can be employed for production of various hydrolase enzymes. *Fusarium* and *Colletotrichum* endophytic species are reported producing amylase, cellulase, lipase, pectinase, and protease. Agar medium method was employed to detect these hydrolase enzymes. *Colletotrichum crassipes* and *Colletotrichum falcatum* fungal endophytes can be employed for production of cellulase, lipase, protease, and amylase and their detection can be done with agar medium (Khan et al., 2017). Proteases produced from *Fusarium* species are detected by agar medium method (Sunitha et al., 2013) while as proteases produced from *Pochonia chlamydosporia* and the detection method employed can be spectrophotometer (Escudero et al., 2016). Various extracellular enzymes like amylase, protease, cellulase, and lipase are produced by various fungal endophytes and are detected through employing either agar medium method or spectrophotometer method (Patil et al., 2015). L-asparaginase produced from some bacterial endophytes like *Bacillus subtilis* and *Bacillus methylotrophicus* can be detected using spectrophotometer (Nongkhlaw and Joshi, 2015). M9 medium can be employed for detection of L-asparaginase produced from bacterial endophytes like *B. licheniformis* and *Paenibacillus senitriformus* (Joshi and Kulkarni, 2016). Some of the common hydrolysases and their detection method are discussed in the following sections.

9.5.1 PECTINASE DETECTION

It is reported that agar medium was employed for detection of pectinase produced from various endophytic strains such as *Fusarium, Aspergillus flavus, Alternaria* (Sunitha et al., 2013); *Acremonium terricola, Phoma tropica* (Bezerra et al., 2012); *Penicillium chrysogenum* (Fouda et al., 2015). Extracellular pectinase are detected from endophytic species like *Pseudomonas hibiscicola, Bacillus anthracis, and Pseudomonas entomophila* through agar diffusion method (Akinsanya et al., 2016). Agar diffusion method is employed for detection of pectinase derived from *Paenibacillus polymyxa* endophyte (Khan et al., 2017). It is suggested that extracellular pectinases derived from fungal endophytes are detected through agar medium and extracellular pectinase derived from bacterial endophytes are detected through agar diffusion method.

9.5.2 AMYLASE DETECTION

Most of the research studies can show extracellular amylases produced from endophytes are detected through agar medium detection method (Maria et al., 2005; Fouda et al., 2015; Kannan et al., 2015; Jurynelliz et al., 2016). Some bacterial endophytes such as *Chryseobacterium indologene*, *Bacillus tequilensis*, and *P. entomophila* are proficient to produce extracellular amylase and their detection through agar diffusion method (Akinsanya et al., 2016). Agar medium method is employed for detection of endophytic *Bacillus* species derived amylase (Carrim et al., 2006; Joe et al., 2016). Agar medium detection method is used for extracellular amylase produced by endophytic fungal strains like *Acremonium, Alternaria, Aspergillus*, and *Fusarium* species (Maria et al., 2005). Agar medium method can be efficient toward extracellular amylase detection derived from fungal endophytes but bacterial endophytes show more sensitivity toward agar diffusion method for amylase detection (Khan et al., 2017). Extracellular amylase produced by some endophytes such as *Cladosporium, Rhizoctonia, Aspergillus, Chaetomium*, and *Curvularia* species can be detected by implementing both agar medium detection method and spectrophotometer (Patil et al., 2015a,b).

9.5.3 CELLULASE DETECTION

Various methods are employed for detection of extracellular cellulase of endophytic origin. Endophytic strains are *Talaromyces emersonii, Pestalotiopsis disseminate, Paecilomyces variotii* (Sunitha et al., 2013); *C. falcatum, C. crassipes, Lasiodiplodia theobromae* (Amirita et al., 2012); *Myrmecridium schulzeri, Trichoderma piluliferum* (Bezerra et al., 2015). Extracellular cellulase production is detected by agar medium method. Extracellular cellulases from bacterial endophytes like *P. polymyxa* and *Bacillus* species are detected by agar diffusion method (Khan et al., 2017). It is claimed that agar diffusion method can be employed for extracellular cellulase detection produced by some bacterial endophytic strains such as *B. anthracis, B. tequilensis*, and *Bacillus aerophilus* (Akinsanya et al., 2016). But, on the other hand, it is reported that cellulase derived from bacterial endophytes of some *Bacillus* and *Acinetobacter* species can be detected by agar medium method (Joe et al., 2016). Cellulase produced extracellularly by some fungal endophytes like *Incertae sedis, Eurotiales, Chaelomiaceae, Nectriaceae*, and *Sporomiaceace* can be detected through employing spectrophotometer and cellulase derived from other fungal endophytes such as *Penicillium citrinum, Thielavia arenaria, Phoma medicaginis*, and *Fusarium proliferatum* can be detected from fluorescence spectrophotometer (Khan et al., 2016).

9.5.4 XYLANASE DETECTION

Extracellular xylanase produced from endophytes can be detected through agar medium method (Bezerra et al., 2012). A study on extracellular xylanase produced from different endophytes suggested their detection by agar medium method (Bezerra et al., 2015). Some fungal endophytes like *Cochliobolus lunatus, Gibberella baccata, M. schulzeri*, and *Acremonium curvulum* can show potential to produce and release xylanase. Xylanases derived from several species of fungal endophytes extracellularly are detected through agar medium method only (Fouda et al., 2015). Xylanases generated from some endophytic bacterial strains can be detected by agar diffusion method (Akinsanya et al., 2016). Xylanase can be produced from *Pseudomonas, Macrococcus*, and *Bacillus* bacterial endophytic strains. It is clarified that xylanase produced from *Bacillus* endophytes can be detected by agar diffusion method only (Khan et al., 2017).

9.5.5 LIPASE DETECTION

Extracellular lipases are reported to be detected by agar medium method extracted from different species of endophytes (Maria et al., 2005; Carrim et al., 2006; Amirita et al., 2012). Agar medium method is considered for detection of extracellular lipase produced by various endophytes of fungal origin, including *P. variotii, Phomopsis longicolla*, and *Fusarium solani* (Sunitha et al., 2013). Agar medium method is preferred for detection of extracellular lipase derived from various fungal endophytic strains like *Myrothecium verrucaria, T. piluliferum, Penicillium commune, Aspergillus niger*, and many more (Bezerra et al., 2015). Agar medium method is used to detect the extracted extracellular lipase from different endophytic bacterial strains of *Bacillus clausii, Bacillus pumilus*, and *B. licheniformis* species (Kannan et al., 2015). Both agar medium and spectrophotometer methods can be implemented for detection of lipase produced by diverse fungal endophytes like *Cladosporium, Rhizoctonia, Aspergillus*, and *Fusarium* species extracellularly (Patil et al., 2015a,b).

9.5.6 PROTEASE DETECTION

Different fungal species of endophytes such as *Umbelopsis isabellina, Hebeloma incarnatulum*, and *Laccaria bicolor* are competent enough to produce extracellular protease (Mayerhofer et al., 2015). Extracellular proteases produced from endophytic species like *Acremonium, Alternaria, Aspergillus, Fusarium*, and *Pestalotiopsis* are detected by agar medium method (Maria et al., 2005). It is altogether supported that agar medium method can be used for detection of extracellular protease produced from endophytes like *Macrophomina phaseolina, Nigrospora sphaerica, F. solani* (Ayob and Simarani, 2016); *Amanita muscaria, Boletus luridus, Cenococcum geophilum* (Nygren et al., 2007); *Monodictys castaneae* (Bezerra et al., 2012); *C. lunatus, A. niger* (Bezerra et al., 2015); *Actinomyces pyogenes, Bacillus circulans* (Carrim et al., 2006); *Methylobacterium, Curtobacterium, Mucilaginibacter* (Chimwamurombe et al., 2016). Extracellular protease produced from endophyte *Colletotrichum gloeosporioides* can be detected using spectrophotometer (Escudero et al., 2016).

9.6 FACTORS AFFECTING ENZYME PRODUCTION

Several factors are involved in affective production of extracellular hydrolases by endophytic microorganisms. Some endophytic strains such as *Bacillus, Pseudomonas, Saccharomyces cerevisiae, Aspergillus*, and *Penicillium* are capable of producing and releasing hydrolase enzymes like amylase, protease, cellulose, and many more (Abd Rahman et al., 2005; Treichel et al., 2010; Sharma et al., 2013; Sundarram and Murthy, 2014). Several factors like pH, temp, etc. influence the growth of microorganisms and enzyme production. It is must to investigate the role of these factors to determine suitable extracellular hydrolase enzymes production.

9.6.1 TEMPERATURE

Temperature plays a critical role in production of extracellular hydrolase enzymes from endophytes as temperature must remain optimum for the growth of microorganism as well as for production of enzyme. The hydrolysis is highly dependent on the effect of temperature. Both temperature ranges influences production and optimum temperature ranges can achieve maximal production (Dhiman et al., 2008; Sharma et al., 2013; Sundarram and Murthy, 2014). Optimum growth temperature is reported in the range of 45–46°C and the optimum temperature of amylase production can be 50°C, in the case of extracellular amylase production by *B. licheniformis* and *B. subtilis* (Sundarram and Murthy 2014). It is further revealed that enzyme production can increase with increasing temperature till optimum temperature is achieved. However, after increasing temperature above optimum value will definitely decrease the enzyme production. To achieve optimal amylase production from endophytic fungal species, the temperature can be optimized in the range of 50–55°C (Jensen and Olsen, 1992). Amylase extracted from *Penicillium* and *Aspergillus* endophytes showed enhanced activity near 30°C (Ramachandran et al., 2004; Thippeswamy et al., 2006). Bacterial endophytes can achieve productive cellulase yield at temperatures 30–45°C (Abou-Taleb et al., 2009). Increased extracellular cellulase production by *Pseudomonas* species is achieved at 35°C (Bakare et al., 2005). Pectinase enzymes produced by *S. cerevisiae, Pseudomonas, Phytophthora, Aspergillus, lactobacillus,* etc., are showing effect on their production through alteration in temperature (Sharma et al., 2013). Temperature is capable of regulating the synthesis and secretion of extracellular protease from microorganisms (Ray et al., 1992; Abd Rahman et al., 2005). Temperature can change the physical properties of cell membrane, so influencing extracellular hydrolases secretion from microbial cells (Abd Rahman et al., 2005). The influence of temperature is assessed on lipase production by *Pseudomonas* species (Cavalcanti et al., 2005).

9.6.2 CARBON SOURCES

The selection of appropriate carbon sources can prove advantageous for enzyme production. It is presented that type of carbon source significantly influences the extracellular hydrolases production (Abou-Taleb et al., 2009; Sundarram and Murthy, 2014). It is inferred that carboxymethyl cellulose can be the best influential carbon source than cellulose in the case of extracellular cellulase enzyme production from endophytic bacterial strains (Narasimha et al., 2006; Niranjane et al., 2007). Cellulase production can be less when carbon source utilized by endophytes is glucose (Muthuvelayudham and Viruthagiri, 2006). It is presented that carboxymethyl cellulose can be employed as a carbon source for enhanced cellulase production by *Bacillus* strains (Abou-Taleb et al., 2009). It is stated that glucose, maltose, wheat bran, sucrose, and banana waste can be employed as carbon source for extracellular amylase production by endophytes like *B. subtilis, B. licheniformis,*

Bacillus amyloliquefaciens, and *A. niger* (Sundarram and Murthy, 2014). Glucose can be best suited as carbon source for enhanced amylases production by *Bacillus* species. Microbial pectinase production can require banana peel, wheat bran and sugarcane bagasse as carbon source to maintain adequacy (Sharma et al., 2013). It is reported that common carbon sources of microbial lipase production are triacylglycerols, fatty acids, hydrolysable esters, tweens, bile salts, and glycerol (Sharma et al., 2001; Gupta et al., 2004). It is claimed that lipidic carbon sources can definitely provide high lipase yield (Treichel et al., 2010).

9.6.3 NITROGEN SOURCES

Nitrogen source implemented for the production of extracellular hydrolase enzymes may be organic or inorganic (Cihangir and Sarikaya, 2004; Abd Rahman et al., 2005; Treichel et al., 2010). Their data is clarified that extracellular enzyme yield can be simulated through supplementation of organic and inorganic nitrogen sources. Most commonly used inorganic nitrogen sources include ammonium sulfate, ammonium chloride and ammonium hydrogen phosphate. On the other hand, most commonly used organic source of nitrogen is peptone, soybean meal, beef extract, urea, and yeast extract (Sundarram and Murthy, 2014). Extracellular production of amylase by endophyte *Aspergillus oryzae* utilized sodium nitrate as inorganic nitrogen source and malt as organic nitrogen source provided highest yield of enzyme (Goto et al., 1998). The use of peptone and sodium nitrate as nitrogen sources can prove serviceable for *Aspergillus* and *Penicillium* endophytic strains (Møller et al., 2004; Kunamneni et al., 2005; Raul et al., 2014). Yeast extract as nitrogen source can be proficient enough for enhancing the cellulase production by *Bacillus* endophytes (Abou-Taleb et al., 2009). Organic nitrogen sources can be competent enough than inorganic sources for optimizing the cellulase production by *B. subtilis* endophytic species (Ray et al., 2007). Mixture of corn steep liquor and ammonium nitrate is the most appropriate source of nitrogen for microbial production of extracellular lipase (Yan and Yan, 2008).

9.6.4 pH

In order to procure maximal production of extracellular hydrolases from endophytic microbes, it is crucial to optimize the pH of the reaction. Studies convince that optimum pH can be pivotal for the stability of extracellular enzymes produced as enzymes are pH sensitive (Ray et al., 1992; Haki and Rakshit, 2003; Chadha et al., 2003; Dhiman et al., 2008). The optimum pH required for efficient production of amylase from *A. oryzae, Aspergillus ficuum*, and *A. niger* can be in the range of 5–6 (Carlsen et al., 1996; Møller et al., 2004; Sundarram and Murthy, 2014). Yield of extracellular amylases produced by *S. cerevisiae* endophytic species can be elevated in the pH range of 5–7 (Knox et al., 2004; Tanyildizi et al., 2005; Ashraf et al., 2005). Production of extracellular lipase by microorganisms can be influenced through pH change (Kar et al., 2008). It is investigated that pH control is having high influence on lipase production by *Acinetobacter* species (Treichel et al., 2010). It is uncovered that highest yield of extracellular protease released by *Pseudomonas aeruginosa* species can be observed at pH 7 (Abd Rahman et al., 2005).

High yield of microbial acid protease can be achieved at pH 3–4 (Ray et al., 1992); neutral proteases can be procured at pH 6.8–7.0 (Moormann et al., 1993) and alkaline protease can be attained at pH 9–10 (Jasvir et al., 1999). Enzymatic processes and components transport across the cell membrane can be altered through pH change (Moon and Parulekar, 1991, Abd Rahman et al., 2005). Cellulase production by *Bacillus* species can be increased by maintaining the pH at 7 (Abou-Taleb et al., 2009). It is stated that maximum production of extracellular cellulase can be obtained from *Bacillus* and *Micrococcus* species through optimizing the pH at 7 (Immanuel et al., 2006; Ray et al., 2007; Abou-Taleb et al., 2009). Optimization of pH can be always valuable for receiving augmented extracellular xylanase and pectinase production from endophytes such as *Pseudomonas, Aspergillus, Penicillium, Bacillus*, and *Streptomyces* species (Dhiman et al., 2008; Sharma et al., 2013).

9.6.5 INOCULUM SIZE

An indirect relationship can be developed based on inoculum size and extracellular hydrolase enzyme yield from endophytes (Abd Rahman et al., 2005; Abou-Taleb et al., 2009). Focus can be laid on optimization of inoculum size for attaining maximal enzyme production. It is elucidated that 3% inoculum size can gradually lead to higher extracellular cellulase yield from endophytes. Cellulase enzyme production by *Bacillus* strains elevated up to 3% inoculum size but decreased thereafter (Abou-Taleb et al., 2009; Ray et al., 2007). Various inoculum sizes of *Pseudomonas* species are investigated for protease production (Abd Rahman et al., 2005). It is indicated that inoculum size of 4% can be used for enhanced extracellular protease yield. Various inoculum sizes optimum for different microbial strains achieve desirable protease yields. It is suggested that 1% inoculum size for *Bacillus* species (Mehrotra et al., 1999) and 5% inoculum size for *B. licheniformis* (Mabrouk et al., 1999) attain improvised yields.

9.6.6 SHAKING RATE

It is claimed that culture-containing flasks need to be kept on rotary shaker with required rpm to achieve proper aeration for maximizing extracellular hydrolase enzymes production by endophytes (Abd Rahman et al., 2005; Abou-Taleb et al., 2009; Treichel et al., 2010). Shaking condition can be advantageous for higher cellulase yield than static condition in the case of extracellular cellulase production by endophytic species of *Bacillus* strain (Khan and Husaini, 2006). To achieve higher yield of cellulase from endophytes like *B. amyloliquefaciens*, the shaking rate applied should be in the range of 150–200 rpm (Abou-Taleb et al., 2009). On the other hand, it is reported that shaking condition can lead to decrease the yield of extracellular protease produced by *Penicillium aeruginosa* species (Abd Rahman et al., 2005). Sometimes excess agitation and aeration can be the reason behind drop in the production rate of extracellular enzymes (Pourrat et al., 1988). However, *B. licheniformis* showed higher protease production with agitation of 250–400 rpm (Mabrouk et al., 1999). Agitation of 150 rpm can show the highest protease production by *Bacillus* species (Razak et al., 1997).

9.7 APPLICATIONS

Endophytes are unexplored source of novel natural products. Researchers are focusing on various endophytic strains such as *Aspergillus terreus, Colletotrichum, Pestalotiopsis, Fusarium*, and many more that produce a large amount of hydrolase enzymes like protease, amylase, xylanase, etc. (Ahmed et al., 2016; Naik et al., 2019). Extracellular enzymes, including hydrolases, are used for many industrial applications (Fig. 9.2). Utility of hydrolases in a particular industry depends on their chemical nature and favorable working conditions (Mandal and Banerjee, 2019). As shown in Table 10.1, different endophytic fungal and bacterial strains can be used for extracellular hydrolases production. Cellulase shows potential for many applications (Sudheep et al., 2017). Fungal endophytes are marvelous in producing broad range of extracellular hydrolase enzymes. Different extracellular hydrolases produced from endophytes and their applications are reported in various literatures. Hydrolase enzymes show their applications in baking, brewing, medicinal, agriculture, wine, animal feedstock, and many more. Extracellular hydrolases can be used for the production and development of medicines, cosmetics, and clinical reagents (Zhang and Kim, 2010; Murray et al., 2013). Extracellular hydrolases such as proteases can be used for production of digestive and anti-inflammatory drugs (Zhang and Kim, 2010). Some researchers focused on biotechnological applications of extracellular hydrolases in chemical, fuel, food, agriculture, paper, and textile sectors (Raghukumar et al., 1994; Sette and Bonuglisantos, 2013). Hydrolases potential in food and beverage sectors is explored (Velmurugan and Lee, 2012). Endophytic extracellular hydrolases can show consistency, easy handling, optimum environmental conditions, and no chemical catalyst requirement (Sudheep et al., 2017). However, due to microbial nature of endophytes, it becomes less tedious to genetically modify them for the optimization of hydrolase enzyme production from a specified endophyte. Roles of endophyte-derived extracellular hydrolase enzymes in different areas and sectors are stated as follows (Table 10.2).

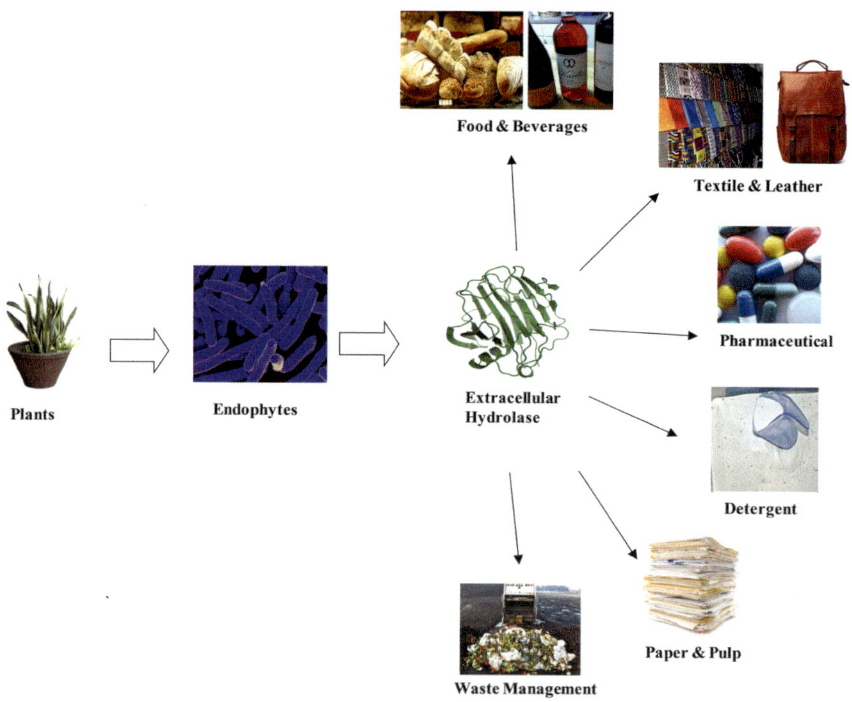

FIGURE 9.2 Diagrammatic representation of application of extracellular hydrolases isolated from endophytic microbes.

TABLE 10.1 Extracellular Hydrolases Produced from Various Endophyte Species.

Hydrolase Enzyme	Endophyte species	References
Pectinase	*Fusarium*	Bezerra et al. (2012)
	Aspergillus flavus	Sunitha et al. (2013)
	Alternaria	Fouda et al. (2015)
	Acremonium terricola	Akinsanya et al. (2016)
	Phoma tropica	
	Penicillium chrysogenum	
	Pseudomonas hibiscicola	
	Bacillus anthracis	
	Pseudomonas entomophila	
Amylase	*Aspergillus*	Carrim et al. (2006)
	Colletotrichum gloeosporioides	Amirita et al. (2012)
	Mycelia sterilia	Sunitha et al. (2013)
	Curvularia brachyspira	Rabha et al. (2014)
	Actinomyces pyogenes	Kannan et al. (2015)
	Bacillus pumilus	Jurynelliz et al. (2016)
Cellulase	*Talaromyces emersonii*	Bezerra et al. (2012)
	Colletotrichum falcatum	Amirita et al. (2012)
	Penicillium glandicola	Sunitha et al. (2013)
	Acinetobacter	Akinsanya et al. (2016)
	Enterobacter ludwigii	Joe et al. (2016)
Lipase	*Paecilomyces variotii*	Carrim et al. (2006)
	Lasiodiplodia theobromae	Amirita et al. (2012)
	Myrothecium verrucaria	Sunitha et al. (2013)
	Bacillus megaterium	Bezerra et al. (2015)
	Pseudomonas stutzeri	
Protease	*Pochonia chlamydosporia*	Carrim et al. (2006)
	P. tropica	Nygren et al. (2007)
	Tetraploa aristata	Bezerra et al. (2012)
	Amanita muscaria	Chimwamurombe et al. (2016)
	Boletus luridus	Escudero et al. (2016)
	Piloderma fallax	
	Corynebacterium renale	
	Methylobacterium	
Xylanase	*Paenibacillus polymyxa*	Khan et al. (2017)
	P. entomophila	Bezerra et al. (2012)
	Alternaria alternate	Fouda et al. (2015)
	A. terricola	Akinsanya et al. (2016)
Phosphatases	*Chaelomiaceae*	Khan et al. (2016)
	Nectriaceae	
	Phoma medicaginis	
	Thielavia arenaria	

TABLE 10.2 Some Other Industrial Applications of Extracellular Hydrolases Produced by Endophytes.

Industry	Enzyme	Function
Cosmetics	Protease, Lipase	Removal of dead skin; Skin care
Organic synthesis	Lipase	Synthesis of pharmaceuticals, polymers, biodiesel, biosurfactants
Polymer	Lipase	Polycondensation, ring-opening polymerization of lactones, carbonates
Animal feed	Phytase, Xylanase	Hydrolyze phytic acid to release phosphorous; Enhanced digestibility of starch

9.7.1 FOOD AND BEVERAGE INDUSTRY

Enzymes produced from various endophytes can be employed in different sectors of food and beverage industry. Extracellular hydrolase enzymes such as lipase, lactase, xylanase, protease, pectinase, amylase, and cellulase produced by endophytes can be systematically implemented for performing various noteworthy functions in dairy industry, baking industry and beverage industry (Yamasaki et al., 1964; El-Zalaki and Hamza, 1979; Taylor and Richardson, 1979; Smart et al., 1985; Jin et al., 1998). Lipase produced from *A. niger* can be used for enhancing cheese ripening and customizing flavor of cheese (Neelakantan et al., 1999). The potential of catalase, xylanase, and lipase generated by *A. niger* and protease produced from *Aspergillus* species is reported (Saxena et al., 2001). Catalases can be used for cheese quality improvement in dairy industry. Xylanases and lipases can be recruited for dough conditioning and stability and proteases to be implemented for improvement of bread aroma in baking industries.

Pectinase, amylase, cellulase, and protease of endophytic origin are documented in various studies for their role in beverage industries. Depectinization involves pectinase, brewing is mainly done by amylase, cellulase is capable for liquefaction of fruits, and protease is mainly focused to restrict haze formation (Yamasaki et al., 1964; El-Zalaki and Hamza, 1979; Jin et al., 1998; Singh et al., 2016). Xylanases can be used for improving nutritional qualities of wheat, clarification of wines and juices, extraction of juice, oils, spices, and pigments and enhancing the texture and volume of bread (Sudheep et al., 2017). Frequent use of pectinase for decreasing astringency and increasing pigmentation is suggested in the wine industry (Tucker and Woods, 1991; Naik et al., 2019). Endophytic (*Aspergillus* and *Bacillus* species) proteases can be used for clarification of protein drinks, alcoholic beverages, fruit juices, and xanthan gum (Mandal and Banerjee, 2019). Two fungal commercial proteases, Kojizyme and Flavourzyme, can be employed in the fermentation of soy sauce and seasoning of fish, meat, casein and gluten like proteinaceous foods. Pectinase enzymes isolated extracellularly from *Aspergillus, Rhizopus, Trichoderma, Pseudomonas, Penicillium*, and many more are directly involved in food industries for fruit ripening, tomato pulp removal, and improvement of protein in baby food and extraction of oil (De Gregorio et al., 2002). Role of various endophytic hydrolase enzymes in some food sectors is categorized in Table 10.3. Cellulase, hemicellulase, and pectinase enzymes are employed in coffee industry (Soccol et al., 2008). The use of extracellular hydrolase enzymes such as amylases, cellulases, pectinases, and proteases for fruit juice production is suggested by various authors (Albersheim, 1966; Kashyap et al., 2001; Mantovani et al., 2005; Semenova et al., 2006; Jalis et al., 2014). Glucanase, peptidase, protease, esterase, lipase, and amylase are successfully implemented in flavor industry (de Souza, 2010; Shahani et al., 1976). Protease and catalase are utilized in egg processing (Singh et al., 2007). Cellulase, glucanase, pectinase, and tannase are employed in tea industry (Pasha and Reddy, 2005). Catalase and lipase can be employed for the production of butter and butter oils (Gupta et al., 2003). Amylase, invertase, pectinase, and protease can prove useful in confectionery industry (Kour et al., 2019). Extracellular hydrolase enzymes produced from endophytes can be useful in brewing industry (Okamura et al., 2001). Amylase, cellulase, glucanase, hemicellulase, lipase, protease, xylanase, and pentosanase are most commonly used hydrolases in brewing industry. Amylase, cellulase, protease, hemicellulase, and pentosanase are mostly used extracellular enzymes of endophytic origin in making of biscuits and breads (Taniwaki et al., 2001). Xylanases help in improving the yield of edible dyes and mineral salts (Polizeli et al., 2005). Lipases can be utilized for the production of chocolates, noodles, and bread (Undurraga et al., 2001; Aravindan et al., 2007).

9.7.2 PAPER AND PULP INDUSTRY

Microbial enzymes are even known for their great role in improving performance in paper and pulp industries. Cellulase derived from fungal endophytes can be substantially used in various processes in paper and pulp industry (Buchert et al., 1996; Wang et al., 2004). Cellulases can be employed in pulp and paper industry for fibrillation, enhancing drainage, and decreasing vessel picking (Walter and Gallatin, 1962; Fuentes and Roberts, 1988; Uchimoto et al., 1988). The use of xylanase and cellulose together for deinking in paper and pulp industries is reported (Dhiman et al., 2008). Endophytic amylases can reflect potential in paper and pulp industry due its efficacy in biological hydrolysis of starch (Gessesse and Mamo, 1999). Xylanases produced by endophytes can be employed for prebleaching the pulps in paper manufacture (Sudheep et al., 2017). Xylanases are capable of showing competency in accelerating the release of lignin from pulp and reducing the quantity of chlorine needed for bleaching in paper and pulp industry (Wong and Saddler, 1993; Beg et al., 2001; Saxena et al., 2015). Pulp depitching can be achieved through inculcation of microbial lipase (Irie et al., 1989).

TABLE 10.3 Application of Some Common Extracellular Hydrolase Enzymes Produced by Endophytes in Different Sectors of Food Industries.

Sector	Enzyme	Function
Dairy	Protease, Lipase, Lactase	Milk coagulation, debittering, enhancing cheese ripening; Faster cheese ripening, flavor customized cheese; Lactose reduced milk, whey products
Baking	Amylase, Xylanase, Lipase	Flour adjustment, bread softness; Dough conditioning; Dough stability and conditioning
Beverage	Pectinase, Cellulase, Amylase, Protease	Depectinization; Fruit liquefaction; Starch hydrolysis; Restrict haze formation

9.7.3 LEATHER AND TEXTILE INDUSTRY

Extracellular enzymes produced by endophytes can show potential in leather and textile industries. Endophyte-derived extracellular hydrolase enzymes like protease, pectinase, cellulase, xylanase, amylase, and lipase can perform important functions in leather and textile industries to improve the quality of the product (Alkorta et al., 1998; Pandey et al., 1999; de Souza et al., 2015; Singh et al., 2016). Cellulase produced from endophytes can be implemented effectively in textile industry (Gusakov et al., 2000). Cellulases act on cotton yarns for indigo dye removal, prevent damage to garment, and reduce effluent load (Kour et al., 2019). The role of cellulases and xylanases in textile industry for fiber treatment processes is suggested (Archna et al., 2015). Lipases can be used for desizing of denim and other cotton fabrics along with removal of lubricants (Handelsman et al., 1998; Hasan et al., 2006). Pectinases are employed along with amylases, cellulases, and lipases for removing sizing agents (Hoondal et al., 2002; Kour et al., 2019). Proteases can be useful for extraction of dull and stiff gum layer of sericin from raw silk to achieve softness and luster (Mandal and Banerjee, 2019; Kour et al., 2019). It is uncovered that continuous treatment to silk and wool fibers with proteases can provide distinctive finishes (Doshi and Shelke, 2001). Amylases can be implemented for desizing process, that is, removal of starch from the fabrics (Mojsov et al., 2018). Extracellular hydrolase enzyme use can improve leather texture and quality (Ward et al., 2005; dos Santos Aguilar and Sato, 2018). *A. niger* and *Aspergillus tamari* endophytes producing extracellular alkaline protease can be utilized for unhairing and bating processes in leather industry (Anandan et al., 2007). Neutral protease produced from endophyte *A. flavus* can be employed for dehairing and soaking processes in leather industry (de Souza et al., 2015). The potential of extracellular lipase and amylase enzymes in leather industry is reported (Singh et al., 2016). Lipase can be utilized for degreasing process and amylase can perform fiber splitting function.

9.7.4 DETERGENT INDUSTRY

In detergent industry, there is a long history behind hydrolases usage. Nowadays, hydrolases can be seen in most of the detergents. Hydrolase enzymes can remove protein, starch oil, and fats present on the fabric in the form of stain (Hasan et al., 2010). Enzyme lipase isolated extracellularly from endophytic strains like *A. oryzae* and *Trichosporon asahii* are capable to perform removal of oil stains (Gerhartz 1990; Kumar et al., 2009). Protease produced from *Trichoderma harzianum* can show effective washing performance (Savitha et al., 2011). Microbial protease can be implemented for digestion of organic stains like grass, blood, egg, and human sweat (Hasan et al., 2010; Kuhad et al., 2011). Amylase produced by *Aspergillus* species can be enrolled for carbohydrate stain removal (de Souza 2010). Lipase isolated from fungal endophytes can be effective against fat stain elimination (Greenough et al., 1996). The role of *A. oryzae*–derived protease is stated for removal of protein stain (Vishwanatha et al., 2009). Cellulase enzyme can be employed to perform enhanced color clarification (Kuhad et al., 2011). Cutinase can be employed in making dish washing and laundry detergents (Flipsen et al., 1998, Pio and Macedo, 2009).

9.7.5 PHARMACEUTICAL INDUSTRY

Enzymes produced by endophytic microbes are gaining valuable appreciation for their extensive use in pharmaceutical industry. In the field of medicines, hydrolase enzymes produced by endophytes can prove useful in many ways. Hydrolase enzymes like tannases can be employed for synthesis of antibacterial drugs (Belmares et al., 2004). Tannases can be produced extracellularly by some fungal endophytes such as *Aspergillus, P. variotii*, and *Penicillium* species (Battestin and Macedo, 2007; González et al., 2017). The role of endophyte-derived extracellular lipase is uncovered for enrichment of polyunsaturated fatty acids (PUFAs) of plant and animal origin, like borage oil, menhaden oil, and tuna oil (Dong et al., 1999). Lipases can resolve racemic mixtures to enhance drug production (Houde et al., 2004). Lipase enzymes can help in production of anti-inflammatory drugs, anticancer drugs, antiviral drugs, antihypertensive drugs, anti-Alzheimer drugs, and vitamin A supplements (Kovac et al., 1996; Akimoto et al., 1999; Kour et al., 2019). Immobilized lipases can be useful in nutraceuticals synthesis (Aravindan et al., 2007). Extracellular protease produced by fungal endophytes can show a variety of applications in pharmaceutical industry (Naik et al., 2019). L-Asparaginase derived from endophytes such as *Pseudomonas, Acinetobacter*, and many more can be utilized for production of antitumor drugs. Collagenase produced by *Clostridium* species can be valuable for making drugs to treat skin ulcers. Amylase and lipase produced by *Bacillus* and *Aspergillus* species can prove proficient for making drugs to prevent digestive disorders (Singh et al., 2016).

9.7.6 BIOLOGICAL WASTE MANAGEMENT

Wastes generated from various industries need to be effectively managed. Nowadays, whole concern is toward efficacious management of waste through biological means. Hydrolase enzymes can prove beneficial in terms of biological waste management. It is reported in various studies that the effective neutralization of toxic and harmful industrial wastes occurs with the utilization of extracellular hydrolases produced by various endophytes (Libra et al., 2003; Casa et al., 2003; Levin et al., 2005). *Phanerochaete chrysosporium* derived ligninases, proteases, and glucanases are reported for assisting waste degradation (Bumpus and Aust, 1987). Their involvement in degradation of diverse organopollutants is highlighted.

Amylase produced by *B. licheniformis* and *Aspergillus* species can be effectively utilized for degradation of vegetable wastes. Lipase released by *A. oryzae* and *Candida tropicalis* can generate potential for degradation of crude oil hydrocarbons. Protease produced by *Chrysosporium* species can help in successful bioremediation of keratinic wastes. Cutinase isolated from *Fusarium* species can prove advantageous for degradation of plastics (Singh et al., 2016; Kour et al., 2019).

9.8 CONCLUSION AND FUTURE PROSPECTS

Scientists are well aware about endophytes (bacteria and fungi) potential of releasing extracellular hydrolase enzyme to hydrolyze variety of polymeric compounds such as chitin, pectin, protein, and cellulose. Endophytes demand special attention particularly toward their competency in producing extracellular hydrolases. Endophytes generate extracellular hydrolases when they colonize different plant tissues. The ability of endophytic microorganisms to produce and release extracellular hydrolase enzymes can prove helpful in numerous ways. It is beneficial to achieve endophyte–host symbiosis and counteract emerging issues such as plant pathogen interaction and tolerance against host defense mechanism. Endophyte derived from broad variety of plants growing under different environmental conditions seizes more propensities to yield higher quantities of extracellular hydrolases. With advancement of technology, most sensitive and proficient methods are required for enzyme detection and quantification. The factors like carbon and nitrogen sources, pH, temperature, inoculum size, and shaking rate play very crucial role in achieving desirable adequate extracellular hydrolase enzyme production from different endophytic strains. It is essential to optimize these factors as they indirectly regulate the growth and nutrition of the endophytes for efficient enzyme yield. Microbial enzymes seem to be more proficient than enzymes from other sources mostly due to their stable nature and broad range of optimum conditions. Enzymes of microbial origin are gaining interest in industrial applications due to their less toxic, economical, and eco-friendly nature. Implementation of endophyte-derived hydrolases in industries like leather, textile, and detergent can reduce the excessive use of harsh chemicals which will ultimately lower the discharge of harmful and toxic chemicals in the environment. Extracellular hydrolase enzymes of endophytic origin can reflect significant potential in waste management. Therefore, endophytes are an engrossing niche for prospecting new natural products that can be used as enzymes with industrial applications.

KEYWORDS

- **endophytes**
- **extracellular hydrolases**
- **enzyme production**
- **enzyme quantification**
- **enzyme detection**
- **industrial applications**

REFERENCES

Abd Rahman, R. N. Z.; Geok, L. P.; et al. Physical Factors Affecting the Production of Organic Solvent-Tolerant Protease by *Pseudomonas aeruginosa* strain K. *Biores. Technol.* **2005,** *96*, 429–436.

Abou-Taleb, K. A.; Mashhoor, W. A.; et al. Nutritional and Environmental Factors Affecting Cellulase Production by Two Strains of Cellulolytic Bacilli. *Aus. J. Basic Appl. Sci.* **2009,** *3*, 2429–2436.

Ahmed, S. A.; Saleh, S. A.; et al. Characterization and Valuable Applications of Xylanase from Endophytic Fungus *Aspergillus terreus* KP900973 Isolated from *Corchorus olitorius*. *Biocatal. Agric. Biotechnol.* **2016,** *7*, 134–144.

Akimoto, M.; Nagashima, Y.; Sato, D. A Kinetic Study on Lipase-Catalyzed Interesterification of Soybean Oil with Oleic Acid in a Continuous Packed-Bed Reactor. *Appl. Biochem. Biotechnol.* **1999,** *81*, 131–142.

Akinsanya, M. A.; Ting, A.; et al. Biodiversity, Enzymatic and Antimicrobial Activities of Bacterial Endophytes in Selected Local Medicinal Plants. *J. Biomed. Pharm. Res.* **2016,** *16*, 1.

Albersheim, P. Pectin Lyase from Fungi. *Meth. Enzyme* **1966,** *8*, 628–631.

Alkorta, I.; Garbisu, C.; et al. Industrial Applications of Pectic Enzymes: A Review. *Process Biochem*. **1998,** *33*, 21–28.

Almeida, M. N.; Guimarães, V. M.; et al. Cellulases and Hemicellulases from Endophytic *Acremonium* Species and Its Application on Sugarcane Bagasse Hydrolysis. *Appl. Biochem. Biotechnol.* **2011,** *165*, 594–610.

Amirita, A.; Sindhu, P.; et al. Enumeration of Endophytic Fungi from Medicinal Plants and Screening of Extracellular Enzymes. *World J. Sci. Technol.* **2012,** *2*, 13–19.

Anandan, D.; Marmer, W. N.; Dudley, R. L. Isolation, Characterization and Optimization of Culture Parameters for Production of an Alkaline Protease Isolated from *Aspergillus tamarii*. *J. Ind. Microbiol. Biotechnol.* **2007,** *34*, 339–347.

Aravindan, R.; Anbumathi, P.; Viruthagiri, T. *Lipase Applications in Food Industry*; 2007.

Archna, S.; Priyank, V.; et al. Bioprospecting for Extracellular Hydrolytic Enzymes from Culturable Thermotolerant Bacteria Isolated from Manikaran Thermal Springs. *Res. J. Biotechnol.* **2015,** *10*, 4.

Ashraf, H.; Qadeer, M. A.; Iqbal, J. Pearl Millet, A. Source of Alpha Amylase Production by *Bacillus licheniformis*. *Biores. Technol.* **2005,** *96*, 1201–1204.

Ayob, F. W.; Simarani, K. Endophytic Filamentous Fungi from a *Catharanthus roseus*: Identification and Its Hydrolytic Enzymes. *Saudi Pharm. J.* **2016,** *24*, 273–278.

Azevedo, J. L.; Maccheroni, W.; et al. Endophytic Microorganisms: A Review on Insect Control and Recent Advances on Tropical Plants. *Electron. J. Biotechnol.* **2000,** *3*, 15–16.

Bakare, M. K.; Adewale, I. O.; et al. Purification and Characterization of Cellulase from the Wild-Type and Two Improved Mutants of *Pseudomonas fluorescens*. *Afr. J. Biotechnol.* **2005,** *4*, 9.

Barz, W.; Daniel, S.; et al. Elicitation and Metabolism of Phytoalexins in Plant Cell Cultures. In *Plant Cell Biotechnol*; 1988; pp 211–230.

Battestin, V.; Macedo, G. A. Effects of Temperature, pH and Additives on the Activity of Tannase Produced by *Paecilomyces variotii*. *Electron. J. Biotechnol.* **2007,** *10*, 191–199.

Beg, Q.; Kapoor, M.; et al. Microbial Xylanases and Their Industrial Applications: A Review. *Appl. Microbiol. Biotechnol.* **2001,** *56*, 326–338.

Belmares, R.; Contreras-Esquivel, J. C.; et al. Microbial Production of Tannase: An Enzyme with Potential Use in Food Industry. *LWT—Food Sci. Technol.* **2004,** *37*, 857–864.

Benhamou, N.; Gagné, S.; et al. Bacterial-Mediated Induced Resistance in Cucumber: Beneficial Effect of the Endophytic Bacterium *Serratia plymuthica* on the Protection against Infection by *Pythium ultimum*. *Phytopathology* **2000;** *90*, 45–56.

Bezerra, J. D. P.; Santos, M. G. S.; Svedese, V. M.; Lima, D. M. M.; et al. Richness of Endophytic Fungi Isolated from *Opuntia ficus-indica* Mill (Cactaceae) and Preliminary Screening for Enzyme Production. *World J. Microbiol. Biotechnol.* **2012,** *28*, 1989–1995.

Bezerra, J. D.; Nascimento, C. C.; et al. Endophytic Fungi from Medicinal Plant *Bauhinia forficata*: Diversity and Biotechnological Potential. *Braz. J. Microbiol.* **2015,** *46*, 49–57.

Bhagobaty, R. K.; Joshi, S. R. Enzymatic Activity of Fungi Endophytic on Five Medicinal Plant Species of the Pristine Sacred Forests of Meghalaya, India. *Biotechnol. Bioprocess. Eng.* **2012,** *17*, 33–40.

Bischoff, K. M.; Wicklow, D. T.; et al. Extracellular Hemicellulolytic Enzymes from the Maize Endophyte *Acremonium zeae*. *Curr. Microbiol.* **2009,** *58*, 499–503.

Boer, W. D.; Folman, L. B.; et al. Living in a Fungal World: Impact of Fungi on Soil Bacterial Niche Development. *FEMS Microbiol. Rev.* **2005,** *29*, 795–811.

Bonugli-Santos, R. C.; dos Santos M. R.; et al. Marine-Derived Fungi: Diversity of Enzymes and Biotechnological Applications. *Front. Microbiol.* **2015,** *6*, 269.

Brem, D.; Leuchtmann, A. Epichloë Grass Endophytes Increase Herbivore Resistance in the Woodland Grass *Brachypodium sylvaticum*. *Oecologia* **2001,** *126*, 522–530.

Buchert, J.; Carlsson, G.; et al. Surface Characterization of Unbleached Kraft Pulps by Enzymatic Peeling and ESCA. *Holzforschung—Intern. J. Biol. Chem. Phys. Technol. Wood.* **1996,** *50*, 69–74.

Bumpus, J. A.; Aust, S. D. Biodegradation of Environmental Pollutants by the White Rot Fungus *Phanerochaete chrysosporium*: Involvement of the Lignin Degrading System. *BioEssays* **1987,** *6*, 166–170.

Burke, R. M.; Cairney, J. W. G. Purification and Characterization of a β-1,4-Endoxylanase from the Ericoid Mycorrhizal Fungus *Hymenoscyphus ericae*. *New Phytol.* **1997,** *135*, 345–352.

Carlsen, M.; Nielsen, J.; Villadsen, J. Growth and α-Amylase Production by *Aspergillus oryzae* during Continuous Cultivations. *J. Biotechnol.* **1996,** *45*, 81–93.

Carrim, A. J. I.; Barbosa, E. C.; Vieira, J. D. G. Enzymatic Activity of Endophytic Bacterial Isolates of *Jacaranda decurrens* Cham (Carobinha-do-campo). *Braz. Arch. Biol. Technol.* **2006,** *49*, 353–359.

Casa, R.; D'Annibale, A.; et al. Reduction of the Phenolic Components in Olive-Mill Wastewater by an Enzymatic Treatment and Its Impact on Durum Wheat (*Triticum durum* Desf.) Germinability. *Chemosphere* **2003,** *50*, 959–966.

Castro, R. A.; Quecine, M. C.; et al. Isolation and Enzyme Bioprospection of Endophytic Bacteria Associated with Plants of Brazilian Mangrove Ecosystem. *SpringerPlus* **2014,** *3*, 382.

Cavalcanti, E. D. A. C.; Gutarra, M. L. E.; et al. Lipase Production by Solid-State Fermentation in Fixed-Bed Bioreactors. *Braz. Arch. Biol. Technol.* **2005,** *48*, 79–84.

Chadha, R.; Kumbhar, B. K.; Sarkar, B. C. Enzymatic Hydrolysis of Carrot for Increased Juice Recovery. *J. Food Sci. Technol.* **2003,** *40*, 35–39.

Chimwamurombe, P. M.; Grönemeyer, J. L.; Reinhold-Hurek, B. Isolation and Characterization of Culturable Seed-Associated Bacterial Endophytes from Gnotobiotically Grown Marama Bean Seedlings. *FEMS Microbiol. Ecol.* **2016,** *92*, fiw083.

Chow, Y.; Ting, A. S. Endophytic L-Asparaginase-Producing Fungi from Plants Associated with Anticancer Properties. *J. Adv. Res.* **2015,** *6*, 869–876.

Cihangir, N.; Sarikaya, E. Investigation of Lipase Production by a New Isolate of *Aspergillus* sp. *World J. Microbiol. Biotechnol.* **2004,** *20*, 193–197.

Correa, A.; Pacheco, S.; et al. Potent and Specific Inhibition of Glycosidases by Small Artificial Binding Proteins (Affitins). *PLoS One* **2014,** *9*, 97438.

De Almeida, M. N.; Guimarães. V. M.; et al. Cellulases and Hemicellulases from Endophytic *Acremonium* species and Its Application on Sugarcane Bagasse Hydrolysis. *Appl. Biochem. Biotechnol.* **2011,** *165*, 594–610.

De Gregorio, A.; Mandalari, G.; et al. SCP and Crude Pectinase Production by Slurry-State Fermentation of Lemon Pulps. *Biores. Technol.* **2002,** *83*, 89–94.

de Souza, P. M. D. Application of Microbial α-Amylase in Industry—A Review. *Braz. J. Microbiol.* **2010,** *41*, 850–861.

de Souza, P. M.; de Assis Bittencourt, M. L.; et al. A Biotechnology Perspective of Fungal Proteases. *Braz. J. Microbiol.* **2015,** *46*, 337–346.

Dhiman, S. S.; Sharma, J.; Battan, B. Industrial Applications and Future Prospects of Microbial Xylanases: A Review. *BioResources* **2008,** *3*, 1377–1402.

Dhouib, H.; Zouari, I.; et al. Potential of a Novel Endophytic *Bacillus velezensis* in Tomato Growth Promotion and Protection against Verticillium Wilt Disease. *Biol. Control* **2019,** *139*, 104092.

Dong, H.; Gao, S.; Han, S. P.; Cao, S. G. Purification and Characterization of a *Pseudomonas* sp. Lipase and Its Properties in Non-Aqueous Media. *Biotechnol. Appl. Biochem.* **1999,** *30*, 251–256.

dos Santos Aguilar, J. G.; Sato, H. H. Microbial Proteases: Production and Application in Obtaining Protein Hydrolysates. *Food Res. Intern.* **2018,** *103*, 253–262.

Doshi, R.; Shelke, V. *Enzymes in Textile Industry—An Environment-Friendly Approach*; 2001.

El-Zalaki, M. E.; Hamza, M. A. Edible Mushrooms as Producers of Amylases. *Food Chem.* **1979,** *4*, 203–211.

Escudero, N.; Ferreira, S. R.; et al. Chitosan Enhances Parasitism of *Meloidogyne javanica* Eggs by the Nematophagous Fungus *Pochonia chlamydosporia*. *Fungal Biol.* **2016,** *120*, 572–585.

Esteves, A. C.; Saraiva, M.; et al. Botryosphaeriales Fungi Produce Extracellular Enzymes with Biotechnological Potential. *Can. J. Microbiol.* **2014,** *60*, 332–342.
Fernandes, E. G.; Pereira, O. L.; et al. Diversity of Endophytic Fungi in Glycine Max. *Microbiol. Res.* **2015,** *181*, 84–92.
Ferreira, A.; Quecine, M. C.; et al. Diversity of Endophytic Bacteria from Eucalyptus Species Seeds and Colonization of Seedlings by *Pantoea agglomerans*. *FEMS Microbiol. Lett.* **2008,** *287*, 8–14.
Flipsen, J. A. C.; Appel, A. C. M.; et al. Mechanism of Removal of Immobilized Triacylglycerol by Lipolytic Enzymes in a Sequential Laundry Wash Process. *Enzyme Microbial. Technol.* **1998,** *23*, 274–280.
Fouda, A. H.; Hassan, S. E. D.; et al. Biotechnological Applications of Fungal Endophytes Associated with Medicinal Plant *Asclepias sinaica* (Bioss). *Ann. Agric. Sci.* **2015,** *60*, 95–104.
Fuentes, J. L.; Robert, M. Process of Treatment of a Paper Pulp by an Enzymic Solution. Eur Pat 262040, 1988.
Garg, G.; Singh, A.; et al. Microbial Pectinases: An Ecofriendly Tool of Nature for Industries. *Biotechnology* **2016,** *6*, 47.
Gazis, R.; Chaverri, P. Diversity of Fungal Endophytes in Leaves and Stems of Wild Rubber Trees (*Hevea brasiliensis*) in Peru. *Fungal Ecol.* **2010,** *3*, 240–254.
Gerhartz, W. Industrial Uses of Enzymes. *Enzyme Ind. Prod. Appl.* **1990,** 77–92.
Gessesse, A.; Mamo, G. High-Level Xylanase Production by an Alkaliphilic *Bacillus* sp. by Using Solid-State Fermentation. *Enzyme Microbial. Technol.* **1999,** *25*, 68–72.
González, M. C.; Buenrostro-Figueroa, J.; et al. *Tannases Current Developments in Biotechnology and Bioengineering: Production, Isolation and Purification of Industrial Products*; Elsevier, 2017; pp 471–489.
Gopinath, S. C.; Anbu, P.; et al. Strategies to Characterize Fungal Lipases for Applications in Medicine and Dairy Industry. *BioMed Res. Intern.* **2013.**
Gopinath, S. C.; Hilda, A.; Anbu, P. Extracellular Enzymatic Activity Profiles in Fungi Isolated from Oil-Rich Environments. *Mycoscience* **2005,** *46*, 119–126.
Goto, C. E.; Barbosa, E. P.; et al. Production of Amylase by *Aspergillus fumigatus* Utilizing Α-Methyl-D-Glycoside, A Synthetic Analogue of Maltose, as Substrate. *FEMS Microbiol. Lett.* **1998,** *167*, 139–143.
Gray, E. J.; Smith, D. L. Intracellular and Extracellular PGPR: Commonalities and Distinctions in the Plant–Bacterium Signaling Processes. *Soil Biol. Biochem.* **2005,** *37*, 395–412.
Greenough, R. J.; Perry, C. J.; Stavnsbjerg, M. Safety Evaluation of a Lipase Expressed in *Aspergillus oryzae*. *Food Chem. Toxicol.* **1996,** *34*, 161–166.
Gupta, R.; Gigras, P.; et al. Microbial α-Amylases: A Biotechnological Perspective. *Process Biochem.* **2003,** *38*, 1599–1616.
Gupta, R.; Gupta, N.; et al. Bacterial Lipases: An Overview of Production, Purification and Biochemical Properties. *Appl. Microbiol. Biotechnol.* **2004,** *64*, 763–781.
Gurung, N.; Ray, S.; et al. A Broader View: Microbial Enzymes and Their Relevance in Industries, Medicine and Beyond. *BioMed Res. Int.* **2013**.
Gusakov, A. V.; Sinitsyn, A. P.; et al. Surface Hydrophobic Amino Acid Residues in Cellulase Molecules as a Structural Factor Responsible for their High Denim-Washing Performance. *Enzyme Microbial. Technol.* **2000,** *27*, 664–671.
Haki, G. D.; Rakshit. S. K. Developments in Industrially Important Thermostable Enzymes: A Review. *Res. Technol.* **2003,** *89*, 17–34.

Hall, T. J.; Schreiber, L. R.; Leben, C. Effects of Xylem-Colonizing *Bacillus* spp. on Verticillium wilt in Maples. *Plant Dis.* **1986,** *70*, 521–524.
Hallmann, J.; Quadt-Hallmann, A.; et al. Bacterial Endophytes in Agricultural Crops. *Can. J. Microbiol.* **1997,** *43*, 895–914.
Handelsman, J.; Rondon, M. R.; et al. Molecular Biological Access to the Chemistry of Unknown Soil Microbes: A New Frontier for Natural Products. *Chem. Biol.* **1998,** *5*, R245–R249.
Hankin, L.; Anagnostakis, S. L. The Use of Solid Media for Detection of Enzyme Production by Fungi. *Mycologia* **1975,** *67*, 597–607.
Harnpicharnchai, P.; Champreda, V.; et al. A Thermotolerant β-Glucosidase Isolated from an Endophytic Fungi, *Periconia* sp., with a Possible Use for Biomass Conversion to Sugars. *Protein Expr. Purif.* **2009,** *67*, 61–69.
Hasan, F.; Shah, A. A.; et al. Enzymes Used in Detergents: Lipases. *Afr. J. Biotechnol.* **2010,** *9*, 4836–4844.
Hasan, F.; Shah, A. A.; Hameed, A. Industrial Applications of Microbial Lipases. *Enzyme Microbial. Technol.* **2006,** *39*, 235–251.
Hegde, S. V.; Ramesha, A.; Srinvas, C. Optimization of Amylase Production from an Endophytic Fungi *Discosia* sp. Isolated from *Calophyllum inophyllum. J. Agric. Technol.* **2011,** *7*, 805–813.
Hoondal, G.; Tiwari, R.; et al. Microbial Alkaline Pectinases and Their Industrial Applications: A Review. *Appl. Microbiol. Biotechnol.* **2002,** *59*, 409–418.
Houde, A.; Kademi, A.; Leblanc, D. Lipases and Their Industrial Applications. *Appl. Biochem. Biotechnol.* **2004,** *118*, 155–170.
Howard, R. L.; Abotsi, E. L. J. R.; et al. Lignocellulose Biotechnology: Issues of Bioconversion and Enzyme Production. *Afr. J. Biotechnol.* **2003,** *2*, 602–619.
Immanuel, G.; Dhanusha, R.; et al. Effect of Different Growth Parameters on Endoglucanase Enzyme Activity by Bacteria Isolated from Coir Retting Effluents of Estuarine Environment. *Int. J. Environ. Sci. Technol.* **2006,** *3*, 25–34.
Irie, Y.; Matsukura, T.; Hata, K. De-Resinification of Mechanical Wood Pulp Using Glycerol Lipase to Decompose Triglyceride Compounds in Pulp and Paper Manufacture. Eur Pat App EP374700, 1989.
Jalgaonwala, R. E.; Mohite, B. V.; Mahajan, R. T. A Review: Natural Products from Plant Associated Endophytic Fungi. *J. Microbiol. Biotechnol. Res.* **2011,** *1*, 21–32.
Jalis, H.; Ahmad, A.; et al. Utilization of Apple Peels for the Production of Plant Cell-Wall Degrading Enzymes by *Aspergillus fumigatus* MS16. *J. Anim. Plant Sci.* **2014,** *24*, 64–67.
Jasvir, S.; Gill, N.; et al. Studies on Alkaline Protease Produced by *Bacillus* sp. NG 312. *Appl. Biochem. Biotechnol.* **1999,** *76*, 57–63.
Jensen, B.; Olsen, J. Physicochemical Properties of a Purified Alpha-Amylase from the Thermophilic Fungus *Thermomyces lanuginosus. Enzyme Microbial. Technol.* **1992,** *14*, 112–116.
Jin, B.; Van Leeuwen, H.; Patel, B.; Yu, Q. Utilisation of Starch Processing Wastewater for Production of Microbial Biomass Protein and Fungal α-Amylase by *Aspergillus oryzae. Biores. Technol.* **1998,** *66*, 201–206.
Joe, M. M.; Devaraj, S.; et al. Isolation of Phosphate Solubilizing Endophytic Bacteria from *Phyllanthus amarus* Schum & Thonn: Evaluation of Plant Growth Promotion and Antioxidant Activity under Salt Stress. *J. Appl. Res. Med. Aroma Plants* **2016,** *3*, 71–77.
Joshi, R. D.; Kulkarni, N. S. Optimization Studies on L-Asparaginase Production from Endophytic Bacteria. *Int. J. Appl. Res.* **2016,** *2*, 624–629.

Jurynelliz, R.-V.; David, P.; et al. Enzymatic and Bacterial Activity of Fungal Strains Isolated from *Alpinia zerumbet* Abs of Pap, 251st ACS Nat Meet & Expos, San Diego, CA, USA, March 13–17, 2016; CHED-1130.

Kalyanasundaram, I.; Nagamuthu, J.; et al. Production, Purification and Characterisation of Extracellular L-Asparaginase from Salt Marsh Fungal Endophytes. *World J. Pharm. Pharm. Sci.* **2015,** *4*, 663–677.

Kannan, R.; Damodaran, T.; Umamaheswari, S. Sodicity Tolerant Polyembryonic Mango Root Stock Plants: A Putative Role of Endophytic Bacteria. *Afr. J. Biotechnol.* **2015,** *14*, 350–359.

Kar, T.; Delvigne, F.; et al. Investigation of the Effect of Different Extracellular Factors on the Lipase Production by *Yarrowia lipolityca* on the Basis of a Scale-Down Approach. *J. Ind. Microbiol. Biotechnol.* **2008,** *35*, 1053.

Kashyap, D. R.; Vohra, P. K.; et al. Applications of Pectinases in the Commercial Sector: A Review. *Biores. Technol.* **2001,** *77*, 215–227.

Kaul, S.; Ahmed, M.; Zargar, K.; et al. Prospecting Endophytic Fungal Assemblage of *Digitalis lanata* Ehrh. (Foxglove) as a Novel Source of Digoxin: A Cardiac Glycoside. *3 Biotech* **2013,** *3*, 335–340.

Kaur, R.; Saxena, A.; et al. Production and Characterization of a Neutral Phytase of *Penicillium oxalicum* EUFR-3 Isolated from Himalayan Region. *Nusantara Biosci.* **2017,** *9*, 68–76.

Khan, A. L.; Al-Harrasi, A.; et al. Endophytic Fungi from Frankincense Tree Improves Host Growth and Produces Extracellular Enzymes and Indole Acetic Acid. *PLoS One* **2016,** *11*, e0158207.

Khan, A. L.; Shahzad, R.; et al. Endophytic Microbes: A Resource for Producing Extracellular Enzymes. *Endophytes: Crop Prod. Prot.* **2017,** 95–110.

Khan, F. A. B. A.; Husaini, A. A. S. A. Enhancing α-Amylase and Cellulase in Vivo Enzyme Expressions on Sago Pith Residue Using *Bacilllus amyloliquefaciens* UMAS 1002. *Biotechnology* **2006,** *5*, 391–403.

Kirk, O.; Borchert, T. V.; Fuglsang, C. C. Industrial Enzyme Applications. *Curr. Opin. Biotechnol.* **2002,** *13*, 345–351.

Knox, A. M.; du Preez, J. C.; Kilian, S. G. Starch Fermentation Characteristics of *Saccharomyces cerevisiae* Strains Transformed with Amylase Genes from *Lipomyces kononenkoae* and *Saccharomycopsis fibuligera*. *Enzyme Microbial. Technol.* **2004,** *34*, 453–460.

Kour, D.; Rana, K. L.; et al. Agriculturally and Industrially Important Fungi: Current Developments and Potential Biotechnological Applications. In *Recent Advancement in White Biotechnology through Fungi;* 2019; pp 1–64.

Kovac, A.; Stadler, P.; et al. Hydrolysis and Esterification of Acylglycerols and Analogs in Aqueous Medium Catalyzed by Microbial Lipases. *Biochim. Biophys. Acta (BBA)—Lipid Metab.* **1996,** *1301*, 57–66.

Kudanga, T.; Mwenje, E. Extracellular Cellulase Production by Tropical Isolates of *Aureobasidium pullulans*. *Can. J. Microbiol.* **2005,** *51*, 773–776.

Kuhad, R. C.; Gupta, R.; Singh, A. Microbial Cellulases and Their Industrial Applications. *Enzyme Res.* **2011,** *2011*, 280696.

Kumar, S. S.; Kumar, L.; et al. A Thiol-Activated Lipase from *Trichosporon asahii* MSR 54: Detergent Compatibility and Presoak Formulation for Oil Removal from Soiled Cloth at Ambient Temperature. *J. Ind. Microbiol. Biotechnol.* **2009,** *36*, 427.

Kumar, V.; Yadav, A. N.; et al. Unravelling Rhizospheric Diversity and Potential of Phytase Producing Microbes. *SM J. Biol.* **2016,** *2*, 1009.

Kumaresan, V.; Suryanarayanan, T. S. Endophytes Assemblages in Young Mature and Senescent Leaves of *Rhizophora apiculata*: Evidence for the Role of Endophytes in Mangrove Litter Degradation. *Fungal Divers* **2002,** *9*, 81–91.

Kunamneni, A.; Permaul, K.; Singh, S. Amylase Production in Solid State Fermentation by the Thermophilic Fungus *Thermomyces lanuginosus*. *J. Biosci. Bioeng.* **2005,** *100*, 168–171.

Larkin, R. P.; Hopkins, D. L.; Martin, F. N. Recovered from a Disease-Suppressive Soil. *Pathology* **1996,** *86*, 812–819.

Leo, V. V.; Passari, A. K.; et al. A Novel Triculture System (CC3) for Simultaneous Enzyme Production and Hydrolysis of Common Grasses through Submerged Fermentation. *Front. Microbiol.* **2016,** *7*, 447.

Levin, L.; Forchiassin, F.; Viale, A. Ligninolytic Enzyme Production and Dye Decolorization by *Trametes trogii*: Application of the Plackett–Burman Experimental Design to Evaluate Nutritional Requirements. *Process Biochem.* **2005,** *40*, 1381–1387.

Libra, J. A.; Borchert, M.; Banit, S. Competition Strategies for the Decolorization of a Textile-Reactive Dye with the White-Rot Fungi *Trametes versicolor* under Non-Sterile Conditions. *Biotechnol. Bioeng.* **2003,** *82*, 736–744.

Lin, E. S.; Wang, C. C.; Sung, S. C. Cultivating Conditions Influence Lipase Production by the Edible Basidiomycete *Antrodia cinnamomea* in Submerged Culture. *Enzyme Microbial. Technol.* **2006,** *39*, 98–102.

Lu, W. J.; Wang, H. T.; et al. Effect of Inoculating Flower Stalks and Vegetable Waste with Ligno-Cellulolytic Microorganisms on the Composting Process. *J. Environ. Sci. Health, Part B* **2004,** *39*, 871–887.

Mabrouk, S. S.; Hashem, A. M.; et al. Optimization of Alkaline Protease Productivity by *Bacillus licheniformis* ATCC 21415. *Biores. Technol.* **1999,** *69*, 155–159.

Mandal, S.; Banerjee, D. Proteases from Endophytic Fungi with Potential Industrial Applications. In *Recent Advancement in White Biotechnology through Fungi;* 2019; pp 319–359.

Mantovani, C. F.; Geimba, M. P.; Brandelli, A. Enzymatic Clarification of Fruit Juices by Fungal Pectin Lyase. *Food Biotechnol.* **2005,** *19*, 173–181.

Maria, G. L.; Sridhar, K. R.; Raviraja, N. S. Antimicrobial and Enzyme Activity of Mangrove Endophytic Fungi of Southwest Coast of India. *J. Agric. Technol.* **2005,** *1*, 67–80.

Marlida, Y.; Delfita, R.; et al. Isolation, Characterization and Production of Phytase from Endophytic Fungus Its Application for Feed. *Pak. J. Nutr.* **2010,** *9*, 471–474.

Mayerhofer, M. S.; Fraser, E.; Kernaghan, G. Acid Protease Production in Fungal Root Endophytes. *Mycologia* **2015,** *107*, 1–11.

Mehrotra, S.; Pandey, P. K.; et al. The Production of Alkaline Protease by a *Bacillus* Species Isolate. *Biol. Res. Technol.* **1999,** *67*, 201–203.

Mendu, D. R.; Ratnam, B. V. V.; et al. Affinity Chromatography of α-Amylase from *Bacillus licheniformis*. *Enzyme Microb. Technol.* **2005,** *37*, 712–717.

Mojsov, K.; Andronikov, D.; et al. Production and Application of α-Amylase Enzyme in Textile Industry. *Tekst. Ind.* **2018,** *66*, 23–28.

Møller, K.; Sharif, M. Z.; Olsson, L. Production of Fungal α-Amylase by *Saccharomyces kluyveri* in Glucose-Limited Cultivations. *J. Biotechnol.* **2004,** *111*, 311–318.

Moon, S. H.; Parulekar, S. J. A Parametric Study of Protease Production in Batch and Fed-Batch Cultures of *Bacillus firmus*. *Biotechnol. Bioeng.* **1991,** *37*, 467–483.

Moormann, M.; Schlochtermeier, A.; Schrempf, H. Biochemical Characterization of a Protease Involved in the Processing of a *Streptomyces reticuli* Cellulase (Avicelase). *Appl. Environ. Microbiol.* **1993,** *59*, 1573–1578.

Murray, P. M.; Moane, S.; et al. Sustainable Production of Biologically Active Molecules of Marine Based Origin. *New Biotechnol.* **2013,** *30*, 839–850.

Muthuvelayudham, R.; Viruthagiri, T. Fermentative Production and Kinetics of Cellulase Protein on *Trichoderma reesei* Using Sugarcane Bagasse and Rice Straw. *Afr. J. Biotechnol.* **2006,** *5*, 20.

Naik, B. S.; Abrar, S.; Krishnappa. M. Industrially Important Enzymes from Fungal Endophytes. In *Recent Advancement in White Biotechnology through Fungi;* 2019; pp. 263–280.

Naik, B. S.; Krishnamurthy, Y. L. Endophytes: The Real Untapped High Energy Biofuel Resource. *Curr. Sci.* **2010,** *98*, 883.

Narasimha, G.; Sridevi, A.; et al. Nutrient Effects on Production of Cellulolytic Enzymes by *Aspergillus niger*. *Afr. J. Biotechnol.* **2006,** *5*, 472–476.

Neelakantan, S.; Mohanty, A. K.; Kaushik, J. K. Production and Use of Microbial Enzymes for Dairy Processing. *Curr. Sci.* **1999,** 143–148.

Nieves, R. A.; Ehrman, C. I.; et al. Survey and Analysis of Commercial Cellulase Preparations Suitable for Biomass Conversion to Ethanol. *World J. Microbiol. Biotechnol.* **1997,** *14*, 301–304.

Niranjane, A. P.; Madhou, P.; Stevenson, T. W. The Effect of Carbohydrate Carbon Sources on the Production of Cellulase by *Phlebia gigantea*. *Enzyme Microbial. Technol.* **2007,** *40*, 1464–1468.

Nongkhlaw, F. M.; Joshi, S. R. L-Asparaginase and Antioxidant Activity of Endophytic Bacteria Associated with Ethnomedicinal Plants. *Ind. J. Biotechnol.* **2015,** *14*, 59–64.

Nygren, C. M.; Edqvist, J.; et al. Detection of Extracellular Protease Activity in Different Species and Genera of Ectomycorrhizal Fungi. *Mycorrhiza* **2007,** *17*, 241–248.

Okamura, T.; Ogata, T.; et al. Characteristics of Wine Produced by Mushroom Fermentation. *Biosci. Biotechnol. Biochem.* **2001,** *65*, 1596–1600.

Pandey, A.; Benjamin, S.; et al. The Realm of Microbial Lipases in Biotechnology. *Biotechnol. Appl. Biochem.* **1999,** *29*, 119–131.

Panuthai, T.; Sihanonth, P.; et al. An Extracellular Lipase from the Endophytic Fungi *Fusarium oxysporum* Isolated from the Thai Medicinal Plant, *Croton oblongifolius* Roxb. *Afr. J. Microbiol. Res.* **2012,** *6*, 2622–2638.

Pasha, C.; Reddy, G. Nutritional and Medicinal Improvement of Black Tea by Yeast Fermentation. *Food Chem.* **2005,** *89*, 449–453.

Patil, M. G.; Pagare, J.; et al. Extracellular Enzymatic Activities of Endophytic Fungi Isolated from Various Medicinal Plants. *Int. J. Curr. Microbiol. Appl. Sci.* **2015a,** *4*, 1035–1042.

Patil, M. P.; Patil, R. H.; Maheshwari, V. L. Biological Activities and Identification of Bioactive Metabolite from Endophytic *Aspergillus flavus* L7 Isolated from *Aegle marmelos*. *Curr. Microbiol.* **2015b,** *71*, 39–48.

Petrini, O.; Sieber, T. N.; et al. Ecology, Metabolite Production, and Substrate Utilization in Endophytic Fungi. *Nat. Toxins* **1993,** *1*, 185–196.

Pio, T. F.; Macedo, G. A. Cutinases: Properties and Industrial Applications. *Adv. Appl. Microbiol.* **2009,** *66*, 77–95.

Pleban, S.; Ingel, F.; Chet, I. Control of *Rhizoctonia solani* and *Sclerotium rolfsii* in the Greenhouse Using Endophytic *Bacillus* spp. *Eur. J. Plant Pathol.* **1995,** *101*, 665–672.

Polizeli, M. L. T. M.; Rizzatti, A. C. S.; et al. Xylanases from Fungi: Properties and Industrial Applications. *Appl. Microbiol. Biotechnol.* **2005,** *67*, 577–591.

Pourrat, H.; Barthomeuf, C.; et al. Production of Semi-Alkaline Protease by *Aspergillus niger*. *J. Ferment. Technol.* **1988,** *66*, 383–388.

Rabha, A. J.; Naglot, A.; et al. In Vitro Evaluation of Antagonism of Endophytic *Colletotrichum gloeosporioides* against Potent Fungal Pathogens of *Camellia sinensis*. *Ind. J. Microbiol.* **2014,** *54*, 302–309.

Raghukumar, C.; Raghukumar, S.; et al. Laccase and Other Lignocellulose Modifying Enzymes of Marine Fungi Isolated from the Coast of India. *Bot. Mar.* **1994,** *37*, 515–524.

Ramachandran, S.; Patel, A. K.; et al. Alpha Amylase from a Fungal Culture Grown on Oil Cakes and Its Properties. *Braz. Arch. Biol. Technol.* **2004,** *47*, 309–317.

Rana, K. L.; Kour, D.; et al. Endophytic Fungi: Biodiversity, Ecological Significance, and Potential Industrial Applications. In *Recent Advancement in White Biotechnology through Fungi*; 2019; pp 1–62.

Rashid, S.; Charles, T. C.; Glick, B. R. Isolation and Characterization of New Plant Growth-Promoting Bacterial Endophytes. *Appl. Soil Ecol.* **2012,** *61*, 217–224.

Raul, D.; Biswas, T.; et al. Production and Partial Purification of Alpha Amylase from *Bacillus subtilis* (MTCC 121) Using Solid State Fermentation. *Biochem. Res. Int.* **2014**.

Ray, A. K.; Bairagi, A.; et al. Optimization of Fermentation Conditions for Cellulase Production by *Bacillus subtilis* CY5 and *Bacillus circulans* TP3 Isolated from Fish Gut. *Acta Ichthyol. Piscat.* **2007,** *1*, 47–53.

Ray, M. K.; Devi, K. U.; et al. Extracellular Protease from the Antarctic Yeast *Candida humicola*. *Appl. Environ. Microbiol.* **1992,** *58*, 1918–1923.

Razak, C. N. A.; Tang, S. W.; et al. Preliminary Study on the Production of Extracellular Protease from a Newly Isolated *Bacillus* sp. (No. 1) and the Physical Factors Affecting Its Production. *Pertanika J. Sci. Technol.* **1997,** *5*, 169–177.

Reddy, P. V.; Lam, C. K.; Belanger, F. C. Mutualistic Fungal Endophytes Express a Proteinase that Is Homologous to Proteases Suspected to be Important in Fungal Pathogenicity. *Plant Physiol.* **1996,** *111*, 1209–1218.

Ruma, K.; Sunil, K.; Prakash, H. S. Antioxidant, Anti-Inflammatory, Antimicrobial and Cytotoxic Properties of Fungal Endophytes from *Garcinia* species. *Int. J. Pharm. Pharm. Sci.* **2013,** *5*, 889–897.

Saikkonen, K.; Helander, M.; et al. In Endophyte-Mediated Interactions between Woody Plants and Insect Herbivores? *Proc of the 9th Internat Sym on Insect-Plant Relat*, 1996; 269–271.

Savitha, S.; Sadhasivam, S.; et al. Fungal Protease: Production, Purification and Compatibility with Laundry Detergents and Their Wash Performance. *J. Taiwan Inst. Chem. Eng.* **2011,** *42*, 298–304.

Saxena, R. K.; Gupta, R.; et al. Role of Fungal Enzymes in Food Processing. *Appl. Mycol. Biotechnol.* **2001,** *1*, 353–386.

Schardl, C. L.; Leuchtmann, A.; Spiering, M. J. Symbioses of Grasses with Seedborne Fungal Endophytes. *Annu. Rev. Plant Biol.* **2004,** *55*, 315–340.

Schulz, B.; Boyle, C. The Endophytic Continuum. *Mycol. Res.* **2005,** *109*, 661–686.

Schulz, B.; Boyle, C.; et al. Endophytic Fungi: A Source of Novel Biologically Active Secondary Metabolites. *Mycolog. Res.* **2002,** *106*, 996–1004.

Semenova, M. V.; Sinitsyna, O. A.; et al. Use of a Preparation from Fungal Pectin Lyase in the Food Industry. *Appl. Biochem. Microbiol.* **2006,** *42*, 598–602.

Sette, L. D.; Santos, R. C. B. Ligninolytic Enzymes from Marine-Derived Fungi: Production and Applications. In *Marine Enzymes for Biocatalysis*; 2013; pp 403–427.

Shahani, K. M.; Arnold, R. G.; et al. Role of Microbial Enzymes in Flavor Development in Foods. *Biotechnol. Bioeng.* **1976,** *18*, 891–907.

Sharma, N.; Rathore, M.; Sharma, M. Microbial Pectinase: Sources, Characterization and Applications. *Rev. Environ. Sci. Biotechnol.* **2013,** *12*, 45–60.

Sharma, R.; Chisti, Y.; et al. Production, Purification, Characterization and Applications of Lipases. *Biotechnol. Adv.* **2001,** *19*, 627–662.

Shukla, L.; Suman, A.; et al. Syntrophic Microbial System for Ex-Situ Degradation of Paddy Straw at Low Temperature under Controlled and Natural Environment. *J. Appl. Biol. Biotechnol.* **2016,** *4*, 30–37.

Singh, R.; Kumar, M.; et al. Microbial Enzymes: Industrial Progress in 21st Century. *Biotechnology* **2016,** *6*, 174.

Singh, S.; Wakeling, L.; Gamlath, S. Retention of Essential Amino Acids during Extrusion of Protein and Reducing Sugars. *J. Agric. Food Chem.* **2007,** *55*, 8779–8786.

Smart, J. B.; Crow, V. L.; Thomas, T. D. Lactose Hydrolysis in Milk and Whey Using Beta-Galactosidase from *Streptococcus thermophilus*. *N.Z. J. Dairy Sci. Technol.* **1985.**

Soccol, C. R.; Dalla Santa, H. S.; et al. Mushrooms—A Promising Source to Produce Nutraceuticals and Pharmaceutical Byproducts. *Curr. Top Bioprocess Food Ind.* **2008,** *1*, 439–448.

Sorgatto, M.; Guimarães, N. C. A.; et al. Purification and Characterization of an Extracellular Xylanase Produced by the Endophytic Fungus, *Aspergillus terreus*, Grown in Submerged Fermentation. *Afr. J. Biotechnol.* **2012,** *11*, 8076–8084.

Strong, P. J.; Claus, H. Laccase: A Review of Its Past and Its Future in Bioremediation. *Crit. Rev. Environ. Sci. Technol.* **2011,** *41*, 373–434.

Sturz, A. V.; Christie, B. R.; Matheson, B. G. Associations of Bacterial Endophyte Populations from Red Clover and Potato Crops with Potential for Beneficial Allelopathy. *Can. J. Microbiol.* **1998,** *44*, 162–167.

Sturz, A. V.; Christie, B. R.; Nowak, J. Bacterial Endophytes: Potential Role in Developing Sustainable Systems of Crop Production. *Crit. Rev. Plant Sci.* **2000,** *19*, 1–30.

Sudheep, N. Á.; Marwal, A.; et al. Fascinating Fungal Endophytes Role and Possible Beneficial Applications: An Overview. In *Plant-Microbe Interactions in Agro-Ecological Perspectives: Volume 1: Fundamental Mechanisms, Methods and Functions;* 2017; pp 255–273.

Sundarram, A.; Murthy, T. P. K. α-Amylase Production and Applications: A Review. *J. Appl. Environ. Microbiol.* **2014,** *2*, 166–175.

Sunitha, V. H.; Nirmala Devi, D.; Srinivas, C. Extracellular Enzymatic Activity of Endophytic Fungal Strains Isolated from Medicinal Plants. *World J. Agric. Sci.* **2013,** *9*, 1–9.

Sunitha, V. H.; Ramesha, A.; et al. Amylase Production by Endophytic Fungi *Cylindrocephalum* sp. Isolated from Medicinal Plant *Alpinia calcarata* (Haw.) Roscoe. *Braz. J. Microbiol.* **2012,** *43*, 1213–1221.

Suto, M.; Takebayashi, M.; et al. Endophytes as Producers of Xylanase. *J. Biosci. Bioeng.* **2002,** *93*, 88–90.

Taechowisan, T.; Peberdy, J. F.; Lumyong, S. Isolation of Endophytic Actinomycetes from Selected Plants and Their Antifungal Activity. *World J. Microbiol. Biotechnol.* **2003,** *19*, 381–385.

Tan, R. X.; Zou, W. X. Endophytes: A Rich Source of Functional Metabolites. *Nat. Prod. Rep.* **2001,** *18*, 448–459.

Taniwaki, M. H.; Silva, N. D.; et al. Comparison of Culture Media, Simplate, and Petrifilm for Enumeration of Yeasts and Molds in Food. *J. Food Prot.* **2001,** *64*, 1592–1596.

Tanyildizi, M. S.; Özer, D.; Elibol, M. Optimization of α-Amylase Production by *Bacillus* sp. Using Response Surface Methodology. *Process Biochem.* **2005,** *40*, 2291–2296.

Taylor, M. J.; Richardson, T. Applications of Microbial Enzymes in Food Systems and in Biotechnology. *Adv. Appl. Microbiol.* **1979,** *25*, 7.

Thippeswamy, S.; Girigowda, K.; Mulimani, V. H. Isolation and Identification of α-Amylase Producing *Bacillus* sp. From Dhal Industry Waste. *Indian J. Biochem. Biophys.* **2006,** *43*, 295–298.

Thomas, L.; Joseph, A.; et al. Industrial Enzymes: Xylanases. *Curr. Dev. Biotechnol. Bioeng.* **2017,** 127–148.

Toghueo, R. M. K.; Ejiya, I. E.; et al. Production of Cellulolytic Enzymes by Endophytic Fungi Isolated from Cameroonian Medicinal Plants. *Int. J. Curr. Microbiol. Appl. Sci.* **2017,** *6*, 1264–1271.

Tomita, F. Endophytes in Southeast Asia and Japan: Their Taxonomic Diversity and Potential Applications. *Fungal Divers* **2003,** *14*, 187–204.

Torres, M.; Dolcet, M. M.; et al. Endophytic Fungi Associated with Mediterranean Plants as a Source of Mycelium-Bound Lipases. *J. Agric. Food Chem.* **2003,** *51*, 3328–3333.

Traving, S. J.; Thygesen, U. H.; et al. A Model of Extracellular Enzymes in Free-Living Microbes: Which Strategy Pays Off? *Appl. Environ. Microbiol.* **2015,** *81*, 7385–7393.

Treichel, H.; de Oliveira, D.; et al. A Review on Microbial Lipases Production. *Food Bioprocess Technol.* **2010,** *3*, 182–196.

Tucker, G. A.; Woods, L. F. J. Enzymes in Production of Beverages and Fruit Juices. *Enzyme Food Process.* **1991,** 201–203.

Uchimoto, I.; Endo, K.; Yamagishi, Y. Improvement of Deciduous Tree Pulp. Japanese Patent 135, 597/88, 1988.

Undurraga, D.; Markovits, A.; Erazo, S. Cocoa Butter Equivalent through Enzymic Interesterification of Palm Oil Mid Fraction. *Process Biochem.* **2001,** *36*, 933–939.

Velmurugan, N.; Lee, Y. S. Enzymes from Marine Fungi: Current Research and Future Prospects. In *Marine Fungi*; 2012; 441–474.

Vijayalakshmi, R.; Kairunnisa, K.; et al. Enzyme Production and Antimicrobial Activity of Endophytic Bacteria Isolated from Medicinal Plants. *Ind. J. Sci. Technol.* **2016,** *9*, 1–8.

Vishwanatha, K. S.; Rao, A. A.; Singh, S. A. Characterisation of Acid Protease Expressed from *Aspergillus oryzae* MTCC 5341. *Food Chem.* **2009,** *114*, 402–407.

Wallenstein, M. D.; Weintraub, M. N. Emerging Tools for Measuring and Modeling the In Situ Activity of Soil Extracellular Enzymes. *Soil Biol. Biochem.* **2008,** *40*, 2098–2106.

Walter, B.; Gallatin, J. C.; Hammermill. Enzymatic Conversion of Cellulosic Fibers. U.S. Pat 3,041,246, 1962.

Ward, O. P.; Qin, W. M.; et al. Physiology and Biotechnology of Aspergillus. *Adv. Appl. Microbiol.* **2005,** *58*, 1–75.

Wipusaree, N.; Sihanonth, P.; et al. Purification and Characterization of a Xylanase from the Endophytic Fungus *Alternaria alternata* Isolated from the Thai Medicinal Plant, *Croton oblongifolius* Roxb. *Afr. J. Microbiol. Res.* **2011,** *5*, 5697–5712.

Wong, K. K.; Saddler, J. N. Trichoderma Xylanases, Their Properties and Application. *Crit. Rev. Biotechnol.* **1992,** *12*, 413–435.

Xiong, X. Q.; Liao, H. D.; et al. Isolation of a Rice Endophytic Bacterium, *Pantoea* sp. Sd-1, with Ligninolytic Activity and Characterization of Its Rice Straw Degradation Ability. *Lett. Appl. Microbiol.* **2014,** *58*, 123–129.

Yadav, A. N.; Sachan, S. G.; Verma, P.; et al. Cold Active Hydrolytic Enzymes Production by Psychrotrophic Bacilli Isolated from Three Sub-Glacial Lakes of NW Indian Himalayas. *J. Basic Microbiol.* **2016,** *56*, 294–307.

Yadav, A. N.; Verma, P.; et al. In Diversity of Culturable Psychrotrophic Bacteria from Leh Ladakh and Bioprospecting for Cold-Active Extracellular Enzymes. *Proc Nat Sem Biotechnol Inter Benefit Man*, Vol. 32, 2012.

Yamasaki, M.; Yasui, T.; Arima, K. Pectic Enzymes in the Clarification of Apple Juice: Part I Study on the Clarification Reaction in a Simplified Model. *Agric. Biol. Chem.* **1964,** *28*, 779–787.

Yan, J. Y.; Yan, Y. J. Optimization for Producing Cell-Bound Lipase from *Geotrichum* sp. and Synthesis of Methyl Oleate in Microaqueous Solvent. *Appl. Microbiol. Biotechnol.* **2008,** *78*, 431–439.

Zhang, C.; Kim, S. K. Research and Application of Marine Microbial Enzymes: Status and Prospects. *Marine Drugs* **2010,** *8*, 1920–1934.

CHAPTER 10

Endophytes: A Hunt for Important Bioactive Compounds

SOFIA SHARIEF KHAN and SHAFAQ RASOOL

School of Biotechnology, Shri Mata Vaishno Devi University, Kakryal, Katra 182320, J & K, India

Corresponding author. E-mail: shafaq.rasool@smvdu.ac.in

ABSTRACT

In recent times, due to various health benefits to humans and plants, bioactive compounds in pharmaceuticals and naturopathy are in high demand. Bioactive compounds are synthesized by microorganisms either alone or in association with the plants. Microorganisms which are found inside the living tissues of plants, known as endophytes, living in a symbiotic/mutualistic association with the host plant though causing no instantaneous or visible adverse effects produce an array of bioactive compounds. Endophytes are found within almost all plants examined to date. Endophytes serve as a treasure for naturally derived and environmentally sustainable products for agricultural, medical, and industrial uses with the least negative environmental effects. The vast biodiversity along with their ability to biosynthesize bioactive metabolites has given the impetus for bioprospecting endophytic microbes with vast biotechnological potential. The need for innovative, more effective and more useful compounds has arisen to provide a better health care system to humans. This chapter focuses on the biodiversity of endophytes producing novel bioactive compounds.

10.1 INTRODUCTION

Since ages, plants have been a vital source of bioactive compounds against various diseases. Off late, microorganisms living inside the plants are reported to produce compounds that have immense medicinal importance (Subbulakshmi et al., 2012). The plant growth-promoting compounds, pest and insect repellents, plant pathogen inhibitors, inducers of stress tolerance, etc are produced by microbes. Therefore, there is a great interest in bioprospection of endophytic microbial organisms from different ecosystems. Endophytes are a group of symbiotic microbes – often bacteria or fungi – that live in symbiotic relationship within plants (Singh and Dubey, 2015). They are ubiquitous in nature and involve a gamut of interactions with their host like mutualism, antagonism, and in some rare cases parasitism as well (Nair and Padmavathy, 2014). These are known to be beneficial for the host plant, for instance, endophytes enhance the plant potential to stand different types of stress conditions and aid plants in fighting against diseases through the production of various bioactive compounds (Joseph and Priya, 2011; Parthasarathi et al., 2012).

Historically medicinal plants hold paramount importance as being a cure to different diseases globally. The formation of the bioactive compounds formed by these medicinal plants differs greatly based on plant species and their relationship with microbes. Medicinal plants play an important role in human health mainly in developing nations and areas affected by poverty. The natural habitats of wild medicinal plants, however, are under threat from overuse and environmental and geopolitical instability (Zhang et al., 2014). Even then, the ability of endophytes inside these bioprospective medicinal plants remains unclear. Endophytes that occupy medicinal plants demonstrate powerful biological activity like antimalarial, antioxidant, anticancer, and antimicrobial activities (Xiao et al., 2014). It should be noted that some endophytic microorganisms obtained from the medicinal plants produce the same compounds as produced by their hosts. If any bioactive product is specific to the endophytic microbe and is not produced by the host, this will not only eliminate the requirement of slow-growing and probably uncommon plants but also maintain the ever-decreasing biodiversity of the earth (Mehanni and Safwat, 2009).

A lot of research is available related to the endophytic diversity and their ability to produce different bioactive compounds. Endophytes are reported from roots, stem, leaf, seeds, buds, fruits, as well as dead plant cells (Specian et al., 2012; Stępniewska and Kuźniar, 2013). The presence of endophytic microbes in a single species of plant varies greatly and relies on multiple factors, like the host developmental stage, host species, environmental conditions, and inoculum density (Dudeja and Giri, 2014; Saikkonen et al., 2010). This chapter brings our attention to different types of endophytes producing novel bioactive compounds. Such studies may result in a better understanding of endophytic microorganisms and address the need for bioprospecting endophytes producing natural bioactive compounds.

10.2 ENDOPHYTIC BIODIVERSITY

It is believed that an individual plant could host a large number of microorganisms as epiphytes (microbial population that grows on the surface of plant) (Afzal et al., 2019) or endophytes (microbial population that is found within the plant tissues) (Andreote et al., 2014, Turner et al., 2013). The existence of endophytes in the plant was first mentioned by De Bary in 1866; he microscopically analyzed plant tissues and found the occurrence of microbial cells within the plant tissues. De Bary primarily defined the endophytic microorganisms as "any organism that grows inside plant tissues", and this explanation still persists with new studies (Wilson, 1995). However, Petrini in 1991 gave the most appropriate definition of endophytic microorganisms, which stated that any microorganism at any stage of its life history is found residing within the plant tissues without inflicting any type of damage to the plant host. Since then, countless efforts have been made to find out the origin of endophytes in various species (Hallmann et al., 1997, Mitter et al., 2013). Initially, the population of the seed-born microorganisms or rhizospheric microbes was considered the main source of endophytic microorganisms (Andreote et al., 2014).

Endophytes are the most abundant group of microorganisms discovered in almost all plants on land. Endophytic microbes have been obtained from a diverse group of plants ranging from large trees (Kaewkla and Franco, 2013), seagrasses (Florea et al., 2015), and also lichens (Ge et al., 2015; Suryanarayanan et al., 2017). Endophytes are linked with plants in the form of bacteria, fungi, mycoplasma, and actinomycetes within the plant tissue (Azevedo et al., 2014). Biodiversity of endophytes is categorized depending on the microorganism and includes primarily endophytic fungi, endophytic bacteria, endophytic actinomycetes, endophytic mycoplasma, and endophytic algae.

10.2.1 ENDOPHYTIC FUNGI

Endophytic fungi are an important group of microbial plant symbionts. They are classified into vast groupings primarily based on their life history and phyletic characters. The basic property for a fungi to be called an endophytic microbe is the presence of a hyphae in tissue at least for a part of the life cycle (Bacon and White Jr, 2000). The rapid visualization of hyphae is performed by acridine orange fluorescence microscopy. Since the identification through the presence of hyphae alone is seldom possible, the said process is carried out through distinct methods like DNA sequencing, immunofluorescence detection, and comparison of the DNA sequences to homologous sequence submitted in the Genbank. The relationship of fungal endophyte differs mainly from mycorrhizae due to the lack of a localized interface of specialized hyphae, nonpresence of coordinated growth of plant fungi, and the absence of plant benefits from nutrient transfer.

The most extensively used anticancer drugs and antibiotics are produced by endophytic fungi. Taxol, which is one of the best known and potential anticancer drugs till date, is obtained from an endophytic fungi *Taxomyces andreanae* (Stierle et al., 1993). Clavatol, sordaricin, jesterone, and javanicin, obtained from different endophytic fungal species, are recognized to own robust antifungal and antibacterial properties (Jalgaonwala et al., 2017). Pestacin, obtained from *Pestalotiopsis microspora*, has enormous antioxidant properties. Similarly, fungal endophytes are known to be the source of many immunosuppressive, anticancer, antidiabetic, and insecticidal compounds. Besides, some compounds providing thermal protection are also known to have been produced by the fungal endophytes. Fungal endophytes are divided into two ecological groups: Clavicipitaceous or grass or balansiaceous endophytes and non-clavicipitaceous or non-balansiaceous endophytes.

10.2.1.1 CLAVICIPITACEOUS OR GRASS OR BALANSIACEOUS ENDOPHYTES

Balansiaceous endophytes are the most extensively studied group owing to their ecological as well as economic significance. These fungal endophytes are found in grass species and grow systematically leading to vertical transmission via the seed. These are from the genera *Balansia*, *Epichloe,* and their anamorphs Ephelis and Neotyphodium. Grass endophytes are estimated to be present in 20–30% of grass species that could play crucial ecological roles in plant groups. It is suggested that these are systemic and often mutualistic for cold season grasses specifically in family Pooideae. These endophytic fungi manufacture a different group of bioactive compounds that are antiinsect alkaloids (lolines and peramine) and toxic in nature (Schardl and Craven, 2003). The fungal partners here avail various nutritional benefits and in return endophytic fungus protect the plant from herbivores through the production of toxic alkaloids.

10.2.1.2 NON-BALANSIACEOUS ENDOPHYTES OR NON-CLAVICIPITACEOUS

Non–clavicipitaceous endophytes are different based on their life cycle as well as phylogenetics. Mostly they belong to the phylum Ascomycota and can be found either intra- or intercellularly (Schulz et al., 2002). They are not only obligate to a particular host but can also adapt to different hosts. The non-balansiaceous endophytes are a widely diversified group and cover a wide variety of fungi from the Ascomycetes and basidiomycetes. The fungal endophytes of this group have a wide range of host and are able of colonizing dicotyledons and monocotyledons. Some non-clavicipitaceous endophytes are specialized and are found inside particular organs or cells, whereas various non-clavicipitaceous endophytic microbes are not specific to a host and a few colonies are found within the host tissue (Schulz et al., 1999).

Rodriguez et al. (2009) reclassified the endophytes; setting apart them into four classes. Class I includes constitutive mutualists (balansiaceous endophytes), Class II and III correspond to inducible mutualists, while Class IV coincides with dark septate root-colonizing fungi. Fungal endophytes are drawing the interest of researchers due to the covert advantages they offer for the host, in many approaches (Amin, 2016; Hartley and Gange, 2009; Le et al., 2009). Fungal endophyte research has indeed outstripped the preliminary phases. Table 10.1 shows fungal endophytes from different medicinal plants producing various bioactive compounds. We can conclude that fungal endophytes are a unique and significant microbial source for the production of bioactive compounds and have drawn interest in their theoretical analysis and potential uses from many researchers. Undoubtedly, the isolation of bioactive compounds from endophytic fungi holds a potential promise.

10.2.2 ENDOPHYTIC BACTERIA

After fungi, the second-most investigated endophytic group are bacterial endophytes. Bacterial endophytes are regarded as a subset of plant

TABLE 10.1 Fungal Endophytes Producing Different Bioactive Molecules.

S. no	Host	Endophyte	Bioactive compound/Activity	References
1	*Garcini dulcis*	*Phomopsis* sp.	Phomoxanthones A and B	Isaka et al. (2001)
2	*Torreya taxifolia*	*Pestalotiopsis microspora*	Pestaloside, caryophyllene sesquiterpenes, pestalotiopsins A and B, 2-hydroxydimeninol	Lee et al. (1995), Pulici et al. (1997), Pulici et al. (1996a, 1996b)
3	*Aegiceras corniculatum*	*Emericella* sp.	Emerimidine A and B and emeriphenolicins A and D	Zhang et al. (2011)
4	*Phlegmariuruscryptomerianus*	*Blastomyces* sp., *Botrytis* sp.	Hupzine A	Ju et al. (2009)
5	*Taxus*	*Metarhizium anisopliae*	Taxol	Visalakchi and Muthumary (2010), Zhang, Zhou, et al. (2009), Jalgaonwala et al. (2017)
6	*Lycopodium serratum chrysogenum*	*Penicillium*	*Huperzine A*	Zhou et al. (2009)
7	*Catharanthus roseus*	*Alternaria* sp.	Vinblastine	Guo et al. (1998)
8	*Iris germanica*	*Rhizopus oryzae*	Irone	Zhang et al. (1999)
9	*Hypericum perforatum*	*Chaetomium globosum*	Hypericin	Kusari et al. (2008)
10	*Salvia miltiorrhiza*	*Trichoderma atroviride* D16	Tanshinone IIA and tanshinone I	Ming et al. (2012)
11	*Ginkgo biloba*	*C. globosum*	Chaetoglobosin G, fumitremorgin C, Gliotoxin,	Li et al. (2011)
12	*Fritillaria unibracteata*	*Fusarium redolens*	Imperialine-3β-d-glucoside and peimisine	Pan et al. (2015)
13	*G. biloba*	*Fusarium oxysporum*	Ginkgolide B	Cui et al. (2012)
14	*Arbutus unedo*	*Talaromyces pinophilus*	Ferrirubin, herquline B,3-*O*-methylfunicone	Vinale et al. (2017)
15	*Markhamia tomentosa*	*Trichoderma longibrachiatum and Syncephalastrum racemosum*	Antifungal and antiproliferative	Ibrahim et al. (2017)

TABLE 10.1 *(Continued)*

S. no	Host	Endophyte	Bioactive compound/Activity	References
16	*Glycyrrhiza glabra Linn.*	Phoma sp.	Thiodiketopiperazine derivatives	Arora et al. (2016)
17	*Nymphaea nouchali*	*C. globosum*	Chaetoglobosin A and C	Dissanayake et al. (2016)
18	*Taxus fauna*	*Chaetomium* sp. and *Penicillium* sp.	Anticancer	Fatima et al. (2016)
19	*Sida acuta*	*Aspergillus sulphureus*	Antibacterial	Murali et al. (2017)
20	Zingiber cassumunar	*Arthrinium* sp. MFLUCC16-1053	β-cyclocitral, 3*E*-cembrene A, laurenan-2-one, sclareol, 2*Z*,6*E*-farnesol, cembrene, β-isocomene and γ-curcumene	Pansanit and Pripdeevech (2018)
21	*Rhizophora mucronata*	*Aspergillus oryzae* BPPTCC 6036, *Emericella nidulans* BPPTCC 6035 *Aspergillus versicolor* BPPTCC 6039	Insecticidal	Abraham et al. (2015)
22	*Aegle marmelos*	*Curvularia australiensis, Cladosporium cladosporioides, and Alternaria alternata*	Antioxidant and antibacterial	Mani et al. (2015)
23	*Camellia oleifera*	*Trichothecium* sp. and *Oidium* sp.	Antifungal	Yu et al. (2018)
24	*Dracaena draco* L.	*Botryodiplodia theobromae* Pat.	Norharman, capsi-amide, coumarin and isocoumarins	Zaher et al. (2015)
25	*C. roseus*	*F. oxysporum*	Vinblastine and vincristine	Kumar et al. (2013)
26	*Erythrina crista-galli*	*Phoma* sp.	Phomol	Weber et al. (2004)
27	*Capsicum annuum*	*A. alternata*	Capsaicin	Devari et al. (2014)
28	*Forsythia suspensa*	*Colletotrichum gloeosporioides*	Phillyrin	Zhang et al. (2012)
29	*Pelargonium sidoides*	*Alternaria* sp.	Linoleic acid and cyclodecasiloxane	Manganyi et al. (2019)

TABLE 10.1 *(Continued)*

S. no	Host	Endophyte	Bioactive compound/Activity	References
30	*Mimusops elengi*	DQ 384608.1 ascomycetes species	Ergoflavin	Deshmukh et al. (2009)
31	*Brassica campestris*	*Mucor* sp.	Biofertilizer	Zahoor et al. (2017)
32	*Simmondsia chinensis*	*Rhodococcus pyridinivorans* and *Oceanobacillus kimchi*	Plant growth promoters	Rosales et al. (2017)
33	*Panax ginseng*	*Plectosphaerella* sp., *Aspergillus* sp., *Penicillium* sp., *Cladosporium* sp., *Ascomycete* sp. *Engyodontium* sp., *Fusarium* sp., and *Verticillium* sp.	Root protection and growth	Wu et al. (2013)
34	*Phaseolus vulgaris* L	*Trichoderma harzianum, Trichoderma polysporum*, and *Trichoderma atroviridae*	Phytostimulation and biocontrol potential	Pierre et al. (2016)
	Flacourtia inermis	*Fusarium decemcellulare*	Anticoagulant; antithrombiotics	Qader et al. (2018)
35	*Leptochloa fusca*	*Enterobacter* sp. *Pantoea stewartii,* and *Microbacterium arborescens*	Bioaugmentation	Ashraf et al. (2018)
36				
37	*Markhamia platycalyx*	*Aspergillus flocculus*	Antimicrobial; Anticancer; antitrypanosomal	Tawfike et al. (2019)
38	*Dendrobium lindleyi*	*F. oxysporum*	Antimutagenic	Bungtongdee et al. (2019)

growth-promoting rhizobacteria generally known as rhizospheric bacteria (Afzal et al., 2019). These are actually a specialized group of bacteria that have developed the potential to penetrate their host plant (Reinhold-Hurek and Hurek, 1998). They show all the significant factors found in rhizobacteria that are compatible with the promotion of host plant growth. (Afzal et al., 2019). Endophytic bacteria are usually found in vascular tissue and intercellular spaces of the plant.

Reportedly, many researchers have isolated endophytic bacteria from a variety of plants, like rice, pea, and sweet corn cultivars (Cho et al., 2007, Kumar et al., 2020; Vasilyeva et al., 2020). A database of all 16S rDNA sequences currently allocated to endophytes was examined along with uncultured and cultured microbes (Hardoim et al., 2015), and it was observed that even though the sequences corresponds to a total of 23 different Phyla of bacteria, four of them (Firmicutes, Actinobacteria, Bacteroidetes, and Proteobacteria) account for 96% of the endophytic sequences of the prokaryotes (Hardoim et al., 2015). More than 50% of the sequences in the database are proteobacteria, and Gammaproteobacteria isolates seem to be the most frequently found as endophytes within this phylum, including genera like *Enterobacter*, *Serratia*, *Pantoea*, *Pseudomonas*, *Stenotrophomonas*, and *Acinetobacter*. *Paenibacillus*, *Mycobacterum*, and *Bacillus* are all well identified amongst the endophytic microbes (Hardoim et al., 2015). It is reported that γ-proteobacteria is the most dominant and diverse endophytic bacterial group in agricultural crops (Miliute et al., 2015). These bacterial endophytes help in plant growth promotion, increased yield, nitrogen fixation, phosphate solubilization, besides protection against plant pathogens. Table 10.2 shows bioactive compounds produced by bacterial endophytes isolated from various plants. In order to explore these rare and promising bacterial endophytes with overall plant beneficial features, a comprehensive approach based on both culture-independent and culture-dependent techniques is required.

10.2.3 ENDOPHYTIC ACTINOMYCETES

Actinomycetes belong to the phylum Actinobacteria and are prokaryotic microorganisms that form spores and own mycelium like fungus (Barka et al., 2016; Chaudhary et al., 2013). There are several benefits to the association of actinomycetes with plants, such as the production of extracellular enzymes, antimicrobial compounds, siderophores, and phytohormones (Gangwar et

TABLE 10.2 Bacterial Endophytes Producing Bioactive Molecules.

S. no	Host	Endophyte	Bioactive compound/Activity	References
1	*Nothapodytes foetida*	*Entrophospora infrequens* SW116	Taxol and taxane	Stierle et al. (1993)
2	*Ageratum conyzoides*	*Shewanella* sp. *and Pseudomonas* sp.	2-amino-3-quinoline carbonitrile and boric acid	(Fitriani et al.)
3	*Cassava (Manihotesculenta)*	*Paenibacillussp. IIRAC-30*	Lipopeptides	Canova et al. (2010)
4	*Balloon flower (Platycodon grandiflorus)*	*Bacillus licheniformis and Bacillus pumilus*	Antifungal activity	Islam et al. (2010)
5	*Bostrychiatenella*	*Penicillium decaturense and Penicillium waksmanii*	Cytochalasin D	Felício et al. (2015)
6	*Forsythia suspensa*	*Colletotrichum gloeosporioides*	Phillyrin	Zhang et al. (2012)
7	*Pelargonium sidoides*	*Alternaria* sp.	Linoleic acid and cyclodecasiloxane	Manganyi et al. (2019)
8	*Moringa peregrina*	*B. licheniformis* MpKL1	Antibacterial	Aljuraifani et al. (2019)
9	*Dracaena cochinchinensis* Lour.	*Pseudonocardia* sp., *Microbacterium* sp., *Streptomyces* sp., *Nocardiopsis* sp., *Brevibacterium* sp., *Arthrobacter* sp., *Tsukamurella* sp., *Brachybacterium* sp., *Rhodococcus* sp., *Nocardia* sp., *Kocuria*sp. *and Nocardioides* sp.	Plant promotion growth and protection	Salam et al. (2017)
10	*Mimusops elengi*	DQ 384608.1 ascomycetes species	Ergoflavin	Deshmukh et al. (2009)
11	*Poplar*	*Pseudomonas putida* W619-TCE	Phytoremediation	Weyens et al. (2015)
12	*Albizia lebbeck*	*Marinomonas* sp., *Bacillus* sp., *Xanthomonas* sp., *Salinococcus* sp. *Pseudomonas* sp., and *Rhizobium* sp.	Reduce heavy metal toxicity	Manikandan et al. (2016)

TABLE 10.2 *(Continued)*

S. no	Host	Endophyte	Bioactive compound/Activity	References
13	*Pyrenacanthavolubilis* Hook	*Bacillus* sp. (KP125955 and KP125956), *Bacillus subtilis* (KY741853), and *Bacillus amyloliquefaciens* (KY741854)	Camptothecin	Soujanya et al. (2017)
14	*Vitis vinifera* cv. Glera	*Pantoagenera, Bacillus,* and *Mirococcus*	Growth promotion and phosphate solubilization	Baldan et al. (2015)
15	*Suaeda japonica*	*Penicillium* sp.	Enhanced seed germination	You et al. (2012)
16	*Grevillea pteridifolia*	Streptomyces NRRl 30566	Kakadumycins	Castillo et al. (2003)
17	*Glycyrrhiza uralensis*	*Bacillus atrophaeus*	Biocontrol agent	Mohamad et al. (2018)
18	*Monstera* sp.	*Streptomyces* sp. (MSU-2110)	Coronamycin	Ezra et al. (2004)
19	*Dendrobium*	*Bacillus megaterium*	Antibacterial	Wang, Liu, et al. (2019)
20	*Hibiscus rosa-sinensis*	*Pseudomonas oryzihabitans*	Asparaginase	Bhagat et al. (2016)
21	*Leptochloa fusca*	*Microbacterium arborescens, Enterobacter* sp. and *Pantoea stewartii*	Bioaugmentation	Ashraf et al. (2018)

al., 2014). In the quest for novel bioactive natural compounds, endophytic actinomycetes have gained attention because new drugs eliminate those to which pathogenic strains have gradually gained resistance. The most commonly identified endophytic actinomycete is *Streptomyces,* which is mainly found in the leaves, stems, and roots of the plant (Golinska et al., 2015). These microbes have been confirmed to establish and facilitate plant growth and induce tolerance to biotic and abiotic stress (Segaran et al., 2017). Actinomycetes as endophytes are popular for unique chemical entities with medicinal significance (Gayathri and Muralikrishnan, 2013; Singh and Dubey, 2015). There are many reports on the production of antimicrobial compounds from different varieties of actinomycetes (Gos et al., 2017). Compounds of biological interest like clethramycin, munumbicins both A and B, kakadumycins, coronamycin, saadamycin, cedarmycin, and naphthomycin (A and K) are reported from *Streptomyces* sp. (El-Gendy and EL-Bondkly, 2010; Zhao et al., 2011). It is suggested that the endophytic actinomycetes are a rich source of herbicidal metabolites (Singh et al., 2018). Table 10.3 shows different endophytic actinomycetes from different medicinal plants and Table 10.4 shows different bioactive compounds along with activity endophytic actinomycetes exhibit.

10.2.4 ENDOPHYTIC MYCOPLASMA

Some Mycoplasma species (phylum Tenericutes) are also found as endophytes. Among self-replicating bacteria, they are the smallest group having a cell membrane, complete translational machinery, and genome containing a minimum set of genes required for replication and growth. However, the mycoplasma lacks a cell wall, unlike other prokaryotes. Some mycoplasma species are reported to exist in symbiosis with few red algae, like *Bryopsis hypnoides, Bryopsis pennata*, and also in *Arcobacter* (Hollants et al., 2011). The occurrence of Mycoplasma species is strictly determined by environmental factors, but there has been no authentic information regarding its utilities (Gouda et al., 2016). Several endophytes that grow within algae and seaweeds are now known (Flewelling et al., 2013). Endophytic microorganisms have been observed in *Bryopsis, Chondrus ocellatus, Ulvellaleptochaete*, etc. (Hollants et al., 2011; Gao et al., 2019; Sahoo et al., 2017; Kamat et al., 2020). Table 10.5 shows different endophytes isolated from from various algal species. One of the most significant sources of research on natural products is the marine environment (de Felício et al., 2015). Even

TABLE 10.3 Endophytic Actinomycetes Isolated from Different Medicinal Plants.

S. no	Host	Endophyte	References
1	*Mangrove medicinal plants of Macao*	*Friedmanniella* sp. *4Q3S-3* and *Nakamurellas* sp. *2Q3S-4-2*	Li et al. (2017)
2	*Terminalia mucronata*	*Micromonospora terminaliae CAP94*	Kaewkla et al. (2017)
3	*Centella asiatica*	*Wenchangensis 234402, Couchioplanes caeruleus SCC 1014, Streptomyces* sp.*, Actinoplanes brasiliensis IFO13938, Micromonospora schwarzwaldensis HKI0641, Gordonia otitidis IFM10032*	Ernawati et al. (2016)
4	*Aloe arborescens*	*Streptomyces zhaozhouensis NEAU-LZS-5*^{T}, *Micrococcus aloeverae AE-6*^{T}	He et al. (2014), Prakash et al. (2014)
5	*Lobelia clavatum*	*Pseudonocardia endophytica YIM 56035*^{T}	Chen et al. (2009)
6	*Elaeagnus angustifolia*	*Micromonospora* sp. *D30401, D30202, D30511C10401 and D30407, Nonomureae* sp. *D10204, Pseudonocardia* sp. *C20201, Planotetraspora* sp. *C10401*	Chen et al. (2011)
7	*Lonicera maackii*	*Allonocardiopsis opalescens*	Du et al. (2013)
8	*Rauwolfia densiflora*	*Streptomyces longisporoflavus*	Akshatha et al. (2014)
9	*Mirabilis jalapa*	*Actinomycete* sp.	Passari et al. (2015)
10	*Achillea millefolium*	*Nocardiopsis* sp., *Micromonospora* sp.,	Machavariani et al. (2014)
11	*Tinospora crispa*	*Streptomyces olivochromogenes*	Pujiyanto et al. (2012)
12	*Spermacoce verticillata*	*Microbispora* sp.	Conti et al. (2012)
13	*Phyllanthus niruri*	*Actinomadura* sp.	Mini Priya (2012)
14	*Catharanthus roseus*	*Streptomyces cavourensis AB184264.1*	Kafur and Basheer (2011)
15	*Psammosilene tunicoides*	*Allostreptomyces psammosilenae YIM DR4008*^{T}	Huang et al. (2017)

TABLE 10.4 Bioactive Compounds and Activity Exhibited by Endophytic Actinomycetes.

S. no	Bioactive compound	Source	Activity	References
1	Indole-3- carbaldehyde, Indole-3-acetic acid, and Indole-3-carboxylic acid	*Microbispora* sp. LGMB259	Antimicrobial	Savi et al. (2015)
2	indole-3-carbaldehyde, cyclo-(L-Pro L-Tyr), brevianamide F, cyclo-(L-Pro-L-Phe) and 1-acetyl-b-carboline	*Aeromicrobium ponti* LGMB491	Antimicrobial	Gos et al. (2017)
3	Lupinacidin C	*Micromonospora lupini*	Anticancer	Igarashi et al. (2011)
4	Camptothecin	*Lysinibacillus* sp.	Anti-cancer	Singh et al. (2017)
5	Ansamitocin	*Nocardia* sp. No. C-15003	Anticancer	Higashide et al. (1977)
6	Anthracycline	*Streptomyces* sp. YIM66403	Antitumor	Li et al. (2015)
7	Miconazole, Ketoconazole, Fluconazole	*Streptomyces olivaceus*, *Streptomyces thermocarboxydus,* *Streptomyces* sp. BPSA 121	Antimicrobial	Passari et al. (2017)
8	Treponemycin	*Streptomyces* strain MS-6-6	Anti-tuberculous	Yassien et al. (2015)
9	Antituberculous	*Streptomyces* sp. MK932-CF8	Anti-prostate cancer	Yamazaki et al. (2015)
10	Dermacozines	*Dermacoccus* sp.	Cytotoxic, radical scavenging	Abdel-Mageed et al. (2010)
11	24-Demethyl-bafilomycin C1	*Streptomyces* sp. CS	Anticancer	Li et al. (2010)
12	Chandrananimycins	*Actinomadura* sp.	Antialgal Antibacterial	Maskey et al. (2003)
13	Thaxtomin A	*Streptomyces scabies*	Cellulose synthesis inhibitor	Francis et al. (2015)
14	Salaceyins A and B	*S. laceyi* MS53	Anticancer	Kim et al. (2006)
15	Saadamycin	*Streptomyces* sp. Hedaya48	Antibiotic	El-Gendy and EL-Bondkly (2010)
16	29-*O*-methylabierixin, Efomycins M, oxohygrolidin and G, abierixin	*Streptomyces* sp.	Antiprotozoal	Supong et al. (2016)

TABLE 10.5 Endophytes Obtained from Different Algae.

S. no	Host	Endophyte	References
1	*Enteromorpha* sp., *Chaetomorpha* sp., *Enteromorpha* sp.	*Aspergillus unguis* AG1.1, *A. unguis* AG1.2, *Perenniporia* sp. AG1.3, *Hypoxylon fragiforme* AG1.4, *A. unguis AG1*.1(G), *Colletotrichum gloeosporioides* PG1.1, *Aspergillus niger* PG1.2, *A. niger* PG1.3, *H. fragiforme* PG1.7, *Aspergillus* sp. VG1.4	Kamat et al. (2020)
2	*Chaetomorpha antennina*	*Penicillium chrysogenum*	Parthasarathy et al. (2020)
3	*Sargassum wightii, Caulerpa peltata, Ulva fasciata, Halymenia venusta, Dictyota dichotoma, Hypnea musiformis, Grateloupia lithophila, Enteromorpha compressa, Ceramium diaphanum, Padina tetrastromatica* and *C.* antennina	*Aspergillus* sp., *Penicillium* sp., and *Engyodontium* sp.	Sarasan et al. (2020)
4	*Carpopeltis affinis*	*Fusarium chlamydosporum* OUPS-N124	Usami et al. (2002)
5	*Sargassum pallidum*	*P. chrysogenum* EN-118	An et al. (2013)
6	*Actinotrichia fragilis*	*Penicillium citrinum* N-059	Tsuda et al. (2004)
7	*Sargassum* sp.	*Aspergillus* sp. SpD081030G1f1	Izumikawa et al. (2010)
8	*Avrainvillea* sp.	*Fusarium* sp. CNL-619	Cueto et al. (2000)
9	*Fucus spiralis*	*Penicillium* sp.	Flewelling, Johnson, et al. (2013)
10	*Pterocladiella tenuis*	*Penicillium chermesinum* EN-480	Liu et al. (2016)

though the macroalgae and fungal association have been described for their ecological significance, there is a lack of research on algal endophytes. Future investigations of algal endophytes in the natural environment are necessary to understand the algal endobiosis.

10.3 SCREENING AND ISOLATION OF ENDOPHYTES WITH BIOACTIVE POTENTIAL

Initially with little knowledge about the endophytes, they were thought of as pathogen-inducing abrasion in the host cell. Various methods were adopted for examining endophytes and their bioactive potential, beginning from the bioassays through which screening of biological molecules was done resulting in the identification, purification, and characterization of the molecules. The fluorescence in situ hybridization-confocal laser scanning microscopy and cultivation-independent assays have confirmed the presence of endophytes in plants (Berg et al., 2014; Mitter et al., 2013). Screening of endophytes using molecular markers and genome mining can determine the desired strains from a large isolated endophytic diversity. Cultivation-dependent and cultivation-independent techniques (metagenomic analysis) can be used to study the diversity of endophytic microbes. Cultivation-based techniques are used for the testing and recovery of endophytic isolates, whereas screening for variations within the general endophytic groups is done by cultivation-independent techniques (Menpara and Chanda, 2013). Microbial or plant growth media, like Rose Bengal agar, Nutrient agar, Potato dextrose agar, Sabouraud dextrose agar, Tryptic soya agar, and any carbon- or nitrogen-containing media, can be used for the isolation of endophytes. The most common technique used for isolating endophytes is isolation from surface-sterilized plant tissue. Genotype, physiological status, growth level, sampling season, habitat, and tissue type are some of the important parameters affecting endophyte colonization in any species of the plant (Gaiero et al., 2013; Golinska et al., 2015). Every endophyte isolated must be grown in suitable conditions for the isolation of the bioactive compounds by extraction with various organic solvents. The extracts obtained were then further processed using different bioassays for different activities (Aly et al., 2010; Garyali et al., 2013; Roopa et al., 2015). Numerous databases like Human Metabolome Database (Wishart et al., 2008), the METLIN database (Smith et al., 2005), and the Madison Metabolomics Consortium Database (Cui et al., 2008) are used for comparing chemical structure with available spectroscopic data in the database.

10.4 ENDOPHYTES PRODUCING DIFFERENT BIOACTIVE COMPOUNDS

The primary resource of bioactive compounds are natural products and the most important bioactive compounds in the world are obtained from microorganisms (Bérdy, 2012). The endophytic microorganisms of various medicinal plants produce several therapeutically active compounds. Many reports are indicating that endophytes isolated from a single plant can produce different bioactive metabolites serving as a splendid source with application in industry, medicine, and agriculture (Strobel and Daisy, 2003; Jalgaonwala et al., 2017; Omojate et al., 2014; Shukla et al., 2014). Isolation of secondary metabolites from endophytic microorganisms is affected by many factors like the period in which the sample was collected, geographical location, and the climate (Shukla et al., 2014). The production of bioactive compounds by endophytic microorganisms have been linked with the host evolution, which also includes higher plants genetic information, permitting microbes to adapt best to the host and do several tasks, like protection from diverse sorts of pathogens, herbivore, and insects (Strobel, 2003). There is a huge scope for the isolation of novel bioactive compounds from endophytes. Several classes of bioactive compounds such as antibacterial, antifungal, antiviral, insecticides, plant growth-promoting compounds, and plant protective agents have been reported to be produced by endophytes. Some important classes of bioactive compounds obtained from medicinal plants are tabulated in Table 10.6.

Several biologically active polyphenolics, terpenoids, and flavonoids have been described from endophytes. Various studies determined that terpenoids are used clinically with great curative effects and have important biological activities. The xylarenic acid and xylarenones A and B, three new sesquiterpenoid obtained from the endophytic fungus *Xylaria* sp. NCY2 isolated from *Torreya jackii* CHUN were evaluated in vitro for antimicrobial and antitumor assays (Hu et al., 2008). Phomanolide obtained from the *Phoma* sp., an endophytic fungus obtained from *Aconitum vilmorinianum* roots, exhibits antiviral activities against influenza virus A (Liu et al., 2019). Paclitaxel obtained from *T. andreanae,* an endophytic fungus, isolated from the *Taxus brevifolia* bark, a diterpene derivative, shows therapeutic positive effect on various types of cancers like rectal, breast, ovarian, colon, lung, and bladder cancer, also on rheumatoid arthritis by affecting mitosis (Stierle et al., 1993). Terpenoids are described as a complex group of natural compounds with a wide variety of biological activities and an enormous potential for the production of drugs. Thus the study of active and novel terpenoids isolated from endophytes is significant.

TABLE 10.6 Shows Different Bioactive Compound Classes Obtained from Medicinal Plants.

Class	Example
Alkyl salicylic acids	Salaceyin A and B
Alkaloids	Taxol, camptothecin, vincristine, nodulisporic acid
Carboxylic acids	Cinnamic acid
Lignans	Podophyllotoxin
Terpenes and terpenoids	Subglutinol A and B, gibberellic acid, xylarenic acid, phomoarcherin A, B and C
Flavonoids and flavonols	Kaempferol
Benzofurans	Pestacin, isopestacin
Cytochalasins	Cytochalasin H, J, and E
Polysaccharides	Mycelial polysaccharide
Polyphenols	Curcumin
Anthraquinones	1,4-Dihydroxyanthraquinone, emodin
Xanthenes	Ergoflavin
Steroids	Beta-sitosterol
Antimicrobials	Cryptocandin, nocardithiocin, javanicin

Alkaloids are quite important molecules, not only just for chemical purposes, but also for their numerous biological properties, like anticancer, antiviral, and antifungal activities. Camptothecin, a potent antineoplastic agent, has been reported from the endophytes *Fusarium solani, Bacillus subtilis* KY741853 and *Entrophospora infrequens* (Ran et al., 2017; Soujanya et al., 2017; Puri et al., 2005). *Fusarium oxysporum* isolated from *Catharanthus roseus produces* vinblastine and vincristine, well-known anticancer alkaloids (Kumar et al., 2013). A natural anthraquinone derivative isolated from endophytic fungus, *Polyporales* sp. isolated from *Rheum emodi,* causes apoptosis in human lung cancer cells (Dar et al., 2017). Ergoflavin, a Xanthene, was obtained from *Mimusops elengi*. It is a dimeric xanthene linked at position-2 having an anticancer activity (Deshmukh et al., 2009). Secalonic acid D, a mycotoxin belonging to ergochrome class, is known to have potent anticancer activities and was isolated from the mangrove endophytic fungus (Zhang et al., 2009). Three novel derivatives of xanthone, isolated from endophytic fungus *Aspergillus sydowii* are 13-*O*-acetylsydowinin B, sydoxanthone A, and sydoxanthone B have moderate immunosuppressive activities (Song et al., 2013).

Wang et al. (2019) have reported a Curtachalasin from endophytic fungus *Xylaria* cf. curta that helps in resistance reversal activity against Fluconazole-Resistant *Candida albicans*. An antiviral compound called Altertoxins, produced by endophyte *Alternaria tenuissima* QUE1Se isolated from *Quercus emoryi,* is effective against HIV-1 virus (Bashyal et al., 2014). Brefeldin A, produced by *Penicillium* sp. FKI-7127, a fungal endophyte, acts as a novel antiviral agent against dengue viruses (Raekiansyah et al., 2017). Podophyllotoxin, a potential natural anticancer compound obtained from endophytic fungi *F. solani* P1 has significant biological and commercial implications (Nadeem et al., 2012). The endophytes related to medicinal plants might provide a broader understanding of plant endophytic evolution and the interaction of symbiotic and mutualistic interactions. The process that allows these microorganisms to communicate with their host plant provides the isolation of several biotechnologically important compounds. Many issues are still not clear, like the combination of plant and endophyte metabolic pathways responsible for bioactivity, the variability of genetic regulation for the synthesis of secondary metabolites between the host plant and the endophyte. To resolve these aspects, the biochemistry and physiology of endophytic microbes must be understood in terms of their role in the development of secondary metabolites. There is obviously a need for further research on the advancement of new technologies and methodologies for their use in the medical, pharmaceutical, and agricultural sectors.

10.5 CONCLUSION

Endophytes are a largely unexplored group capasble of synthesizing bioactive natural compounds having a broad range of biological activities and a huge degree of structural complexity. Bioactive compounds of endophytic origin have exhibit huge potential and application in different fields like agriculture, industry, and medicine. With the help of biotechnology, like microbial fermentation process, genetic engineering, and metabolic engineering, we can better manipulate and understand endophytes. Endophytes have received more attention with the passage of time as they can produce bioactive compounds similar to their host plant. The latest developments, ongoing studies, and previous success related to endophytic microorganisms are sufficient to attract the scientific community's attention towards this new field and to exploit its potential therapeutic uses in medical, pharmaceutical, food, and cosmetics fields.

KEYWORDS

- **endophytes**
- **biodiversity**
- **bioactive compounds**
- **pathogenic microorganisms**
- **bioprospection**

REFERENCES

Afzal, I.; Zabta, K. S.; Shomaila, S.; Shaheen; S. Plant Beneficial Endophytic Bacteria: Mechanisms, Diversity, Host Range and Genetic Determinants. *Microbiol. Res.* **2019**, *221*, 36–49.

Aly, A. H.; Abdessamad, D.; Julia, K.; Peter, P. Fungal Endophytes from Higher Plants: A Prolific Source of Phytochemicals and Other Bioactive Natural Products. *Fungal Divers* **2010**, *41* (1), 1–16.

Amin, N. Endophytic Fungi to Control of Cocoa Pod Borer (*Conopomorpha cramerella*) on Cocoa Plantation. *Res. J. Pharm. Biol. Chem. Sci.* **2016**, *7* (6),1496–1501.

Andreote, F. D.; Thiago, G.; Ademir, D. Exploring Interactions of Plant Microbiomes. *Sci. Agrícola* **2014**, *71* (6), 528–539.

Bacon, C. W.; White, J. F. Jr. Physiological Adaptations in the Evolution of Endophytism in the Clavicipitaceae. In *Microbial Endophytes*; CRC Press, 2000; pp. 251–276.

Barka, E. A.; Vatsa, P.; Sanchez, L., et al. Taxonomy, Physiology, and Natural Products of Actinobacteria. *Microbiol. Mol. Biol. Rev.* **2016**, *80* (1), 1–43.

Bashyal, B. P.; Wellensiek, B. P.; Ramakrishnan, R., et al. Altertoxins with Potent Anti-HIV Activity from *Alternaria tenuissima* QUE1Se, A Fungal Endophyte of *Quercus emoryi*. *Bioorg. Med. Chem.* **2014**, *22* (21), 6112–6116.

Bérdy, J. Thoughts and Facts About Antibiotics: Where We Are Now and Where We Are Heading. *J. Antibiot.* **2012**, *65* (8), 385.

Berg, G.; Grube, M.; Schloter, M.; Smalla, K. Unravelling the Plant Microbiome: Looking Back and Future Perspectives. *Front. Microbiol.* **2014**, *5*, 175–175.

Chaudhary, H. S.; Soni, B.; Shrivastava, A. R.; Shrivastava, S. Diversity and Versatility of Actinomycetes and Its Role in Antibiotic Production. *J. Appl. Pharm. Sci.* **2013**, *3* (8), 83–94.

Cho K. M.; Hong S. Y.; Lee S. M., et al. Endophytic Bacterial Communities in Ginseng and Their Antifungal Activity Against Pathogens. *Microbial. Ecol.* **2007**, *54* (2), 341–351.

Cui, Q.; Lewis, I. A.; Hegeman, A. D., et al. Metabolite Identification Via the Madison Metabolomics Consortium Database. *Nat. Biotechnol.* **2008**, *26* (2), 162.

Dar, R. A.; Majeed, R.; Sheikh, A. A.; et al. Emodin, Isolated and Characterized from An Endophytic Fungus *Polyporales* sp., Induces Apoptotic Cell Death in Human Lung Cancer Cells Through the Loss of Mitochondrial Membrane Potential. *J. Phytopharma.* **2017**, *6* (5), 288–292.

de Felício, R.; Pavão, G. B.; de Oliveira, A. L. L.; et al. Antibacterial, Antifungal and Cytotoxic Activities Exhibited by Endophytic Fungi from the Brazilian Marine Red Alga *Bostrychia tenella* (Ceramiales). *Braz. J. Pharmacog.* **2015**, *25* (6), 641–650.

Deshmukh, S. K.; Mishra, P. D.; Almeida, A. K.; et al. Anti-Inflammatory and Anticancer Activity of Ergoflavin Isolated from An Endophytic Fungus. *Chem. Biodivers.* **2009**, *6* (5), 784–789.

Dudeja, S. S.; Giri, R. Beneficial Properties, Colonization, Establishment and Molecular Diversity of Endophytic Bacteria in Legumes and Non Legumes. *Afr. J. Microbiol. Res.* **2014**, *8* (15), 1562–1572.

El-Gendy, M. M. A.; EL-Bondkly, A. M. A. Production and Genetic Improvement of A Novel Antimycotic Agent, Saadamycin, Against Dermatophytes and Other Clinical Fungi from Endophytic *Streptomyces* sp. Hedaya48. *J. Ind. Microbiol. Biotechnol.* **2010**, *37* (8), 831–841.

Flewelling, A. J.; Ellsworth, K. T.; Sanford, J.; et al. Macroalgal Endophytes from the Atlantic Coast of Canada: A Potential Source of Antibiotic Natural Products? *Microorganisms* **2013**, *1* (1), 175–187.

Florea, S.; Schardl, C. L.; Hollin, W. Detection and Isolation of Epichloë Species, Fungal Endophytes of Grasses. *Curr. Protocol Microbiol.* **2015**, *38* (1), 19A. 1.1–19A. 1.24.

Gaiero, J. R.; McCall, C. A.; Thompson, K. A.; et al. Inside the Root Microbiome, Bacterial Root Endophytes and Plant Growth Promotion. *Am. J. Bot.* **2013**, *100* (9), 1738–1750.

Gangwar, M.; Dogra, S.; Gupta, U. P.; Kharwar, R. N. Diversity and Biopotential of Endophytic Actinomycetes from Three Medicinal Plants in India. *Afr. J. Microbiol. Res.* **2014**, *8* (2), 184–191.

Gao, X.; Ogandaga, C. A. M.; Park, S. K.; et al. Algal Endophytes of Commercial *Chondrus ocellatus* (Gigartinaceae, Rhodophyta) from Different Wild Populations in Korea. *J. Appl. Phycol.* **2019**, *32*, 697–703.

Garyali, S.; Kumar, A.; Reddy, M. S. Taxol Production by An Endophytic Fungus, *Fusarium redolens*, Isolated from Himalayan Yew. *J. Microbiol. Biotechnol.* **2013**, *23* (10), 1372–1380.

Gayathri, P.; Muralikrishnan, V. Isolation and Characterization of Endophytic Actinomycetes from Mangrove Plant for Antimicrobial Activity. *Int. J. Curr. Microbiol. App. Sci.* **2013**, *2* (11), 78–89.

Ge, C.; Liu, B.; Che, J.; et al. Diversity of Bacillus species Inhabiting on the Surface and Endophyte of Lichens Collected from Wuyi Mountain. *Acta Microbiol. Sin.* **2015**, *55* (5), 551–563.

Golinska, P.; Wypij, M.; Agarkar, G.; et al. Endophytic Actinobacteria of Medicinal Plants: Diversity and Bioactivity. *Anton. Van Leeuwen* **2015**, *108* (2), 267–289.

Gos, F. M. W. R.; Savi, D. C.; Shaaban, K. A.; et al. Antibacterial Activity of Endophytic Actinomycetes Isolated from the Medicinal Plant *Vochysia divergens* (Pantanal, Brazil). *Front. Microbiol.* **2017**, *8*, 1642.

Gouda, S.; Das, G.; Sen, S. K.; et al. Endophytes: A Treasure House of Bioactive Compounds of Medicinal Importance. *Front. Microbiol.* **2016**, *7*, 1538.

Hallmann, J.; Quadt-Hallmann, A.; Mahaffee, W. F.; Kloepper, J. W. Bacterial Endophytes in Agricultural Crops. *Can. J. Microbiol.* **1997**, *43* (10), 895–914.

Hardoim, P. R.; Overbeek, L. S. V.; Berg, G.; et al. The Hidden World Within Plants: Ecological and Evolutionary Considerations for Defining Functioning of Microbial Endophytes. *Microbiol. Mol. Biol. Rev.* **2015**, *79* (3), 293–320.

Hartley, S. E.; Gange, A. C. Impacts of Plant Symbiotic Fungi on Insect Herbivores: Mutualism In A Multitrophic Context. *Ann. Rev. Entomol.* **2009**, 54, 323–342.

Hollants, J.; Leroux, O.; Leliaert, F.; et al. Who Is In There? Exploration of Endophytic Bacteria Within the Siphonous Green Seaweed *Bryopsis* (Bryopsidales, Chlorophyta). *PLoS One* **2011**, *6* (10):e26458.

Hu, Z. Y.; Li, Y. Y.; Huang, Y. J.; et al. Three New Sesquiterpenoids from *Xylaria* sp. NCY2. *Helv. Chim. Acta* **2008**, *91* (1), 46–52.

Jalgaonwala, R. E.; Mohite, B. V.; Mahajan, R. T. A Review: Natural Products from Plant Associated Endophytic Fungi. *J. Microbiol. Biotechnol. Res.* **2017**, *1* (2), 21–32.

Joseph, B.; Priya, R. M. Bioactive Compounds from Endophytes and their Potential in. *Am. J. Biochem. Mol. Biol.* **2011**, *1*, 291–309.

Kaewkla, O.; Franco, C. M. M. Rational Approaches to Improving the Isolation of Endophytic Actinobacteria from Australian Native Trees. *Microbial Ecol.* **2013**, *65* (2), 384–393.

Kamat, S.; Kumari, M.; Taritla, S.; Jayabaskaran, C. Endophytic Fungi of Marine Alga from Konkan Coast, India—A Rich Source of Bioactive Material. *Front. Mar. Sci.* **2020**, *7*, 31.

Kumar, A.; Patil, D.; Rajamohanan, P. R.; Ahmad, A. Isolation, Purification and Characterization of Vinblastine and Vincristine from Endophytic Fungus *Fusarium oxysporum* Isolated from *Catharanthus roseus*. *PLoS One* **2013**, *8* (9):e71805.

Kumar, V.; Jain, L.; Jain, S. K.; et al. Bacterial Endophytes of Rice (*Oryza sativa* L.) and Their Potential for Plant Growth Promotion and Antagonistic Activities. *S. Afr. J. Bot.* **2020**, *134*: 50–63.

Le, H. T. T.; Padgham, J. L.; Sikora, R. A. Biological Control of the Rice Root-Knot Nematode *Meloidogyne graminicola* on Rice, Using Endophytic and Rhizosphere Fungi. *Intl. J. Pest. Manag.* **2009**, *55* (1), 31–36.

Liu, S. S.; Jiang, J. X.; Huang, R.; et al. A new antiviral 14-nordrimane sesquiterpenoid from an endophytic fungus *Phoma* sp. *Phytochem. Lett.* **2019**, *29*, 75–78.

Menpara, D.; Chanda, S. Endophytic Bacteria-Unexplored Reservoir of Antimicrobials for Combating Microbial Pathogens. In *Microbial Pathogens and Strategies for Combating them: Science, Technology and Education*, 2013, pp 1095–1103.

Miliute, I.; Buzaite, O.; Baniulis, D.; Stanys, V. Bacterial Endophytes in Agricultural Crops and Their Role in Stress Tolerance: A Review. *Zemdirbyste-Agri* **2015**, *102* (4), 465–478.

Mitter, B.; Petric, A.; Shin, M. W.; et al. Comparative Genome Analysis of *Burkholderia phytofirmans* PsJN Reveals A Wide Spectrum of Endophytic Lifestyles Based on Interaction Strategies with Host Plants. *Front. Plant Sci.* **2013**, *4*, 120.

Nadeem, M.; Ram, M.; Alam, P.; et al. *Fusarium solani*, P1, A New Endophytic Podophyllotoxin-Producing Fungus from Roots of *Podophyllum hexandrum*. *Afr. J. Microbiol. Res.* **2012**, *6* (10), 2493–2499.

Omojate, G. C.; Felix, O. E.; Augustina, O. J.; Christopher, O. E. Mechanisms of Antimicrobial Actions of Phytochemicals Against Enteric Pathogens–A Review. *J. Pharm. Chem. Biol. Sci.* **2014**, *2* (2), 77–85.

Parthasarathi, S.; Sathya, S.; Bupesh, G.; et al. Isolation and Characterization of Antimicrobial Compound from Marine *Streptomyces hygroscopicus* BDUS 49. *World J. Fish. Mar. Sci.* **2012**, *4* (3), 268–277.

Puri, S. C.; Verma, V.; Amna T.; et al. An Endophytic Fungus from *Nothapodytes foetida* that Produces Camptothecin. *J. Nat. Prod.* **2005,** *68*, (12), 1717–1719.

Raekiansyah, M.; Mori, M.; Nonaka, K.; et al. Identification of Novel Antiviral of Fungus-Derived Brefeldin A Against Dengue Viruses. *Trop. Med. Health* **2017**, *45* (1), 32.

Ran, X.; Zhang, G.; Li, S.; Wang, J. Characterization and Antitumor Activity of Camptothecin from Endophytic Fungus *Fusarium solani* Isolated from *Camptotheca acuminate*. *Afr. Health Sci.* **2017**, *17* (2), 566–574.

Reinhold-Hurek, B.; Hurek, T. Life in Grasses: Diazotrophic Endophytes. *Trends Microbil.* **1998**, *6* (4), 139–144.

Roopa, G.; Madhusudhan, M. C.; Sunil K. C. R.; et al. Identification of Taxol-producing Endophytic Fungi Isolated from *Salacia oblonga* Through Genomic Mining Approach. *J. Genet. Eng. Biotechnol.* **2015**, *13* (2), 119–127.

Sahoo, S.; Sarangi, S.; Kerry, R. G. Bioprospecting of Endophytes for Agricultural and Environmental Sustainability.In *Microbial Biotechnology;* J. K. Patra, et al. Eds.; Springer 2017, pp 429–458.

Saikkonen, K.; Saari, S.; Helander, M. Defensive Mutualism Between Plants and Endophytic Fungi? *Fungal Divers.* **2010**, *41* (1), 101–113.

Schardl, C. L.; Craven, K. D. Interspecific Hybridization in Plant-associated Fungi and Oomycetes: A Review. *Mol. Ecol.* **2003**, *12* (11), 2861–2873.

Schulz, B.; Boyle, C.; Draeger, S.; et al. Endophytic Fungi: A Source of Novel Biologically Active Secondary Metabolites. *Mycol. Res.* **2002**, *106* (9), 996–1004.

Schulz, B.; Römmert, A. K.; Dammann, U.; et al. The Endophyte–Host Interaction: A Balanced Antagonism? *Mycol. Res.* **1999**, *103* (10), 1275–1283.

Segaran, G.; Sundar, R. D. V.; Settu, S.; et al. A Review on Endophytic Actinomycetes and Their Applications. *J. Chem. Pharm. Res.* **2017**, *9* (10), 152–158.

Shukla, S. T.; Habbu, P. V.; Kulkarni, V. H.; et al. Endophytic Microbes: A Novel Source for Biologically/Pharmacologically Active Secondary Metabolites. *Asian J. Pharmacol. Toxicol.* **2014**, *2* (3), 1–6.

Singh, H.; Naik, B.; Kumar, V.; Bisht, G. S. Screening of Endophytic Actinomycetes for Their Herbicidal Activity. *Ann. Agrar. Sci.* **2018**, *16* (2), 101–107.

Singh, R.; Dubey, A. K. Endophytic Actinomycetes As Emerging Source for Therapeutic Compounds. *Indo Global J. Pharm. Sci.* **2015**, *5*, 106–116.

Smith, C. A.; O'Maille, G.; Want, E. J.; et al. METLIN: A Metabolite Mass Spectral Database. *Ther. Drug Monit.* **2005**, *27* (6), 747–751.

Song, X. Q.; Zhang, X.; Han, Q. J.; et al. Xanthone Derivatives from *Aspergillus sydowii*, An Endophytic Fungus from the Liverwort *Scapania ciliata* S. Lac and Their Immunosuppressive Activities. *Phytochem. Lett.* **2013**, *6* (3), 318–321.

Soujanya, K. N.; Siva, R.; Kumara, P. M.; et al. Camptothecin-Producing Endophytic Bacteria from *Pyrenacantha volubilis* Hook.(Icacinaceae): A Possible Role of a Plasmid in the Production of Camptothecin. *Phytomedicine* **2017**, *36*, 160–167.

Specian, V.; Sarragiotto, M. H.; Pamphile, J. A.; Clemente, E. Chemical Characterization of Bioactive Compounds from the Endophytic Fungus *Diaporthe helianthi* Isolated from Luehea Divaricata. *Braz. J. Microbiol.* **2012**, *43* (3), 1174–1182.

Stępniewska, Z.; Kuźniar, A. Endophytic Microorganisms—Promising Applications in Bioremediation of Greenhouse Gases. *Appl. Microbiol. Biotechnol.* **2013**, *97* (22), 9589–9596.

Stierle, A.; Strobel, G.; Stierle, D. Taxol and Taxane Production by *Taxomyces andreanae*, An Endophytic Fungus of *Pacific yew*. *Science* **1993**, *260* (5105), 214–216.

Strobel, G. A. Endophytes as Sources of Bioactive Products. *Microbes Infect.* **2003** *5* (6), 535–544.

Strobel, G. A.; Daisy, B. Bioprospecting for Microbial Endophytes and Their Natural Products. *Microbiol. Mol. Biol. Rev.* **2003**, *67* (4), 491–502.

Subbulakshmi, G. K.; Thalavaipandian, A.; Ramesh, V. Bioactive Endophytic Fungal Isolates of *Biota orientalis* (L) Endl., *Pinus excelsa* Wall. and *Thuja occidentalis* L. *Intl. J. Adv. Life Sci.* **2012**, *4*, 9–15.

Suryanarayanan, T. S.; Govindarajulu, M. B.; Rajamani, T.; et al. Endolichenic Fungi in Lichens of Champawat District, Uttarakhand, Northern India. *Mycol. Progress* **2017**, *16* (3), 205–211.

Turner, T. R.; James, E. K.; Poole, P. S. The Plant Microbiome. *Genome Biol.* **2013**, *14* (6), 209.

Vasilyeva, E. N.; Akhtemova, G. A.; Afonin, A. M.; et al. Cultivated Endophytic Bacteria of the Stems and Leaves of the Common Pea (*Pisum sativum* L.). *Environ. Genet.* **2020**, *18* (2), 169–184.

Wang, W. X.; Lei, X.; Ai, H. L.; et al. Cytochalasans from the Endophytic Fungus *Xylaria* cf. curta with Resistance Reversal Activity Against Fluconazole-Resistant *Candida albicans*. *Org. Lett.* **2019**, *21* (4), 1108–1111.

Wilson, D. Endophyte: The Evolution of A Term, and Clarification of Its Use and Definition. *Oikos* **1995**, *73*, 274–276.

Wishart, D. S.; Knox, C.; Guo, A. C.; et al. HMDB: A Knowledgebase for the Human Metabolome. *Nucleic. Acids Res.* **2008**, *37*, 603–610.

Xiao, J.; Zhang, Q.; Gao, Y. Q.; et al. Secondary Metabolites from the Endophytic *Botryosphaeria dothidea* of *Melia azedarach* and Their Antifungal, Antibacterial, Antioxidant, and Cytotoxic Activities. *J. Agric. Food Chem.* **2014**, *62* (16), 3584–3590.

Zhang, G.; Sun, S.; Zhu, T.; et al. Antiviral Isoindolone Derivatives from An Endophytic Fungus *Emericella* sp. Associated with *Aegiceras corniculatum*. *Phytochemistry* **2011**, *72* (11–12), 1436–1442.

Zhang, H.; Ying, C.; Bai, X. Advancement in Endophytic Microbes from Medicinal Plants. *Intl. J. Pharm. Sci. Res.* **2014**, *5* (5), 1589.

Zhang, J. Y.; Tao L. Y.; Liang, Y. J.; et al. Secalonic Acid D Induced Leukemia Cell Apoptosis and Cell Cycle Arrest of G1 with Involvement of GSK-3β/β-catenin/c-Myc pathway. *Cell Cycle* **2009**, *8* (15), 2444–2450.

Zhao, K.; Penttinen, P.; Guan, T.; et al. The Diversity and Anti-Microbial Activity of Endophytic Actinomycetes Isolated from Medicinal Plants in *Panxi plateau*, China. *Curr. Microbiol.* **2011**, *62* (1), 182–190.

CHAPTER 11

Overview of Endophytic Microbial Community of *Cannabis sativa* and Its Metabolites

PUSHPA RUWALI[1], JIYA CHANOTIYA[2], NIHARIKA PANDEY[1], KRISHIKA SAMBYAL[3], and RAHUL VIKRAM SINGH[4*]

[1]*Department of Biotechnology, M. B. Govt. P. G. College, Haldwani-263139, Uttarakhand, India*

[2]*Department of Biochemistry, College of Basic Sciences and Humanities, GBPUAT, Pantnagar-263145, Uttarakhand, India*

[3]*University Institute of Biotechnology, Chandigarh University, Gharuan, Punjab, India*

[4]*Academy of Scientific and Innovative Research (AcSIR), Ghaziabad-201002, Uttar Pradesh, India*

[*]*Corresponding author. E-mail: rahul.negi121@gmail.com*

ABSTRACT

Cannabis sativa is a potent pharmaceutically important medicinal plant that produces approximately 500 compounds and around 113 cannabinoids. In the Indian traditional medicine system, *C. sativa* has been used as a hypnotic, sedative, hallucinogenic, analgesic, and anti-inflammatory agent. Various microbial communities are associated with this plant contributing to its growth, resistance against pathogens, mineral nutrient uptake and production of plant secondary metabolites. Researchers are paying attention to investigating the major role of mutual partnership between plants and microbes to increase the production of metabolites. Therefore, there is a need to explore the clear role of endophytes in the plant. This chapter aims to summarize the relation between endophytic microbial communities and host *Cannabis sativa*.

11.1 INTRODUCTION

A group of microorganisms that aid in providing numerous benefits to their host plants while residing within them in a mutualistic association without any noticeable symptom of infection are known as endophytes (Botella and Diez, 2011; Scott et al., 2018). Both fungal and bacterial strains are the dominant microbial communities in alliance with plants, the importance of which can be elucidated by their capability to provide physical and chemical defense responses to the host (Kusari et al., 2014a) through induction of host's tolerance to various environmental stress and diseases (Porras-Alfaro and Bayman, 2011; Kusari et al., 2013), by balancing the ecosystem (Hamilton et al., 2012), by production of diverse range of biologically active secondary metabolites (Debbab et al., 2012), etc. A comparable plant possessing similar multifunctional properties with a number of advantageous and beneficial endophytes is *Cannabis sativa,* which originated from Central Asia and domesticated for over 5000 years now (Booth and Bohlmann, 2019). It belongs to the family Cannabaceae (Fig. 11.1) with well-known ethnobotanical and medicinal properties. A few critical factors influencing the growth of *C. sativa* include soil fertility, light intensity, temperature, photoperiod, humidity, phytohormones, and microbiome (Winston et al., 2014; Jalali et al., 2019; Eichhorn Bilodeau et al., 2019).

Kingdom	Plantae
Division	Magnoliophyta
Class	Magnoliopsida
Order	Rosales
Family	Cannabaceae
Genus	*Cannabis*
Species	*Cannabis sativa*

FIGURE 11.1 Morphology and biosystematics of *Cannabis sativa.*

The diversity and role of *C. sativa*-associated endophytes have not been studied previously to a large extent. There is an abundance of microbes associated with the plant but the most common ones include bacterial genera, such *as Bacillus*, *Pseudomonas*, *Pantoea*, and *Staphylococcus, and fungal genera, such as Alternaria, Aureobasidium, Cochliobolus, Aspergillus*, and *Penicillium* (Gautam et al., 2013; Scott et al., 2018). The plant is also widely acknowledged for the occupancy of various nitrogenous compounds, terpenes, flavonoids, etc. among which Cannabinoids are unique to this genus with active drug ingredients (Gautam et al., 2013). There are approximately more than 560 phytochemicals been identified from *Cannabis* representing variant chemical classes (Gould, 2015; Radwan et al., 2017), such as fatty acids, steroids, and amino acids represent the primary metabolites, while stilbenoids, terpenoids, flavonoids, alkaloids, lignans, and cannabinoids belong to secondary metabolites (Flores-Sanchez and Verpoorte, 2008; Andre et al., 2016). Cannabinoids accumulate as cannabinoid acids in *Cannabis* plant and get nonenzymatically decarboxylized into their neutral forms during storage. Most of the metabolites are present in *Cannabis* resin with Δ9-tetrahydrocannabinolic acid (THCA) and cannabidiolic acid (CBDA) as the predominant cannabinoids. Others include cannabinol (CBN), cannabidiol (CBD), cannabichromene (CBC), and cannabigerol (CBG). (Taghinasab and Jabaji, 2020). The terpenes myrcene, β-caryophyllene, and α-humulene are commonly present in most of the strains of *Cannabis.* Some other terpenes such as α-pinene, linalool, limonene, bisabolol, and (*E*)-β-farnesene are also found in the resin and isoprenoid moiety of the cannabinoid structure. These originate from the mevalonic acid pathway in the cytosol and the methylerythritol phosphate pathway in plastids through the isoprenoid biosynthetic system (Booth and Bohlmann, 2019). The common members of *C. sativa* include marijuana (drug type) with 1.0–20% of Δ9-tetrahydrocannabinol (THC), intermediate type with 0.3–1.0% of THC, and fiber-rich hemp with less than 0.3% of THC. These varieties are cultivated for medicinal purposes, fiber, seed, oil, etc. (Guo et al., 2017). They provide resistance against insects and parasites, which is assured by the bioactive secondary metabolites secreted by them (Booth et al., 2017). There is a recent growth in the hemp cultivation due to effective use as a source of green pesticides active against ectoparasites (Tabari et al., 2020). The traditional use of *Cannabis* phytocannabinoids for the treatment of headache, menstrual irregularities, insomnia, panting, gout, cough, etc. has been observed in the Chinese medicine system, while the Indian traditional medicine portrays the plant to be useful as nervous system stimulant, digestion stimulant, analgesic, diuretic, aphrodisiac, antiviral, antiparasitic, sedative, and is good for skin (Nuutinen, 2018). All these beneficiaries have paved the path toward the development of various *Cannabis*-based medicines such as Dronabinol (Marinol®, Solvay Pharmaceuticals, Belgium), Sativex (GW

Pharmaceuticals, UK), and Nabilone (Cesamet®, Valeant Pharmaceuticals International, USA). *Cannabis* is a depot of compounds that serve various purposes with respect to its legality for medicinal, recreational, and other usages depending country-wise (Gutierrez and Hohmann, 2011).

11.2 ROLE OF MICROBES IN METABOLITE PRODUCTION FROM *CANNABIS SATIVA*

C. sativa has a partnership with microbial communities both inside and outside its structure contributing toward the betterment of plant's life cycle by providing nutrients, defense, protection, fitness, etc. However, the main factors that affect agricultural plant microbiota are the plant species and genotype itself, followed by environmental factors, developmental stages, etc. (Nunes da Rocha et al., 2009). But the complex relation between *Cannabis* and microbial families has not been unfolded to its maximum yet (Liste and Prutz, 2006; Taghinasab and Jabaji, 2020). Therefore, the role of endophytes (Fig. 11.2) as cannabinoid yielders and contributors toward other ecological services must be explored.

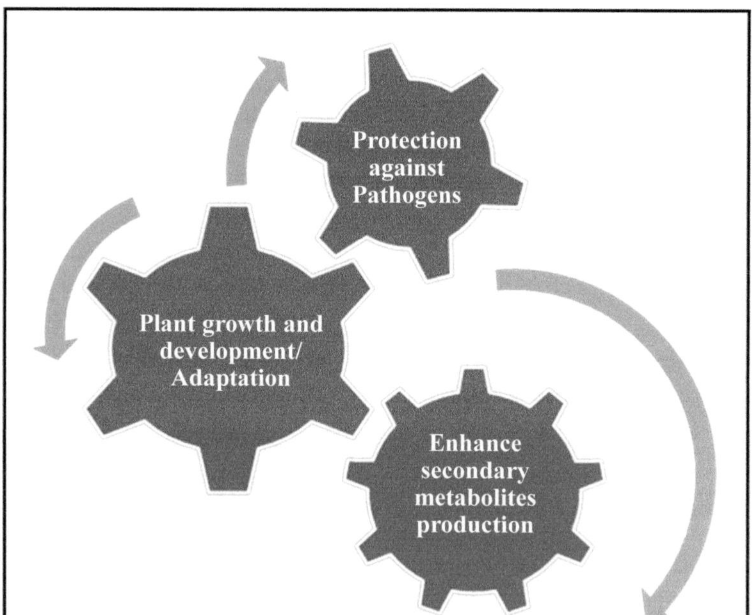

FIGURE 11.2 Role of endophytes in *Cannabis sativa*.

There is an abundance of secondary metabolites present in *C. sativa*, which are specifically produced as an adaptation to defense or stress response, thereby improving plant growth (Gonçalves et al., 2019). Presently, these metabolites are prominent for their diverse use in human medicine making under pharmaceutical industries (Deshmukh et al., 2018). Agricultural practices of *C. sativa* can be improved to a higher extent due to the microbial partnership benefitting the plant (Winston et al., 2014). Both fungal and bacterial endophytic flora is found in leaf, petiole, seed, bud, and root of the *C. sativa* (Tables 11.1 and 11.2). Some fungal genera contributing toward the plant fitness and metabolite production includes *Cochliobolus*, *Aureobasidium*, *Phoma*, *Rhizopus*, *Colletotrichum*, *Cladosporium*, *Alternaria*, *Aureobasidium*, *Cochliobolus*, *Aspergillus*, and *Penicillium* and few bacterial genera are *Pseudomonadaceae*, *Chryseobacterium*, *Oxalobacteraceae*, *Xanthomonadaceae*, *Sphingobacteriales*, *Bacillus*, *Acinetobacter*, *Pantoea*, *Staphylococcus*, *Pseudomonas*, and *Enterobacter* (Gautam et al., 2013; Afzal et al., 2015; Scott et al., 2018; Taghinasab and Jabaji, 2020). All these *Cannabis*-associated microbial communities help to yield a high amount of hemp and marijuana, promote infection and disease resistance, as well as modulate secondary metabolite production. Among the microbial communities, genera in relation to the roots and their surroundings are designated as rhizospheric microbes that act as the first line of defense in the host plant against root pathogens. Hence, they majorly contribute to the activation of secondary metabolite biosynthetic gene clusters for enhancing levels of defense metabolites (Berg et al., 2014; Carrión et al., 2019). Both gram-positive and gram-negative endophytic bacteria are responsible for suppressing phytopathogens (Lugtenberg and Kamilova, 2009). Some important phytochemical combinations such as quercetin and flavonoids are worth making *Cannabis* a potent antioxidant plant (McPartland and Russo, 2001). Some bacterial endophytes such as *Bacillus megaterium*, *Brevibacillus borstelensis*, and other *Bacillus* sp. defend plant against pathogens and prevent them from developing resistance to bioactive metabolites secreted by plant/endophytes through a quorum quenching strategy, which disrupts the quorum sensing signals in the cells of target organism (Kusari et al., 2014b). There are fungal species such as *Curvularia* sp., which enhance abiotic stress tolerance by providing thermal protection to host (Redman et al., 2002). The arbuscular mycorrhizal fungi such as *Diversispora* sp., *Funneliformis mosseae*, and *Glomus caledonium* help the host in tolerating contaminated soil with sewage sludge and phosphogypsum enabling the plant to respond accurately for biomass production (Zielonka et al., 2019).

11.3 ENDOPHYTES OF *CANNABIS SATIVA*

11.3.1 BACTERIAL

Different bacterial communities get associated with *C. sativa*, which mostly belong to a category of γ-proteobacteria and α-proteobacteria. These endophytes get connected with different tissues/parts of the plant from root to shoot, including seeds. The presence of endophytes varies in different parts of the plant (Table 11.1). It has been observed that the same bacteria can associate in a number of ways to different parts of the plant as commonly seen in genera of *Pseudomonas,* which forms the relationship with leaf, petiole, and seed of the plant (Taghinasab and Jabaji, 2020) (Table 11.1).

TABLE 11.1 Few Bacterial Endophytes in Association with *Cannabis sativa*.

S. no.	Tissue/Part	Bacteria	References
1.	Leaf	*Pseudomonas fulva*	Taghinasab and Jabaji (2020)
		Bacillus licheniformis	Taghinasab and Jabaji (2020)
		Bacillus subtilis	Taghinasab and Jabaji (2020)
		Bacillus pumilus	Taghinasab and Jabaji (2020)
		Bacillus megaterium	Taghinasab and Jabaji (2020)
		Achromobacter	Iqbal et al. (2018)
		Alcaligenes	Iqbal et al. (2018)
		Acinetobacter pittii	Iqbal et al. (2018)
		Acinetobacter gyllenbergii	Iqbal et al. (2018)
		Acinetobacter nosocomialis	Iqbal et al. (2018)
		Acinetobacter parvus	Iqbal et al. (2018)
2.	Petiole	*Pantoea*	Taghinasab and Jabaji (2020)
		Staphylococcus	Taghinasab and Jabaji (2020)
		Pseudomonas putida	Taghinasab and Jabaji (2020)
		Pseudomonas fluorescens	Taghinasab and Jabaji (2020)
		Paenibacillus tundrae	Iqbal et al. (2018)
		Paracoccus marcusii	Iqbal et al. (2018)
		Planomicrobium chinense	Afzal et al. (2015)
		Streptomyces eurocidicus	Afzal et al. (2015)
		Agrobacterium	Taghinasab and Jabaji (2020)
		Brevibacterium	Scott et al. (2018)
		Curtobacterium	McPartland (2000)
		Nocardioides albus	Afzal et al. (2015)
		Nocardioides kongjuensis	Afzal et al. (2015)

TABLE 11.1 *(Continued)*

S. no.	Tissue/Part	Bacteria	References
3.	Seed	*Enterobacter*	Taghinasab and Jabaji (2020)
		Herbaspirillum seropedicae	McPartland (2000)
		Exiguobacterium indicum	Iqbal et al. (2018)
		Pseudomonas orientalis	Taghinasab and Jabaji (2020)
		P. putida	Taghinasab and Jabaji (2020)
		P. fluorescens	Taghinasab and Jabaji (2020)
4.	Root	*Acinetobacter*	Taghinasab and Jabaji (2020)
		Azospirillum brasilense	McPartland (2000)
		Gluconacetobacter diazotrophicus	McPartland (2000)
		Chryseobacterium	Taghinasab and Jabaji (2020)
		Enterobacter asburiae	Afzal et al. (2015)
		Enterobacter casseliflavus	Afzal et al. (2015)
		Microbacterium	Taghinasab and Jabaji (2020)
		Rhizobiales	Taghinasab and Jabaji (2020)
		Cedecea	Scott et al. (2018)
		Erwinia	Scott et al. (2018)
5.	Bud	*Mycobacterium*	Taghinasab and Jabaji (2020)
		Acinetobacter	Scott et al. (2018)
		Paenibacillus	Scott et al. (2018)
		Xanthomonas gardneri	Afzal et al. (2015)
		Stenotrophomonas	Scott et al. (2018)
		Serratia marcescens	Iqbal et al. (2018)

11.3.2 FUNGAL

Alike bacterial endophytes, fungal endophytes are also associated with different tissues/parts of the plant (Table 11.2). The major fungal endophytes belong to the *Ascomycetes* fungal class, while Basidiomycetes class such as *Irpex* and *Schizophyllum commune* has also been observed (Taghinasab and Jabaji, 2020). The dominant fungal endophytes are related to leaf followed by petiole of the plant.

11.4 ENDOPHYTIC BENEFITS TO *CANNABIS SATIVA*

Endophytic microorganisms, including bacterial and fungal genera associated with plants, promise diverse potentials of providing fitness benefit to *C. sativa.*

TABLE 11.2 Few Fungal Endophytes in Association with *Cannabis sativa.*

S. no	Tissue/Part	Fungal endophytes	References
1.	Leaf	*Aspergillus*	Taghinasab and Jabaji (2020)
		Penicillium	Taghinasab and Jabaji (2020)
		Phoma	Taghinasab and Jabaji (2020)
		Rhizopus	Taghinasab and Jabaji (2020)
		Colletotrichum	Taghinasab and Jabaji (2020)
		Cladosporium	Taghinasab and Jabaji (2020)
		Curvularia	Taghinasab and Jabaji (2020)
		Cochliobolus	Taghinasab and Jabaji (2020)
		Pezizomycetes	Scott et al. (2018)
		Sordariomycetes	Scott et al. (2018)
		Cryptococcus	Scott et al. (2018)
		Dothideomycetes	Scott et al. (2018)
		Chaetomium globosum	Kusari et al. (2013)
		Sebacina vermifera	Lubna et al. (2019)
		Piriformospora	Lubna et al. (2019)
		Bipolaris	Lubna et al. (2019)
		Porostereum spadiceum	Lubna et al. (2019)
		Penicillium citrinum	Lubna et al. (2019)
		Gibberella fujikuroi	Lubna et al. (2019)
		Neurospora crassa	Lubna et al. (2019)
		Aureobasidium	Taghinasab and Jabaji (2020)
2.	Twig	*Penicillium rubidurum*	Taghinasab and Jabaji (2020)
		C. globosum	Taghinasab and Jabaji (2020)
		Paecilomyces lilacinus	Taghinasab and Jabaji (2020)
3.	Stem	*Rhizopus stolonifer*	Taghinasab and Jabaji (2020)
		Alternaria alternate	Taghinasab and Jabaji (2020)
		Cladosporium	Taghinasab and Jabaji (2020)
4.	Petiole	*Cryptococcus*	Taghinasab and Jabaji (2020)
		Sclerotinia sclerotiorum	Scott et al. (2018)
		Botrytis cinerea	Scott et al. (2018)
		Rhizoctonia	Scott et al. (2018)
		Trichoderma virens	Scott et al. (2018)
		Colletotrichum gloeosporioides	Scott et al. (2018)
		Stachybotrys	Scott et al. (2018)
		Helminthosporium solani	Scott et al. (2018)
		Fusarium solani	Scott et al. (2018)
		Fusarium graminearum	Scott et al. (2018)
		Alternaria	Kusari et al. (2013)

TABLE 11.2 *(Continued)*

S. no	Tissue/Part	Fungal endophytes	References
5.	AM (Arbuscular mycorrhizal)	*Diversispora*	Taghinasab and Jabaji (2020)
		Funneliformis mosseae	Taghinasab and Jabaji (2020)
		Funneliformis geosporum	Taghinasab and Jabaji (2020)
		Glomus caledonium	Taghinasab and Jabaji (2020)
		Glomus occultum	Taghinasab and Jabaji (2020)
6.	Seed	*Aureobasidium*	Taghinasab and Jabaji (2020)
7.	Bud	*Penicillium copticola*	Taghinasab and Jabaji (2020)
		P. lilacinus	Kusari et al. (2013)
		Paecilomyces sumatrense	Kusari et al. (2013)
		Paecilomyces meleagrinum	Kusari et al. (2013)
		Aspergillus versicolor	Kusari et al. (2013)

11.4.1 ENHANCES SECONDARY METABOLITES PRODUCTION

Ample evidences reveal that endophytes have a high potential of triggering plant responses resulting in the production of secondary metabolites in the host (Fig. 11.3) (Pandey et al., 2016). These secondary metabolites are said to mimic the action of the plant's metabolites or generate the host plant compounds that activate signaling pathways targeting the transformation of secondary plant metabolites (Kusari et al., 2017). Plant microbiomes persuade the production of plant hormones and regulators such as abscisic acid, ethephon, cycocel, salicylic acid, γ-aminobutyric acid (GABA), mevinolin, and gibberellins, which further enhance the productivity and functionality of *Cannabis* secondary metabolites (Mansouri and Salari, 2014; Jalali et al., 2019). The cannabinoids, including THC, CBN, and CBD, express their potentials during *Cannabis* stress responses (Mansouri et al., 2013). Cannabinoids are predominantly produced and accumulated in glandular trichomes, which are the hair-like epidermal projections heavily concentrated in the flowers of *Cannabis* plants (Grof, 2018).

11.4.2 BOOSTS PLANT TOLERANCE

In association with endophytes of both bacteria and fungi, the tolerance level of plant gets boosted several folds to bear various adverse environmental or pathological conditions. Different mechanisms operate to withstand these changes, which are as follows:

i) Abundant increase in nitrogen fixation: Nitrogen metabolism is suitably increased, which can be utilized for the formation of various proteins and phyto-compounds in plant.
ii) Siderophore formation: The plant root is enriched mainly with iron through chelating process, which may further be required for various physiological metabolisms of the plant.
iii) Mineral Solubilization: An increase in phosphorus and calcium solubilization occurs, which is a necessity for physiological and biochemical processes of plant, resulting in the profound enhancement of the tolerance level of plant (Berg and Smalla, 2009; Compant et al., 2019).

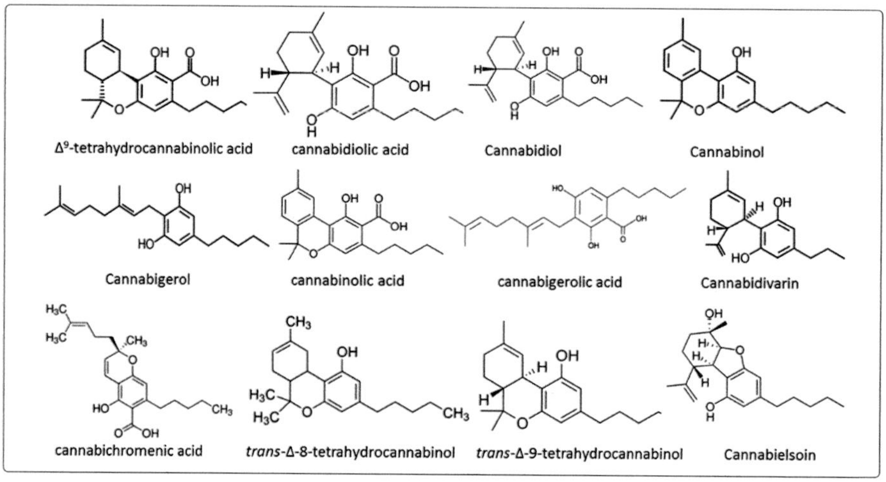

FIGURE 11.3 Secondary metabolites of Cannabis sativa.

11.4.3 ENHANCES PHYTOHORMONE PRODUCTION

There are various evidences that state the increased phytohormone concentration in plant via relation to endophytes. The production of auxin (IAA), Gibberellic acid, cytokinin, and abscisic acid majorly speeds up due to endophytic association (Koberl et al., 2013). Gibberellic acid when applied at a concentration of 100 μM enhances the amount of THC and CBD in the plant tissue. One reasonable hypothesis for this action is that the application of Gibberellic acid adds to the regulation of 1-aminocyclopropane-1-carboxylic acid, ramping the level of ethylene, thus leading to increased THC and CBD content (Mansouri et al., 2011).

11.4.4 PROMOTES GROWTH VIA MULTISPECIES CONSORTIUM

Various consortium formed by multiple bacterial species positively extrapolate the plant growth such as *Gluconacetobacter diazotrophicus*, *Azospirillum brasilense*, *Burkholderia ambifaria*, and *Herbaspirillum seropedicae*, which extremely raises the hemp biomass and height. Consortium not only enhances plant growth but remarkably increases the accumulation of secondary metabolite level, most often seen in CBD, THC, and terpene contents (Botta et al., 2013).

11.4.5 OLERANCE TO HEAVY METALS

In association with AM fungal endophytes, the plant increases tolerance against different heavy metals like Cd, Ni, and Cr, reducing the intoxication (Citterio et al., 2005).

11.4.6 DEFENSE SYSTEM AGAINST PATHOGENS

Plant microbiomes linked with hemp and marijuana display antagonistic activity against the invading pathogens (Scott et al., 2018; Afzal et al., 2015). In dual confrontation assays conducted by Scott et al. (2018), the hemp-linked strains of *Pseudomonas fulva* (BTC6-3 and BTC8-1) and *Pseudomonas orientalis* (BTG8-5 and BT14-4) displayed antifungal action against *Botrytis*. *Pseudomonas* species employ a number of strategies for acting antifungal species, such as

- Release HCN
- Release cellulose, lipopeptides
- Release diffusible antibiotics like PCA, DPAG, pyocyanine, pyoluteorin
- Release volatile compound alike HCN (Haas and Keel, 2003).

All these characters make *Pseudomonas* species act like an effective biocontrol agent. Other *Cannabis* endophytes such as *Paecilomyces lilacinus* A3, *Penicillium* sp. T6, and *Penicillium copticola* L3 effectively hold back the growth of *Cannabis* pathogens such as *Botrytis cinerea* and *Trichothecium roseum* (Kusari et al., 2013).

11.4.7 *QUENCH QUORUM SENSING STRATEGY*

The bacterial endophytes of *Bacillus* seem to employ astonishing machinery to provide defense in plant. It targets the quorum signaling in pathogen microbe, which is the primary mechanism for cell-to-cell communication and determines the target site in the host cell. Common bacterial endophytes that target quorum signaling are *B. megaterium* B4, *B. borstelensis* B8, *Bacillus* sp. B11, *and Bacillus* sp. B3 (Kusari et al., 2014). In a nutshell, these studies unwind the potentials of endophytes for being promising biocontrol agents. These are an ideal substitute for the chemical-based fertilizers and, hence, maintain low pesticide residue levels in the *Cannabis* plant (Nate, 2019).

11.5 CONCLUSION

In the present scenario, biochemical and pharmacological characteristics of *C. sativa* have been acknowledged in various fields of medicine making as an economically valuable crop. The endophytic association in *C. sativa* has been determined, which shows a mutualistic relationship providing a prominent advantage to the plant in terms of increased stress tolerance, plant growth, increased yield, abundant biomass, and positive response in the secondary metabolite profile of the plant, thus, forming the basis for maximum biochemical and pharmacological properties by the activation of numerous phyto-compounds. However, only basic and preliminary knowledge regarding the endophyte mechanism is known till date; therefore, an elaborated aspect of various molecular and biochemical studies can be furthermore a beneficial task.

KEYWORDS

- ***Cannabis sativa***
- **secondary metabolites**
- **endophytes**
- **cannabinoids**
- **traditional medicine system**

REFERENCES

Afzal, I. M. R. A. N.; Shinwari, Z. K.; Iqrar, I. Selective Isolation and Characterization of Agriculturally Beneficial Endophytic Bacteria from Wild Hemp Using Canola. *Pak. J. Bot.* **2015,** *47* (5), 1999–2008.

Andre, C. M.; Hausman, J. F.; Guerriero, G. *Cannabis sativa*: The Plant of the Thousand and One Molecules. *Front. Plant Sci.* **2016,** *7*, 19.

Berg, G.; Smalla, K. Plant Species and Soil Type Cooperatively Shape the Structure and Function of Microbial Communities in the Rhizosphere. *FEMS Microbiol. Ecol.* **2009,** *68* (1), 1–13.

Berg, G.; Grube, M.; Schloter, M.; Smalla, K. Unraveling the Plant Microbiome: Looking Back and Future Perspectives. *Front. Microbiol.* **2014,** *5*, 148.

Booth, J. K.; Bohlmann, J. Terpenes in *Cannabis sativa* – From Plant Genome to Humans. *Plant Sci.* **2019,** *284*, 67–72.

Booth, J. K.; Page, J. E.; Bohlmann, J. Terpene Synthases from *Cannabis sativa*. *PLoS One* **2017,** *12* (3), 0173911.

Botella, L.; Diez, J. J. Phylogenic Diversity of Fungal Endophytes in Spanish Stands of *Pinus halepensis*. *Fungal Diversity* **2011,** *47* (1), 9–18.

Botta, A. L.; Santacecilia, A.; Ercole, C.; Cacchio, P.; Del Gallo, M. *In Vitro* and *In Vivo* Inoculation of Four Endophytic Bacteria on *Lycopersicon esculentum*. *New Biotechnol.* **2013,** *30*, 666–674.

Carrión, V. J.; Perez-Jaramillo, J.; Cordovez, V.; et al. Pathogen-Induced Activation of Disease-Suppressive Functions in the Endophytic Root Microbiome. *Science* **2019,** *366* (6465), 606–612.

Citterio, S.; Prato, N.; Fumagalli, P.; et al. The Arbuscular Mycorrhizal Fungus *Glomus mosseae* Induces Growth and Metal Accumulation Changes in *Cannabis sativa* L. *Chemosphere* **2005,** *59*, 21–29.

Compant, S.; Samad, A.; Faist, H.; Sessitsch, A. A Review on the Plant Microbiome: Ecology, Functions, and Emerging Trends in Microbial Application. *J. Adv. Res.* **2019,** *19*, 29–37.

Debbab, A.; Aly, A. H.; Proksch, P. Endophytes and Associated Marine Derived Fungi—Ecological and Chemical Perspectives. *Fungal Diversity* **2012,** *57* (1), 45–83.

Deshmukh, S. K.; Gupta, M. K.; Prakash, V.; Reddy, M. S. Mangrove-Associated Fungi: A Novel Source of Potential Anticancer Compounds. *J. Fungi* **2018,** *4* (3), 101.

Eichhorn Bilodeau, S.; Wu, B. S.; Rufyikiri, A. S.; MacPherson, S.; Lefsrud, M. An Update on Plant Photobiology and Implications for Cannabis Production. *Front. Plant Sci.* **2019,** *10*, 296.

Flores-Sanchez, I. J.; Verpoorte, R. Secondary Metabolism in Cannabis. *Phytochem. Rev.* **2008,** *7* (3), 615–639.

Gautam, A. K.; Kant, M.; Thakur, Y. Isolation of Endophytic Fungi from *Cannabis sativa* and Study Their Antifungal Potential. *Arch. Phytopathol. Plant Prot.* **2013,** *46* (6), 627–635.

Gonçalves, J.; Rosado, T.; Soares, S.; et al. Cannabis and Its Secondary Metabolites: Their Use as Therapeutic Drugs, Toxicological Aspects, and Analytical Determination. *Medicines* **2019,** *6* (1), 31.

Gould, J. The Cannabis Crop. *Nature* **2015,** *525* (7570), S2–S3.

Grof, C. P. L. Cannabis, from Plant to Pill. *Br. J. Clin. Pharmacol.* **2018,** *84* (11), 2463–2467.

Guo, T. T.; Zhang, J. C.; Zhang, H.; et al. Bioactive Spirans and Other Constituents from the Leaves of *Cannabis sativa* f. Sativa. *J. Asian Nat. Prod. Res.* **2017,** *19* (8), 793–802.

Gutierrez, T.; Hohmann, A. G. Cannabinoids for the Treatment of Neuropathic Pain Are They Safe and Effective?. *Future Neurol.* **2011,** *6* (2), 129–133.

Haas, D.; Keel, C. Regulation of Antibiotic Production in Root-Colonizing *Pseudomonas* spp. and Relevance for Biological Control of Plant Disease. *Ann. Rev. Phytopathol.* **2003,** *41*, 117–153.

Hamilton, C. E.; Gundel, P. E.; Helander, M.; Saikkonen, K. Endophytic Mediation of Reactive Oxygen Species and Antioxidant Activity in Plants: A Review. *Fungal Diversity* **2012,** *54* (1), 1–10.

Iqbal, A.; Arsad, M.; Hashmi, I. Biodegradation of Phenol and Benzene by Endophytic Bacterial Strains Isolated from Refinery Waste Water-Fed *Cannabis sativa*. *Environ. Technol.* **2018,** *39* (13), 1–35.

Jalali, S.; Salami, S. A.; Sharifi, M.; Sohrabi, S. Signaling Compounds Elicit Expression of Key Genes in Cannabinoid Pathway and Related Metabolites in Cannabis. *Ind. Crop Prod.* **2019,** *133*, 105–110.

Koberl, M.; Schmidt, R.; Ramadan, E. M.; Bauer, R.; Berg, G. The Microbiome of Medicinal Plants: Diversity and Importance for Plant Growth, Quality and Health. *Front. Microbiol.* **2013,** 4, 400.

Kusari, P.; Kusari, S.; Spiteller, M.; Kayser, O. Endophytic Fungi Harbored in *Cannabis sativa* L.: Diversity and Potential as Biocontrol Agents against Host Plant-Specific Phytopathogens. *Fungal Diversity* **2013,** *60* (1), 137–151.

Kusari, P.; Spiteller, M.; Kayser, O; Kusari, S. Recent Advances in Research on Cannabis sativa L. Endophytes and Their Prospect for the Pharmaceutical Industry. In *Microbial Diversity and Biotechnology in Food Security*; Springer: New Delhi, 2014a; pp 3–15.

Kusari, P.; Kusari, S.; Lamshöft, M.; Sezgin, S.; Spiteller, M.; Kayser, O. Quorum Quenching Is an Antivirulence Strategy Employed by Endophytic Bacteria. *Appl. Microbiol. Biotechnol.* **2014b,** *98* (16), 7173–7183.

Kusari, P.; Kusari, S.; Spiteller, M.; Kayser, O. Cannabis Endophytes and Their Application in Breeding and Physiological Fitness. In *Cannabis sativa L.—Botany and Biotechnology*, 2nd ed.; Chandra, S., Lata, H., El Sohly, M.A., Eds.; Springer International Publishing: Cham, Switzerland, 2017; pp 419–437.

Liste, H. H.; Prutz, I. Plant Performance, Dioxygenase-Expressing Rhizosphere Bacteria, and Biodegradation of Weathered Hydrocarbons in Contaminated Soil. *Chemosphere* **2006,** *62* (9), 1411–1420.

Lubna, S.; Khan, S.; Latif, A. L.; et al. Growth-Promoting Bioactivities of *Bipolaris* sp. CSL-1 Isolated from *Cannabis sativa* Suggest a Distinctive Role in Modifying Host Plant Phenotypic Plasticity and Functions. *Acta Physiol. Plant.* **2019,** *41* (65), 1–16.

Lugtenberg, B.; Kamilova, F. Plant-Growth-Promoting Rhizobacteria. *Ann. Rev. Microbiol.* **2009,** *63*, 541–556.

Mansouri, H.; Salari, F. Influence of Mevinolin on Chloroplast Terpenoids in *Cannabis sativa*. *Physiol. Mol. Biol. Plants* **2014,** *20* (2), 273–277.

Mansouri, H.; Asrar, Z.; Amarowicz, R. The Response of Terpenoids to Exogenous Gibberellic Acid in *Cannabis sativa* L. at Vegetative Stage. *Acta Physiol. Plant.* **2011,** *33*, 1085–1091.

Mansouri, H.; Salari, F.; Asrar, Z. Ethephon Application Stimulates Cannabinoids and Plastidic Terpenoids Production in *Cannabis sativa* at Flowering Stage. *Ind. Crop Prod.* **2013,** *46*, 269–273.

McPartland, J. M., Russo, E. B. Cannabis and Cannabis Extracts: Greater than the Sum of Their Parts?. *J Cannabis Therapeutics* **2001,** *1* (3–4), 103–132.

Nate, S. Into the Weeds: Regulating Pesticides in Cannabis. *Environ. Health Perspect.* **2019,** *127* (4), 042001.

Nunes da Rocha, U.; Van Overbeek, L.; Van Elsas, J. D. Exploration of Hitherto-Uncultured Bacteria from the Rhizosphere. *FEMS Microbiol. Ecol.* **2009,** *69* (3), 313–328.

Nuutinen, T. Medicinal Properties of Terpenes Found in *Cannabis sativa* and *Humulus lupulus*. *Eur. J. Med. Chem.* **2018,** *157*, 198–228.

Pandey, S. S.; Singh, S.; Babu, C. S. V.; et al. Fungal Endophytes of *Catharanthus roseus* Enhance Vindoline Content by Modulating Structural and Regulatory Genes Related to Terpenoid Indole Alkaloid Biosynthesis. *Sci. Rep.* **2016,** *6*, 26583.

Porras-Alfaro, A.; Bayman, P. Hidden Fungi, Emergent Properties: Endophytes and Microbiomes. *Ann. Rev. Phytopathol.* **2011,** *49*, 291–315.

Radwan, M. M.; Wanas, A. S.; Chandra, S.; ElSohly, M. A. Natural Cannabinoids of Cannabis and Methods of Analysis. In *Cannabis sativa L.–Botany and Biotechnology*, 2017; pp. 161–182.

Redman, R. S.; Sheehan, K. B.; Stout, R. G.; Rodriguez, R. J.; Henson, J. M. Thermotolerance Generated by Plant/Fungal Symbiosis. *Science* **2002,** *298* (5598), 1581–1581.

Scott, M.; Rani, M.; Samsatly, J.; Charron, J. B.; Jabaji, S. Endophytes of Industrial Hemp (*Cannabis sativa* L.) Cultivars: Identification of Culturable Bacteria and Fungi in Leaves, Petioles, and Seeds. *Can. J. Microbial.* **2018,** *64* (10), 664–680.

Tabari, M. A.; Khodashenas, A.; Jafari, M.; et al. Acaricidal Properties of Hemp (Cannabis sativa L.) Essential Oil against *Dermanyssus gallinae* and *Hyalomma dromedarii*. *Ind. Crop Prod.* **2020,** *147*, 112238.

Taghinasab, M.; Jabaji, S. Cannabis Microbiome and the Role of Endophytes in Modulating the Production of Secondary Metabolites: An Overview. *Microorganisms* **2020,** *8* (3), 355.

Winston, M. E.; Hampton-Marcell, J.; Zarraonaindia, I.; et al. Understanding Cultivar-Specificity and Soil Determinants of the *Cannabis* Microbiome. *PLoS One* **2014,** *9* (6), 99641.

Zielonka, D.; Sas-Paszt, L.; Derkowska, E.; Lisek, A.; Russel, S. Occurrence of Arbuscular Mycorrhizal Fungi in Hemp (*Cannabis sativa*) Plants and Soil Fertilized with Sewage Sludge and Phosphogypsum. *J. Nat. Fibres* **2019,** 1–11.

CHAPTER 12

Endophytes: A Big Boon to Agriculture

AROOSA JAN MATTOO and SKARMA NONZOM*

Department of Botany, University of Jammu, J&K 180006, India

**Corresponding author. E-mail: skarmanunzom@yahoo.com*

ABSTRACT

Global population is increasing exponentially, however resources grow arithmetically. The world's population is expected to reach 9.7 billion by 2050. This projection increases the concerns related to feeding and managing the natural resources. In view of enhancing the productivity and fulfilling the basic nutritional demands of rising population, efficient and sustainable agricultural techniques must be employed. On a large scale, agricultural production is dependent on chemical fertilizers and pesticides to ameliorate the nutrient acquisition of plants and to control the pests and pathogens. However, these intensive agricultural techniques are environmentally hazardous, costly and unsustainable. Therefore, search for eco-friendly and sustainable alternatives is essential to meet the pursuit of global food security. Plant microbiomes can be such alternatives as they play a major role in plant growth promotion without exhibiting detrimental effects on the environment. Endophytic microbiomes are regarded as more important as their effects on plant improvement are direct and efficient on account of their residence inside the plant itself. Endophytes directly enhances the photosynthetic efficiency of host plants, improves plant growth and development by phytohormone biosynthesis, nitrogen fixation, siderophore production, phosphate, zinc and potassium solubilization. In addition, they indirectly promote plant growth by augmenting their tolerance to biotic factors, like, pathogens, pests and nematodes along with conferring resistance to various abiotic factors including drought, salinity, temperature extremes, reactive oxygen species and soil contaminants. By using these multi-lucrative microbes as bioinoculants, crop production can be intensified manifold in a sustainable manner.

12.1 INTRODUCTION

Agriculture is the world's largest economic sector involved in crop production on a large scale (Rai et al., 2014). However, it has been facing various challenges in terms of maintaining soil fertility as well as decreasing stress on crops. To increase the soil fertility and productivity, chemical fertilizers are being applied (Majeed, 2018). In addition, crop productivity faces many challenges in terms of pathogens, pests, and nematodes (Rai et al., 2014). This problem leads to the excessive application of pesticides, namely, fungicides, insecticides, and nematicides. Unfortunately, these agricultural practices pose a major threat to soil and water ecology as well as to human health (Laabs et al., 2000; Foley et al., 2005; Aktar et al., 2009). Moreover, these agrochemicals can also adversely affect both the rhizospheric and endospheric communities (Stuart et al., 2010; Nettles et al., 2016; Stuart et al., 2018). Besides, abiotic stress factors, like water stress, temperature stress, salinity, and heavy metals, also adversely affect the crops in diverse ways (Waqas et al., 2017; Zafar et al., 2018; Yadav et al., 2019; Zörb et al., 2018). Therefore, so as to minimize the harmful effects of the conventional methods of agriculture, innovative approaches based on microbial inoculation have recently gained interest.

Plant–microbe interactions are highly diverse in nature and fundamental for maintaining the ecological stability on earth. Endophytes are the endosymbiotic microorganisms, including bacteria, fungi, and actinomycetes, which reside within the plant tissues without exhibiting overt disease symptoms (Bacon and White, 2000). These endophytes, after colonizing the host tissues, are known to be indulged in complex metabolic interactions and further enhance the fitness of their host while acquiring food and shelter in return (Aly et al., 2013; Wani et al., 2015). However, this association is highly dynamic and regarded as endophytic continuum, which ranges from mutualism to pathogenism (Saikkonen et al., 1998; Schulz and Boyle, 2005). Although endophytic microbes are thought to be associated and evolved with the plants since their establishment on the earth, their recognition was highly delayed (Arora and Ramawat, 2017).

Endophytes play an important role in the functioning of agroecosystems. The modulation of plant growth and development is the reflection of coevolutionary significance of these microbes with their host plants (Zhao et al., 2010). Plant–endophyte associations influence the plant growth and effectively improve the agricultural traits, nutrient cycling, and soil quality (Karthik et al., 2016; Puri et al., 2016). Further, these symbionts are the essential components of ecosystem, which play an immense role in decomposition and energy flow (Doran and Zeiss, 2000). In addition, certain endophytes have been observed to bestow protection against various biotic and abiotic stress factors (Baltruschat et al., 2008; Hubbard et al., 2014). To endure stable association, endophytes elicit host plant metabolism that promotes the plant growth and help in their survival in diverse environmental conditions (Das and Varma, 2009). Apart from enhancing photosynthetic ability, phytohormone production, and nutrient acquisition, these symbionts have evolved multiple strategies to cope with biotic and abiotic stresses by reprogramming physiological responses (Wang et al., 2020a). Endophytes, therefore, offer a novel and eco-friendly approach for expanding agriculture with multiple plant growth-promoting (PGP) benefits. They are expected to play a crucial role in integrated pest management in the future (Dey and Pal, 2020). Endophytic inoculants offer the best alternative to intensive agriculture techniques and hence can be used to achieve sustainable agriculture.

The advantages conferred by the endophytes to the plants can be direct or indirect.

12.2 DIRECT ADVANTAGES

In direct advantages, the endophytes are involved directly in the growth and development of plants, such as elevation of photosynthetic capability, production of certain phytohormones, and nutrient acquisition (Fig. 12.1). Some of the endophytes conferring advantages directly to the plants are mentioned in Tables 12.1 and 12.2. The ability of these symbionts to provide direct benefits has immense importance in agriculture as these processes are fundamental to the plant growth and development. Certain advantages directly bestowed by endophytic microbes to the plants are briefly discussed below.

12.2.1 ELEVATION OF PHOTOSYNTHETIC EFFICIENCY

Photosynthetic efficiency can be determined mostly by two parameters—chlorophyll content and photochemical efficiency (Weatherby and Carter, 2013). While chlorophyll content signifies the efficacy of light-harvesting complexes, photochemical efficiency depicts the efficacy of electron transport during photochemical reactions between photosystem-II and photosystem-I. Both the parameters have been observed to get enhanced in plants after endophytic colonization (Xia et al., 2016). In addition, dark reaction of photosynthesis has also been observed to be significantly improved by endophytic interactions (Rho and Kim, 2017). The role of various microbes in improving photosynthetic efficiency has been established in diverse studies. For instance, inoculation of bacterial endophytes, *Bacillus pumilus*, *Chryseobacterium indologene,* and *Acinetobacter johnsonii,* intensified the photosynthetic efficiency in *Beta vulgaris* by increasing their chlorophyll content (Shi et al., 2010). Further, endophyte-free plants of *B. vulgaris* exhibited saturation at lower irradiance than endophyte-inoculated plants. It has also been observed that the genes involved in photosynthesis are upregulated in crop plants when their roots are inoculated with *Trichoderma* species (Harman et al., 2019). Similarly, a pathogenic fungus *Blumeria graminis* is known to adversely depress the photosynthetic parameters in *Achnatherum inebrians*. However, in this plant, association of *Epichloe* endophyte has been observed to decrease the disease index and significantly improve the photosynthesis by increasing the intercellular carbon dioxide, chlorophyll content, and net photosynthetic rate (Xia et al., 2016). In addition to disease protection, this endophyte has also been observed to improve carbon assimilation efficiency and photosystem-II phytochemistry in *Dactylis glomerata* (Rozpądek et al., 2015). Moreover, quantities of LHCI, LHCII, and chlorophyll-b have been observed to enhance in *Epichloe*-infected plants, enhancing their light capturing ability and hence photosynthesis (Rozpądek et al., 2015).

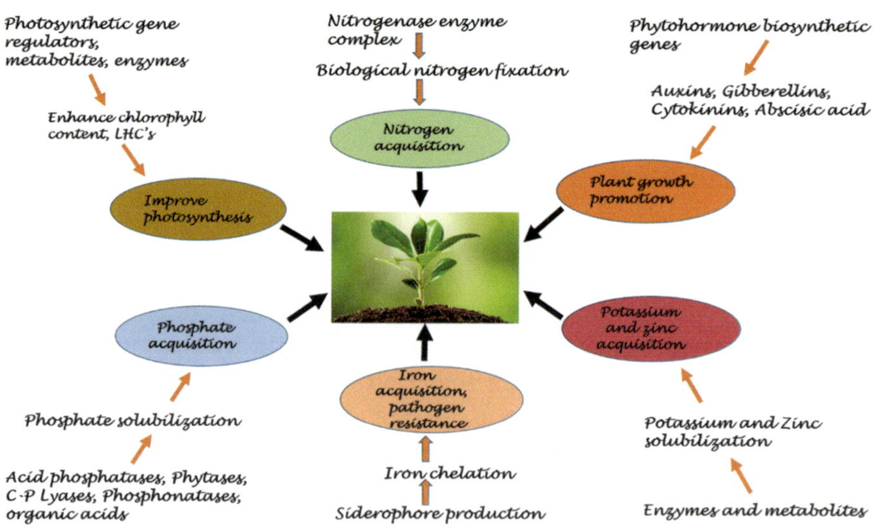

FIGURE 12.1 Schematic representation of various advantages conferred by endophytes which directly promote plant growth and development. (LHCs: Light harvesting complexes.)

TABLE 12.1 Role of Endophytic Bacteria in Direct Plant Growth Improvement.

Crop	**Endophyte**	**Benefits**	Reference
Pinus contorta var. *latifolia* (original host), *Brassica napus*, *Solanum lycopersicum*, *Zea mays* (target hosts)	*Paenibacillus polymyxa*	Nitrogen fixation	Bal et al. (2012), Puri et al. (2015), Padda et al. (2016)
Oryza sativa	*Penibacillus* sp., *Microbacterium* sp., *Bacillus* spp., and *Klebsiella* spp.	Nitrogen fixation	Ji et al. (2014)
Saccharum officinarum	*Herbaspirillum seropedicae*, *Herbaspirillum rubrisubalbicans*, *Azospirillum*, *Gluconacetobacter diazotrophicus*, and *Burkholderia tropica*	Nitrogen fixation	Schultz et al. (2014)
Z. mays	*Bacillus* sp. and *Enterobacter* sp.	Nitrogen fixation	Szilagyi-Zecchin et al. (2014)
Glycine max and *Vigna radiata*	*Klebsiella* sp. and *Pseudomonas* sp.	Zn solubilization	Sharma et al. (2014)
Wheat cultivars (Fritz and Puelche)	*Bacillus*, *Paenibacillus*, *Klebsiella*, and *Acinetobacter*	Auxin and siderophore production, phytate mineralization, and tricalcium phosphate solubilization	Duran et al. (2014)
S. officinarum	*Novosphingobium sediminicola* and *Ochrobactrum intermedium*	Nitrogen fixation	Muangthong et al. (2015)
Sedum plumbizincicola	*Bacillus pumilus* E2S2, *Bacillus* sp. E1S2, *Bacillus* sp. E4S1, *Achromobacter* sp. E4L5 and *Stenotrophomonas* sp. E1L	Siderophore production, phosphate solubilization, IAA production	Ma et al. (2015)
O. sativa	*Burkholderia* spp., *Sphingobium* spp., *Novoshingobium* spp., *Klebsiella* spp.	Nitrogen fixation	Rangjaroen et al. (2015)
Pisum sativum	*Pseudomonas fluorescens*	Phosphate solubilization	Oteino et al. (2015)
Phyllostachys edulis	*Alcaligenes* sp., *Enterobacter* sp., and *Bacillus* sp.	K and P solubilization	Yuan et al. (2015)

TABLE 12.1 *(Continued)*

Crop	Endophyte	Benefits	Reference
S. lycopersicum	*Pseudomonas*, *Rhizobium*, *Rhodococcus*, and *Agrobacterium*	Siderophore production	Abbamondi et al. (2016)
Moringa peregrine (host) *S. lycopersicum* (target)	*Bacillus, Methylobacterium*, and *Sphingomonas*	ACC deaminase activity, indole-3-acetic acid production, phosphate solubilization	Khan et al. (2016)
Prunus persica and *Pyrus communis*	*Leifsonia shinshuensis*, *Sphingomonas parapaucimobilis*, *Brevundimonas vesicularis*, *Agrobacterium tumefaciens*, *Brevundimonas diminuta*, *Stenotrophomonas rhizophilia*, and *Pseudoxanthomonas mexicana*	IAA production, nitrogen fixation, and phosphate solubilization	Liaqat and Eltem (2016)
Pteris vittata	*Bacillus*, *Enterobacter*, *Stenotrophomonas*, and *Rhizobium*	IAA and siderophore production	Tiwari et al. (2016)
Phyllanthus amarus	*Acinetobacter* sp. and *Bacillus* sp.	Phosphate solubilization	Joe et al. (2016)
S. officinarum	*Enterobacter cloaceae*, *B. pumilus*, and *Pseudomonas* sp.	K and Zn solubilization	Pirhadi et al. (2016)
Z. mays	*Bacillus* sp.	Phosphate solubilization	de Abreu et al. (2017)
O. sativa	*Acinetobacter* sp. and *Serratia* sp.	Zn solubilization	Othman et al. (2017)
Sedum alfredii	*P. fluorescens* Sam05	IAA production	Chen et al. (2017)
Teucrium polium	*Bacillus cereus*, *Bacillus subtilis*	IAA production, phosphate solubilization	Hassan (2017)
Musa paradisiaca	*Aneuinibacillus* sp. and *Lysinibacillus* sp.	Phosphate solubilization	Matos et al. (2017)
O. sativa	*Flavobacterium*, *Pseudomonas*	IAA and siderophore production, phosphate solubilization	Walitang et al. (2017)

TABLE 12.1 *(Continued)*

Crop	Endophyte	Benefits	Reference
Triticum aestivum	*Pseudomonas fragi*, *Pantoea dispera*, *Pantoae agglomerans*, *Rhizobium* sp.	Zn solubilization, HCN, extracellular enzyme production	Kamran et al. (2017)
Z. mays	*Bacillus* sp.	Oxidase activity, Zn and phosphate solubilization	Mumtaz et al. (2017)
Pennisetum glaucum	*Bacillus sp.*	N, P, K solubilization, IAA, and siderophore production	Ribeiro et al. (2018)
T. aestivum	*Pseudomonas* sp. MN12	Zn acquisition, protein synthesis	Rehman et al. (2018)
G. max	*Enterobacter, Acinetobacter, Pseudomonas, Bacillus*, and *Acinetobacter calcoaceticus* DD161	Nitrogen fixation, siderophore and IAA production	Zhao et al. (2018)
Pistacia atlantica	*Pseudomonas*, *Stenotrophomonas*, *Bacillus*, *Pantoea*, and *Serratia*	Auxin and gibberellin biosynthesis, siderophore production, phosphate solubilization, atmospheric nitrogen fixation, protease, and hydrogen cyanide production	Etminani and Harighi (2018)
O. sativa	*Achromobacter xylosoxidans, Enterobacter ludwigii, Bacillus aerius, B. cereus, Rhizobium nepotum, Bacillus amyloliquenfaciens, P. fragi,* and *P. agglomerans*	IAA production, enhanced dry weight, increased plant height	Susilowatia et al. (2018)
Camellia sinensis	*Herbaspirillum* sp. and *Brevundimonas* sp.	Nitrogen fixation, P-solubilization, IAA and siderophore production, ACC-deaminase activity	Yan et al. (2018)

TABLE 12.1 *(Continued)*

Crop	Endophyte	Benefits	Reference
Ipomoea batatas	*Klebsiella, Enterobacter,* and *Herbaspirillum*	IAA production	Dhungana et al. (2019)
S. lycopersicum	*Leclercia adecarboxylata*	IAA production, ACC deaminase activity	Kang et al. (2019)
Suaeda maritima	Zhihengliuella halotolerants and *Brachybacterium* sp.	Nitrogen fixation	Alishahi et al. (2020)
Z. mays	*Burkholderia* sp. FDN2-1	K and P solubilization	Baghel et al. (2020)
Capsicum annuum	*Bacillus megaterium*	Zn solubilization	Bhatt and Maheshwari (2020)
Cicer arietinum	*Enterobacter* sp.	Improved productivity, Zn solubilization	Ullah et al. (2020)
Leucaena leucocephala	*Rhizobium radiobacter* LB2	Zn and P-solubilization	Verma et al. (2020)
Lilium lancifolium	*P. polymyxa*	Organic acid production, ACC deaminase activity, IAA and siderophore production, phosphate solubilization, nitrogen fixation	Khan et al. (2020)

TABLE 12.2 Role of Endophytic Fungi in Direct Plant Growth Improvement.

Crop	Endophyte	Benefits	Reference
Saccharum officinarum	*Epicoccum nigrum*	Growth improvement, biomass production	Fávaro et al. (2012)
Cucumis sativus	*Paecilomyces formosus*	Gibberellin and IAA production	Khan et al. (2012)
Solanum *lycopersicum* and *Brassica campestris*	*Scolecobasidium humicola*	Improves plant growth under organic nitrogen conditions	Mahmoud and Narisawa (2013)
S. officinarum	*Alternaria* sp., *Aspergillus niger*, *Annulohypoxylon stygium*, *Talaromyces wortmannii*, and *Trichoderma atroviride*	Secretes hydrolytic enzymes, biomass production	Robl et al. (2013)
Lens esculenta	*Trichoderma gamsii* NFCC1 2177	Phosphate solubilization, chitinase activity, ammonia	Rinu et al. (2014)
Hordeum brevisubulatum	*Epichloë*	N, P, K acquisition	Song et al. (2014)
Oryza sativa and *S. officinarum*	*Rhodotorula paludigenum* DMKU-RP301 and *Torulaspora globosa* DMKU-RP31	IAA and siderophore production	Nutaratat et al. (2014)
Dactylis glomerata	*Epichloe typhina*	Enhanced photosynthesis	Rozpądek et al. (2015)
O. sativa	*Phomopsis liquidambari*	Improves nitrogen acquisition	Yang et al. (2015)
Triticum durum	*Glomus intraradices* BEG72, *Glomus mossae*, and *T. atroviride* MUCL 45632	P, Zn, K, Fe acquisition and protein content elevation	Colla et al. (2015)
S. lycopersicum	*Trichoderma pseudokoningii*	Phosphate solubilization, IAA, and siderophore production	Chadha et al. (2015)
S. lycopersicum	*Fusarium fusarioides* and *Trichoderma peudokoningii*	IAA production, phosphate solubilization	Neha et al. (2015)
Aromatic rice	KN2, KN6, KN9, KN11, KN13	Phosphate solubilization	Syamsia et al. (2015)
Solanum nigrum	*Fusarium tricinctum* RSF-4L *Alternaria alternata* RSF-6L	IAA and biomass production, chlorophyll	Khan et al. (2015)

TABLE 12.2 *(Continued)*

Crop	Endophyte	Benefits	Reference
Camellia sinensis	*A. niger Penicillium sclerotium*	IAA production Phosphate and Zn solubilization	Nath et al. (2015)
Terminalia bellerica	*Acremonium sclerotigenum*	Siderophore production	Prathyusha et al. (2015)
Ocimum sanctum and *Aloe vera*	AVR1, T2S1, TL1, AVR3, TL2	Siderophore production, phosphate solubilization	Yadav et al. (2016)
Clethra barbinervis	*Phialocephala fortinii* and Rhizodermea *veluwensis*	Siderophore production, seedling growth promotion	Yamaji et al. (2016)
S. lycopersicum, O. sativa, Zea mays, and *C. sativus*	*Trichoderma* sp.	Improve photosynthesis	Palma et al. (2016), Fu et al. (2018), Harman and Uphoff (2019), Segarra et al. (2007)
Teucrium polium	*Penicillium chrysogenum Penicillium crustosum*	Phosphate solubilization. IAA	Hassan (2017)
Brassica napus	*Fusarium* sp. CBRF44 *Alternaria* sp. CBSF68 *Penicillium* sp. CBRF65	IAA and siderophore production Phosphate solubilization	Shi et al. (2017)
Z. mays	*Gaeumannomyces cylindrosporus*	Improve growth and photosynthesis	Ban et al. (2017)
Miscanthus sinensis	*Chaetomium cupreum*	Siderophore production, Fe and Al uptake	Haruma et al. (2017)
Glycine max	*Galactomyces geotrichum* WLL1	Nutrient uptake and assimilation	Waqas et al. (2017)
S. lycopersicum	DSE fungi (A101, A104, and A105)	N, P, K acquisition	Vergara et al. (2017)
O. sativa Z. mays	ENF3, ENF5, ENF22 (*Penicillium simplicissimum*) ENF22, ENF44, ENF53, ENF32, ENF49, ENF31, ENF52	IAA production Phosphate solubilization, siderophore production	Potshangbam et al. (2017)

TABLE 12.2 *(Continued)*

Crop	Endophyte	Benefits	Reference
Oxalis corciniculata	*Aspergillus fumigatus*, *Fusarium proliferatum*	GA, siderophore and IAA production, phosphate solubilization	Bilal et al. (2018a,b)
Triticum aestivum	*Penicillium roqueforti*	IAA production, nutrient uptake, stress tolerance	Ikram et al. (2018)
Chenopodium album (native host) *O. sativa* and *G. max* (target hosts)	*Aspergillus flavus* CSH1	Siderophore, IAA and GA production, plant growth promotion	Lubna et al. (2018)
T. aestivum	581PDA1, 582PDA6, 582PDA7	Siderophore production	Ripa et al. (2019)
O. sativa	A101, A103	Nutrient acquisition	Vergara et al. (2019)
Arachis hypogaea	*Phomopsis liquidambaris*	Increased nodulation and nitrogen fixation	Xie et al. (2019)
S. lycopersicum	*Periconia macrospinosa* and *Cadophora* sp.	Improvement in N and P uptake	Yakti et al. (2019)
A. vera	*Trichoderma harzanium*	IAA and siderophore production, phosphate solubilization, HCN production	Chowdhary and Sharma (2020)
Cymbidium aloifolium	*Penicillium crysogenum*, *Aspergillus sydowii*, and *Aspergillus terreus*	Siderophore production	Chowdappa et al. (2020)
Sorghum bicolor	EU006 (*Fusarium* sp.), EU009 (*Fusarium* sp.), EU012 (*Curvularia* sp.), EU016 (*E. nigrum*), EU020 (*Trichoderma asperellum*), and EU025 (*Alternaria* sp.)	Siderophore production	Rajini et al. (2020)
Grasses	*Epichloe*	Improve soil fertility, nutrient acquisition	Wang et al. (2020a,b)
Lolium perenne	*Epichloë*	Improvement in N, P, Mn, and K uptake	Chen et al. (2020)

Similarly, chlorophyll-a, chlorophyll-b, and total chlorophyll content of drought-stressed paddy plants increased after their inoculation with endophytic fungi in comparison to the noninfected ones (Syamsia et al., 2018). Endophytic fungi *Aspergillus fumigatus* TS1 and *Fusarium proliferatum* BRL1 isolated from *Oxalis corniculata* have been analyzed for PGP activities of mutant rice Waito-C. Interestingly, it was observed that the endophyte-associated plants exhibited high root/shoot length, biomass production, and chlorophyll content apart from other PGP potencies (Bilal et al., 2018a). Similarly, photosynthesis was observed to improve in *Zizania latifolia* after their inoculation with endophytic fungus *Ustilago esculenta* (Yan et al., 2013). In this study, the infected plants exhibited lower light compensation point but elevated light saturation point and quantum yield. In addition, rate of carboxylation and regeneration of Rubisco and RuBP have also been observed to be enhanced (Yan et al., 2013). Interestingly, in C3 plants, endophytes have been reported to have their role in enhancing photosynthetic efficacy at high CO_2 concentrations. In most of these plants, higher concentrations of atmospheric CO_2 downregulate the photosynthesis (Makino and Mae, 1999). However, in endophyte-inoculated rice seedlings, this effect has been observed to be reversed. This has been attributed to the elevated photosynthetic electron transport and mesophyll conductance rates in endophytic plants in comparison to nonendophytic plants under high CO_2 concentrations (Rho et al., 2019).

12.2.2 PHYTOHORMONE PRODUCTION

Phytohormones are small organic signal molecules, which act at very low concentrations for normal plant growth and development (Davies, 1995). They belong to diverse chemical groups; however, the major classes include auxins, gibberellins, cytokinins, abscisic acid (ABA), ethylene, and brassinosteroids (Han et al., 2018). Interestingly, besides plants, the endophytic microbes have also been reported to synthesize some of these chemicals and further promote the host plant growth and development (Waqas et al., 2015; Ali et al., 2017).

A. Auxins

Auxins or indole-3-acetic acid (IAA) represents one of the major hormones, which help in stem elongation, promotion of root growth, photosynthesis, resistance to diverse stressful conditions, cell division stimulation, and differentiation (Duca et al., 2014; Tivendale et al., 2014). Numerous studies have revealed that endophytes are capable of producing IAA (Hallmann et al., 1997). For example, an endophytic bacterium *Bacillus subtilis* isolated from *Solanum lycopersicum* is able to produce IAA (Khan et al., 2016). In soybean and corn, 39.6% of endophytic bacterial isolates produced IAA (Yu et al., 2016). In rice, endophytic bacteria, including *Flavobacterium* and *Pseudomonas*, showed the potential of synthesizing IAA along with siderophore production and phosphate solubilization (Walitang et al., 2017). Moreover, *Klebsiella*, *Enterobacter*, and *Herbaspirillum* endophytic to sweet potato have also been shown to produce IAA (Dhungana and Itoh, 2019). Similarly, many other crop plants harbor IAA-producing bacteria (Fuentes-Ramirez et al., 1993; Phetcharat and Duangpaeng, 2012; Khan et al., 2016). In addition to bacteria, there are reports on fungal endophytes producing IAA. For example, 16 endophytic fungal isolates associated with aromatic rice exhibited significant IAA-synthesizing potential (Syamsia et al., 2015). Similarly, endophyte *Colletotrichum fructicola* associated with *Coffea arabica* was found to stimulate the plant growth by biosynthesizing IAA (Numponsak et al., 2018). Fungal endophytes of cucumber also produced IAA under abiotic stress, hence augmenting their host's health in adverse conditions (Khan et al., 2012).

B. Gibberellins

Gibberellin or gibberellic acid (GA) is another phytohormone crucial for cell elongation, seed germination, floral initiation, and fruiting (Albermann et al., 2013). Endophytic fungi are identified as potent growth mediators of crops (Khan et al., 2012). A few studies have been carried out on the GA production by endophytic bacteria, and it is not a well-known trait among them (Bastián et al., 1998; Khan et al., 2014). Tomato plants have been observed to show significant increase in growth attributes, like shoot elongation, chlorophyll content, and overall dry weight on account of gibberellin and IAA production by endophytic bacterium *Sphingomonas* sp. (Khan et al., 2014). Certain endophytes are known to promote plant growth through phytohormone biosynthesis under various abiotic stress factors. For instance, a GA-synthesizing endophytic fungus *Porostereum spadiceum* revives the growth of soybean inhabiting saline environment (Hamayun et al., 2017). Similarly, *A. fumigatus* TS1 and *F. proliferatum* BRL1 isolated from *O. corniculata* have been shown to possess the ability to produce gibberellins and mediate endogenous plant hormone production (Bilal et al., 2018a). Similarly, certain endophytic fungi promote the host plant growth under multiple abiotic stress factors via production of gibberellins and IAA (Waqas et al., 2012).

C. Cytokinins and ABA

Cytokinin is an essential regulator of cell cycle in plants. It also influences seed dormancy, leaf senescence, and apical dominance (Kieber, 2002). These are mostly synthesized in the roots and are translocated to other parts of the plant via xylem (Greene, 1980). Approximately, 200 natural and synthetic cytokinins are currently known. Endophytic bacteria belonging to the genera *Bacillus*, *Pseudomonas*, *Serratia*, *Flavobacterium,* and *Micrococcus* of various legume crops have been shown to biosynthesize cytokinins (UmaMaheswari et al., 2013). Production of cytokinin-like compounds has also been reported in *Pseudomonas resinovorans* and *Paenibacillus polymyxa*, evaluated by cucumber cotyledon greening bioassay (Bhore et al., 2010). Similarly, ABA, also known as stress hormone, is not directly associated with plant growth promotion. However, it indirectly influences crop productivity by combating abiotic stress factors. For example, a bacterial endophyte *Bacillus amyloliquefaciens* RWL-1 was found to mitigate salinity stress by synthesizing ABA (Shahzad et al., 2017). Similarly, endophytic species of *Azospirillum* promoted plant growth and increased water stress tolerance in maize (Cohen et al., 2009). These PGP bacteria have also been reported to synthesize IAA and gibberellins along with ABA.

12.2.3 ENDOPHYTES AS BIOFERTILIZERS

Plant nutrients represent one of the important constituents for growth and development and play a key role in sustainable agriculture (Yadav et al., 2017a). In the past decade, in order to minimize the harsh impact of agrochemicals, microbes are being used to enhance the nutrient uptake and maintain the health of soil ecosystem (Vyas et al., 2017). Biofertilizers are defined as the microbes, which enhance the availability of nutrients to the crop plants (Yadav, 2018). Nowadays, the use of PGP endophytic consortia as bioinoculants along with a low dose of chemical fertilizers is practiced (Kumar et al., 2017a). These microbes have gained attention due to their eco-friendly and economical nature (Sahoo et al., 2018). Endophytes have the advantage to penetrate and colonize the internal plant tissues, thereby establishing an intimate and fruitful association with the host. Plant growth improvement via diverse processes is the outcome of this optimistic association (Saravanakumar and Samiyappan, 2007). Apart from enhancing the nutrient uptake, endophytes also maintain the soil microbial dynamics, thereby playing a role in rejuvenating the plant microbiomes (Kumar et al., 2017b). Owing to the negative impacts imposed by the use of chemical fertilizers, endophytic inoculants can be employed as biofertilizers to increase the soil fertility and improve plant growth.

12.2.4 BIOLOGICAL NITROGEN FIXATION

Nitrogen is an essential constituent of enzymes, proteins, nucleic acids, and chlorophyll, the principal photosynthetic pigment (Laghari et al., 2016). Among different sources, atmosphere represents the main reservoir of nitrogen. It is found in the form of dinitrogen (N_2), which is inert and hence unavailable to the plants (Aczel, 2019). Dinitrogen has to be fixed or reduced into usable forms, such as, nitrates (NO_3^-) and ammonium ions (NH_4^+) to make them feasible for uptake by the plants (Gupta et al., 2012). Biological nitrogen fixation offers a proficient alternative to these agronomic chemicals. The ability of fixing atmospheric nitrogen is attributed to those microbes that possess nitrogenase, the key enzyme complex for biological nitrogen fixation. Many bacteria possess this capability and are classically termed diazotrophs. Endophytic diazotrophic bacteria have also been worked out, which mostly belong to the genera *Pseudomonas, Arthrobacter, Azospirillum, Bacillus, Herbaspirillum, Azotobacter, Gluconacetobacter, Serratia, Enterobacter, Azoarcus*, and *Klebsiella* (Wei et al., 2014; White Jr et al., 2014; Verma et al., 2014). For example, *Gluconacetobacter diazotrophicus* endophytic to sugarcane was found to fix nitrogen in vitro (*Suman* et al., 2005, 2008). Similarly, in rice, *Azospirillum*, *Herbaspirillum*, *Burkholderia,* and *Rhizobium leguminosarum* bv. *trifolii* constitute the main endophytic diazotrophs (Elbeltagy et al., 2001; Govindarajan et al., 2008; Estrada et al., 2013; Aon et al., 2015).

On the other hand, fungal endophytes have not been reported to fix the atmospheric dinitrogen due to the absence of dinitrogenase enzyme complex. However, nitrogen acquisition properties have effectively been attributed to many of them. For example, the endophytic fungus *Phomopsis liquidambari* assists the host *Oryza sativa* in acquiring soil nitrogen (Yang et al., 2015). After inoculating this symbiont, nitrate and ammonium content of the rhizospheric soil of infected plants have been observed to get enhanced under diminished nitrogen conditions in comparison to noninfected plants. Furthermore, nitrification rate was also enhanced in these plants. In another study, *P. liquidambari* was found to enhance the nodulation and nitrogen fixation in *Arachis hypogea* (Xie et al., 2019). Detailed investigation revealed the upregulated expression of genes related to phenol and flavonoid synthesis, which increases the chemotaxis and nodulation-related processes of *Bradyrhizobium* (a diazotrophic bacterium).

12.2.5 PHOSPHATE SOLUBILIZATION

Phosphorus (P) is a crucial macronutrient for plant growth and development and is involved in diverse processes, including photosynthesis, mineral uptake, cell division, biological oxidation, and protein synthesis (Illmer and Schinner, 1999). It also promotes root development and early flowering. The concentration of bioavailable P in soil is extremely low as the bigger fraction of it exists as insoluble minerals salts, rocks, or organic compounds (Sharma et al., 2013). Most of the P has poor mobility as it forms complexes with calcium, iron, or aluminum and thus is not able to support plant growth (Chhabra et al., 2013; Ramanuj and Shelat, 2018).

Both bacterial and fungal endophytes are able to convert the insoluble P into soluble and accessible forms. This is accomplished mostly by releasing organic acids, such as acetate, citrate, oxalate, tartarate, lactate, gluconate, ketogluconate, glycolate, and succinate (Khan et al., 2009; Stella and Halimi, 2015; Yadav et al., 2015). However, endophytes are also reported to solubilize the organic-bound phosphates via introduction of certain enzymes, namely, nonspecific acid phosphatases, phytases, C-P lyases, and phosphonatases (Illmer et al., 1995; Ngwene et al., 2016; Adhikari and Pandey, 2019). Phosphate solubilization is a prevalent attribute in endophytic bacteria. The majority of the endophytic bacteria colonizing agricultural crops, including rice, maize, wheat, peanut, and legumes, showed phosphate solubilization in plate assay (Puente et al., 2009; Verma et al., 2014, 2015). For example, endophytic *B. subtilis, Bacillus megaterium, Pseudomonas putida, Pantoea ananatis,* and *Brevibacillus agri* isolated from O. sativa solubilized phosphate from tricalcium phosphate, iron phosphate, and aluminum phosphate (Borah et al., 2017). Similarly, endophyte *Burkholderia* sp. has been reported to solubilize iron phosphate, rock phosphate, aluminum phosphate, and tricalcium phosphate (Baghel et al., 2020). Interestingly, certain species of *Bacillus* were found more efficient phosphate solubilizers in maize (de Abreu et al., 2017). Survey of literature also reveals the role of endophytes in stress-exposed plants, which showed enhanced phosphate-solubilizing potential and plant growth, thereby helping them in their survival (Forchetti et al., 2007).

For example, *Acinetobacter* sp. and *Bacillus* sp. endophytic to *Phyllanthus amarus* solubilized phosphate under salt stress (Joe et al., 2016).

Endophytic fungi also bear the potential to solubilize the phosphates, thereby augmenting the nutrient acquisition of host plants. For example, significant phosphate solubilization activity has been attributed to an endophytic fungus *Fusarium verticillioides* (Radhakrishnan et al., 2015). Many dark septate endophytic fungi (DSE) are also able to solubilize calcium, iron, and aluminum phosphates (Spagnoletti et al., 2017). However, the endophytic fungi colonizing the roots of *Taxus wallichiana* exhibited the production of phosphatase and phytase enzymes, correlating the potential of phosphate solubilization to these enzymes (Adhikari and Pandey, 2019). Similarly, the production of many organic acids, like malic acid, succinic acid, oxalic acid, lactic acid, and citric acid, has also been well established in many fungal isolates, which are known to degrade the phosphate complexes.

12.2.6 POTASSIUM AND ZINC SOLUBILIZATION

Potassium is an essential macronutrient for normal growth and development of plants (Marschner, 1986). It increases the rate of photosynthesis via carbon assimilation and enhances carbon mobility and resistance against diseases (Sangakkara et al., 2000; Rehm and Schmitt, 2002). Although soil contains a large amount of potassium, yet most of it is unavailable for plant uptake (Etesami et al., 2017). Approximately 90–98% of soil potassium is present as feldspar, mica, biotite, illite, muscovite, and orthoclase (Sparks, 1987; Andrist-Rangel et al., 2010). Similarly, zinc is an important micronutrient and acts as a cofactor of over 300 enzymes and proteins involved in photosynthesis, respiration, cell cycle, nucleic acid metabolism, and translation (Marschner, 1986). Many endophytic bacteria are able to solubilize K and Zn. K solubilizing endophytic bacteria mostly belong to *Bacillus edaphicus*, *Bacillus circulans*, *Bacillus mucilaginosus*, *Acidithiobacillus ferrooxidans*, *Burkholderia* sp., *Alcaligenes* spp., and *Paenibacillus* spp. (Yuan et al., 2015; Suman et al., 2016; Baghel et al., 2020; Kushwaha et al., 2020). However, zinc solubilization is carried out by many bacterial endophytes, including *Achromobacter, Arthrobacter, Pseudomonas, Pantoea, Bacillus, Bordetella, Staphylococcus, Exiguobacterium, Flavobacterium, Kocuria, Klebsiella,* and *Providencia* (Sharma et al., 2014; Kamran et al., 2017; Singh et al., 2018; Ullah et al., 2020). For instance, three endophytic bacterial genera, namely, *Enterobacter cloacae, B. pumilus,* and *Pseudomonas* sp., from sugarcane-solubilized potassium and zinc in a plate assay (Pirhadi et al., 2016). Further, inoculation of *E. cloacae* into the pot containing wheat seedlings exhibited enhanced K uptake. In *Phoenix dactylifera,* 19 out of 85 endophytic bacterial strains solubilized zinc from zinc oxide (Yaish et al., 2015).

In addition, endophytic fungi also bear the potential of solubilizing the abovementioned elements from their corresponding minerals. For example, Nath et al. (2015) reported the potential of *Aspergillus niger, Aspergillus terreus,* and *Penicillium sclerotiorum* in solubilizing K and Zn. Similarly, *Glomus intraradices* BEG72, *Glomus mossae,* and Trichoderma atroviride MUCL 45632 have been found to assist in the acquisition of K and Zn by *Triticum durum* apart from acquiring other essential nutrients (Colla et al., 2015). Besides, certain DSE have also been reported to solubilize K (Vergara et al., 2017).

12.2.7 SIDEROPHORE PRODUCTION

Iron is an essential micronutrient involved in DNA synthesis, photosynthesis, and respiration (Rout and Gyana, 2015). It plays a crucial role in metabolic processes by acting as a cofactor of various enzymes (Balk and Schaedler, 2014). Inspite of being the fourth most abundant element in the earth's crust, iron remains largely unavailable for both the plants and microbes due to its occurrence as ferric ion (Fe^{3+}), which is insoluble at physiological pH (Bou-Abdallah, 2010; Saha et al., 2016). Therefore, many microorganisms have adapted a strategy to acquire iron by releasing siderophores. The latter are low molecular weight, high-affinity iron-chelating secondary metabolites which arrest ferric ions from diverse environments to solubilize them for efficacious uptake (Lacava et al., 2008).

Numerous endophytes have the ability to synthesize siderophores that enhance the availability of iron to their host plants. This in turn proves beneficial to the host plants in that they deprive the phytopathogens of iron and inhibit their growth (Whipps, 2001; Suman et al., 2016). Therefore, siderophore-producing endophytes not only exhibit the direct PGP activities but also provide the competitive advantage to their host plants (Ribeiro and Simões, 2019). Many endophytic bacteria, including *Streptomyces, Methylobacterium,* and *Pseudomonas,* from agricultural crops were found active in siderophore production (Rungin et al., 2012; Lacava and *Azevedo,* 2013; Walitang et al., 2017). Similarly, as many as 14 endophytic bacteria isolated from *Cicer arietinum* and *Pisum sativum* have been shown to produce siderophores, especially of hydroxamate and carboxylate type (Maheshwari et al., 2019).

Siderophore production and their subsequent growth-promoting attributes have also been reported from the fungal endophytes, particularly of hydroxamate type (Bartholdy et al., 2001; Kajula et al., 2010). For example, endophytic fungi belonging to genera *Acremonium, Aspergillus, Colletotrichum, Fusarium, Penicillium, Arthrinium, Phaeotheca* isolated from different crop plants are mainly reported to produce siderophores (Ohra et al., 1995; Shi et al., 2017; Chowdapa et al., 2020). Similarly, a study revealed siderophore production, antimicrobial and anti-oxidant activity of fungal endophytes recovered from Pinus sylvestris and Rhododendron tomentosum (Kajula et al., 2010). Intriguingly, only those endophytes that showed antimicrobial properties exhibited siderophore-producing ability. However, few reports suggest that siderophores of endophytic fungi have lower affinity to sequester iron in comparison to bacterial siderophores (Whipps, 2001; Loper and Henkels, 1999).

12.3 INDIRECT ADVANTAGES

Although biotic and abiotic stress tolerance of plants do not represent direct growth enhancement, yet they indirectly affect their growth and development (Fig. 12.2). On account of being sedentary creatures, plants have to cope with an array of unfavorable conditions to adapt to a particular habitat. Biotic stressors involve pathogens, pests, and nematodes, which can cause heavy losses in agriculture. However, abiotic stress factors include drought, salinity, temperature extremes, oxidative stress, and heavy metal stress. Abiotic stresses represent the critical growth-limiting stressors affecting the survival of plants in changing or extreme environmental conditions. Interestingly, endophytes are known to augment their host plants in tolerating both these stressors by secreting diverse bioactive metabolites. Some of the indirect known benefits conferred by endophytes are mentioned below.

12.3.1 BIOTIC STRESS TOLERANCE

Every year, agriculture faces huge loss due to pathogens, pests, and nematodes, which signify a considerable trammel on global food security. Plant health is an integral part of sustainable agriculture. However, chemical pesticides employed for the eradication of these entities pose diverse environmental problems (Aktar et al., 2009). Therefore, search for the alternative to these chemicals has become imperative. Endophytes are widely known for their antimicrobial, anti-insecticidal, and antinematicidal properties (Selim et al., 2011; Sudha et al., 2016; Abraham et al., 2015; Khan et al., 2019). It is believed that endophytes have evolved different mechanisms to adapt themselves inside the plant tissues and make themselves safe from the external harmful organisms. Therefore, in order to compete with these organisms, they produce various toxic bioactive secondary metabolites (Hallmann and Sikora, 1996). These metabolites belong to diverse chemical groups ranging from alkaloids to glycosides (Mousa and Raizada, 2013; Patil et al., 2016). The endophytic metabolites may either be similar to the ones produced by the corresponding host or altogether different from them (Kusari et al., 2008; Kusari and Spiteller, 2012). By producing all the antagonistic metabolites, endophytes enhance the defensive properties of their hosts.

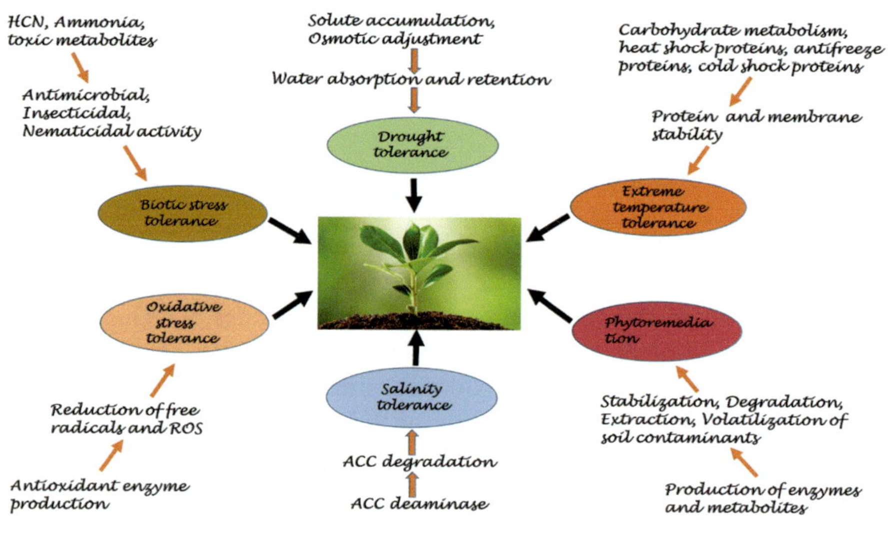

FIGURE 12.2 Schematic representation of endophytes in bestowing biotic and abiotic stress tolerance to host plants. (ROS: Reactive oxygen species; ACC: 1-aminocyclopropane-1-carboxylic acid.)

A) Pathogens

Endophytes have been reported to act as biocontrol agents against many bacterial and fungal pathogens that cause diseases in plants. For instance, endophytes of different cereals are found to exhibit antifungal activity against *Gaeumannomyces graminis,* the causal agent of many diseases in wheat (Coombs et al., 2004). Similarly, *B. subtilis* acts as promising antagonist of *Xanthomonas oryzae,* causal agent of leaf blight of rice

(Nagendran et al., 2013). A maize endophyte *Pseudomonas aeruginosa* served as an efficient biocontrol agent against six phytopathogens, namely, *Sclerotium rolfsii, Rhizoctonia solani, Pythium aphanidermatum, Fusarium oxysporum, Macrophomina phaseolina,* and *Alternaria* sp. (Sandhya et al., 2017). Further, 10 endophytic fungal isolates of wheat showed antagonism against *Phytophthora infestans* and *F. oxysporum* f.sp. *albedinis* (Sadrati et al., 2013). Similarly, *Epicoccum nigrum,* an important endophytic fungus of sugarcane inhibited various phytopathogens, such as *Ceratocystis paradoxa, F. verticillioides,* and *Xanthomonas albilineans* (Fávaro et al., 2012). On the other hand, endophytic *Pseudomonas aurantiaca* exhibits strong antifungal activity against *Colletotrichum falcatum,* the causal organism of red rot of sugarcane (Mehnaz et al., 2014). A plethora of endophytic microbes possess the similar activities and, therefore, can effectively be used as inoculants for biocontrol agents of agricultural crops.

B) Pests

Apart from pathogens, insects are one of the potent destroyers of crop plants and can account for huge losses either directly or indirectly by acting as vectors of numerous infectious agents (Culliney, 2014; Heck, 2018). Although insecticides are being used to control these losses, yet their use has not been appreciated due to the range of environmental hazards they cause (Mahmood et al., 2016). Another disadvantage of using these chemicals is that many insects have developed resistance to the frequently used insecticides. However, novel metabolites produced by the endophytes are expected to overcome these drawbacks. Numerous endophytes have been worked out against many pests. For example, various endophytic isolates associated with *A. inebrians* (Drunken Horse grass) showed more than 90% of mortality rates in *Aphis gossypii* (Shi et al., 2013). Similarly, *Spodoptera litura,* a polyphagous pest showing resistance to known insecticides, was killed by an endophytic fungus *Cladosporium uredinicola* (Thakur et al., 2013).

C) Nematodes

Phytoparasitic nematodes are ubiquitous in nature and an expensive burden on agriculture. This burden can be relieved by employing endophytes as biocontrol agents. Endophytic isolates from several plants have been observed to exhibit nematicidal activities against various nematodes, such as *Caenorhabditis elegans, Bursaphelenchus xylophilus,* and *Meloidogyne incognita* (Zheng et al., 2008; Ponpandian et al., 2019). Nematicidal activity of endophytic microbes is attributed to certain enzymes and metabolites secreted by them. For example, Chaetomium globosum produced chaetoglobosin A, B and flavipin effective against Meloidogyne javanica (Khan et al., 2019). Similarly, endophytic bacterium Bacillus cereus exhibited significant biocontrol against *M. incognita.* This activity was ascribed to enzymes like chitosanase, alkaline serine protease, and neutral protease (Hu et al., 2017).

12.3.2 ABIOTIC STRESS TOLERANCE

Plants have evolved several strategies to perceive stress signals and accordingly respond to them at varied degrees. During these stressed conditions, interactive regulatory mechanisms mediated by signaling molecules or cofactors are involved (Dombrowski, 2003). Temperature extremes, drought, salt stress, nutrient stress, and heavy metal stress represent the major abiotic stress factors that downgrade the plant survival (Chaves and Oliveira, 2004). Endophytes have been observed to promote plant growth by conferring tolerance to these stresses (Eid et al., 2019). Abiotic stress tolerance conferred by plant–endophyte interactions either involve the activation of stress response after being exposed to stress or secretion of antistress metabolites by endophytic symbionts (Redman et al., 1999; Schulz et al., 2002). The main abiotic stress factors are described below.

A) Drought

Drought is the most critical limiting factor for plant growth and development in agriculture. However, endophytic associations are reported to enhance the drought tolerance of crops manifold (Ullah et al., 2019). For example, *Curvularia*-inoculated rice seedlings were found to grow for a longer period without water in comparison to endophyte-free plants (Redman et al., 2011). In addition, the seed yield of endophyte colonized plants was also observed to be increased. The mechanisms underlying this capability include accumulation and translocation of assimilates, osmotic adjustments, and maintenance of cell wall elasticity (Nieves-Cordones et al., 2019). In plants, enzymes particularly, H^+-ATPase and H^+-PPase, confer drought tolerance. It was observed that root endophytic bacteria colonizing *Capsicum annuum* enhanced vacuolar H^+-PPase (H^+-pumping pyrophosphatase) activity, thereby helping the host to grow under water-limiting conditions (Vigani et al., 2018). Also, *Cladosporium cladosporioides* and an unidentified fungal endophyte improved water stress tolerance in *Nicotiana benthamiana* (Dastogeer et al., 2018). Recently, Morsy et al. (2020) tested various endophytes in conferring tolerance to abiotic stress conditions. Interestingly, they found that a fungal endophyte *Ampelomyces* sp. effectively mitigated the drought tolerance in tomato plants. Besides, *Azospirillum lipoferum* secretes ABA, which is found to reduce the water stress in maize (Cohen et al., 2009). Similarly, *Azospirillum*-inoculated plants showed enhanced biomass, soluble sugar, soluble protein, antioxidant potential, proline content, and the elevated expression of drought-responsive genes.

B) Salinity

Salinity is the major abiotic stress factor in plants as around 20% of the world's cultivated lands are saline (Glick et al., 2007; Shrivastava and Kumar, 2015). Higher salt concentrations in soil cause ion imbalances, which hurdles water absorption by plants (Sheldon et al., 2004). However, many endophytes have been reported to promote plant growth under high salt conditions. While checking endophyte-soybean interactions, Khan et al. (2011) observed that endophytic fungus *Penicillium minioluteum* effectively promoted plant growth by enhancing biomass, chlorophyll content, and shoot elongation. Interestingly, low endogenous ABA and high salicylic acid (SA) contents were detected in endophyte-infected plants in response to salinity stress. Recently, *Fusarium* sp., an endophyte of salt-adapted Pokkali rice, was found to confer drought tolerance to salt-sensitive rice variety IR-64 by promoting its growth under high salinity conditions (Sampangi-Ramaiah et al., 2020). Similarly, *Penicillium* sp. significantly enhanced the tolerance of tomato plants to salinity when placed under 300 mM salt concentration in comparison to nonendophytic plants (Morsy et al., 2020). Salinity is observed to affect crop yield in majority of plants via biosynthesis of ethylene, which has a negative effect on root growth (Feng and Barker, 1992). However, various studies unveiled that the level of ethylene can be reduced by some endophytes that secrete 1-aminocyclopropane-1-carboxylic acid (ACC) deaminase (Afridi et al., 2019). This enzyme breaks the immediate precursor of ethylene in its biosynthetic pathway. Various endophytes, like *Enterobacter, Azospirillum, Bacillus, Serratia, Burkholderia, Acinetobacter, Pseudomonas, Achromobacter, Ralstonia, Alcaligenes, Rhizobium,* and *Agrobacterium,* are known to produce ACC deaminase (Zhang et al., 2011; Xu et al., 2014; Verma et al., 2015).

C) Temperature

Temperature extremes (low and high) represent another stress factor detrimental for plant growth and survival. High temperature impairs plant growth, photosynthesis and reproduction by protein denaturation, alteration of enzyme activity, and generation of reactive oxygen species (ROS) (Ihsan et al., 2019). Endophytes help the plants inhabiting xeric environments by stabilizing heat shock proteins and secretion of anti-oxidant enzymes, such as catalases (CATs) and peroxidases (PODs) (Ismail et al., 2018; Khan et al., 2020). For example, *Aspergillus japonicus* EuR-26 mitigated the thermal stress in soybean and sunflower by negotiating various potencies of CAT, ascorbic acid oxidase, and ABA (Ismail et al., 2018). Similarly, *B. cereus* SA1 conferred thermotolerance to soybean by modulating the phytohormone production (increased ABA and decreased SA), anti-oxidant enzyme secretion [superoxide dismutase (SOD) and ascorbic acid POD] and further enhanced the quantity of heat shock proteins (Khan et al., 2020).

In addition, an endophytic fungus *Thermomyces* sp. was isolated from the roots of a hot desert-adapted plant *Cullen plicata* (Ali et al., 2018). This symbiont was then evaluated for conferring heat-stress tolerance in cucumber plants grown in field during summer in Egypt. Surprisingly, it was observed to eliminate the adverse effects of heat stress by elevating the amount of several anti-oxidant enzymes, sugars, soluble proteins, saponins, and flavonoids. Moreover, water use efficiency, photosynthesis rate, and root elongation were found to improve in endophyte-inoculated plants in comparison to the untreated ones (Ali et al., 2018). Similarly, low temperature affects photosynthesis, respiration, and yield of plants (Hendrickson et al., 2004). It slows down the cell growth and stiffens the cell membrane by changing its fluidity (Kumar et al., 2018). Liquid inside the cells begins to freeze and form crystals that may pierce the plasma membrane and kill the cells (Pearce, 2001). Interestingly, this does not happen to most of the plants harboring endophytes growing in natural habitats, which help them to survive under such adverse conditions (Barka et al., 2006; Yadav et al., 2017b). It has been observed that endophytes improve cold tolerance by modulating photosynthetic activity and carbohydrate metabolism that results in the deposition of cold stress–related metabolites, such as, starch, trehalose, phenolics, proline, and polyols (Barka et al., 2006; Fernandez et al., 2012). Other strategies include production of antifreeze proteins and cold-active enzymes (Hashim et al., 2013; Yadav et al., 2017b; Furhan, 2020). A bacterial endophyte *Burkholderia phytofirmans* PsJN has been reported to confer cold stress tolerance to grapevine (Barka et al., 2006). Further, it enhanced the physiological activity and biomass of the host plant manifold. Similarly, fungal endophytes have also been reported to help plants to adapt well to extreme cold climates. This can be observed by the fact that many plants harboring endophytes grow well in Arctics and Antarctica. For example, *Cadophora, Geomyces, Fusarium, Gyoerffyella, Aspergillus, Davidiella, Entrophospora, Microdochium, Mycocentrospora,* and *Phaeospaeria* isolated from the leaves of an Antarctic plant *Colobanthus quitensis* have been indicated in helping the host plant in adapting to such extreme conditions (Rosa et al., 2010). Similarly endophytes *Cryptococcus, Mycopappus, Melampsora, Rhizosphaera, Phaeosphaeria, Mrakia, Venturia, Leptosphaeria,* and *Tetracladium* have been reported to confer cold stress tolerance to various Arctic plants (Zhang and Yao, 2015).

D) Oxidative stress

Oxidative stress of the plants in response to environmental stress factors involves the production of ROS, such as superoxide anions, hydroxyl radicals, and hydrogen peroxide (Lata et al., 2018). ROS production can bring oxidative damage to lipids, proteins, and nucleic acids. However, it has been observed that colonization of endophytic bacteria upregulates the expression of ROS-degrading genes, such as gene-coding SOD and glutathione reductase (GR) (Lata et al., 2018). In an experiment, *Brassica rapa* infected with endophytic *Piriformospora indica* was treated with polyethylene glycol to mimic drought stress. Surprisingly, an upregulation of anti-oxidant enzymes, namely, CATs, PODs, and SODs was observed in this plant within 24 hours (Sun et al., 2010). Similarly, an endophytic bacterium *Sphingomonas* SaMR12 was isolated from Sedum alfredii (Wang et al., 2020b). In order to evaluate its PGP activities, SaMR12 was inoculated in nonhost plant *Brassica juncea*. It was observed that, apart from various advantages endowed by this endophyte, it also relieved the oxidative stress in *B. juncea.* It was accomplished by decreasing the concentration of hydrogen peroxide (H2O2) and improving the activities of SOD, CAT, POD, and GR. In this way, endophytes help the plants to overcome various abiotic stresses and further improve their growth.

E) Phytoremediation

Phytoremediation involves the process of using plants and associated microbes to reduce the toxic contaminants in the environment (Prasad, 2004; Dinckinson et al., 2009). Many contaminants, including heavy metals, halogenated hydrocarbons, and polycyclic aromatic hydrocarbons, are phytotoxic. Majority of the plants are incapable of degrading these chemicals. However, certain endophytes have shown the ability to degrade many toxic chemicals and, therefore, play a crucial role in phytoremediation of contaminated soils (Newman and Reynolds, 2005). For this, they have been observed to employ their metabolites, enzymes, siderophores, organic acids, etc. (Soleimani et al., 2010; Yousaf et al., 2010). Endophytes enhance the phytoremediation by improving phytostabilization, phytodegradation, phytovolatilization, and phytoextraction (Germaine et al., 2009; Chen et al., 2010; Weyens et al., 2010; Zhang et al., 2011).

A number of studies have been carried out on both bacterial and fungal endophytes for assessing their role in phytoremediation. For instance, *Bacillus thuringiensis* isolated from *Pinus* roots was found to enhance the heavy metal accumulation and growth promotion of *Alnus firma.* It was found efficient in removing heavy metals, namely, As, Ni, Pb, Zn, and Cu. In addition, this symbiont also possessed IAA production, phosphate solubilization, siderophore production, and ACC deaminase potential (Babu et al., 2013). Similarly, *Methylobacterium oryzae* and *Burkholderia* sp. endophytic to rice showed reduced Ni and Cd toxicity in tomato plants in controlled conditions (Madhaiyan et al., 2007). An endophytic bacterial consortium has been observed to enhance the arsenic phytoremediation in *Solanum nigrum* (Mukherjee et al., 2018).

Similarly, *Mucor* strains endophytic to *Brassica campestris* and *Brassica* napus accomplished phytoremediation against multiple pollutants, including Cr^{6+}, Mn^{2+}, Co^{2+}, Cd, and Pb (Deng et al., 2013; Zahoor et al., 2017). *P. liquidambari,* endophytic to *O. sativa* and *A. hypogea,* has been observed to improve the host tolerance to phenanthrene and phenoloic acids (Chen et al., 2013; Xie et al., 2015; Fu et al., 2018). Certain endophytes act synergistically to improve the plant growth as well as ecophysiological responses of plants to stress conditions. For example, in zinc and aluminum stressed *Glycine max,* two fungal endophytes *Paecilomyces formosus* LHL10 and *Sphingomonas* sp. LK11 were observed to confer higher plant growth (biomass, root/shoot length, and chlorophyll content) in comparison to the noninoculated ones (Bilal et al., 2018b). In this way, these symbionts help the host plants to decrease phytotoxicity and thereby their adaptation to polluted environment (Chen et al., 2010; Zhang et al., 2011; He et al., 2020). Hence, they should be employed as bioinoculants in agricultural crops to improve their survival in contaminated soils.

12.4 CONCLUSION AND FUTURE PERSPECTIVES

Endophytes offer an engrossing environment-friendly resource. Employing them as inoculants can reduce our dependency on chemical fertilizers and pesticides which otherwise pose various threats to different ecosystems. Upon direct inoculation into the seeds or aerial parts, they can ameliorate the crop productivity of target hosts with less competition as compared to rhizospheric or phyllospheric microbes. They provide an economical and sustainable alternative to traditionally practiced agricultural techniques. By using them in agriculture, basic nutritional demands of over-increasing population, especially those of developing countries can be met without causing pollution and other health hazards. On account of furnishing multiple direct and indirect PGP benefits in an eco-friendly approach, endophytic symbionts represent efficient bioinoculants for attaining sustainable agriculture.

In addition, they can be potent source of novel metabolites which may display bounteous PGP activities. Specifically, the endophytic microbes colonizing the plants inhabiting extreme environmental conditions can be screened for novel and desired PGP traits. Crop varieties with phytopathogen-resistant endophytes can be designed to counter the huge losses caused by these pathogenic counterparts every year. Further, genetic engineering of these microbes with specific and maximum growth promotion–related genes should be executed. It would be easier to engineer the microbes in comparison to that of plants. Additionally, by using engineered or nonengineered endophytes, ethical-related issues to genetically modified crops can be eliminated. Intriguingly, the consideration of endophyte-like biological tools would be beneficial in attaining sustainable development goals proposed by United Nations as a universal call to eradicate poverty, hunger and protect the environment which is supposed to be achieved by 2030.

KEYWORDS

- **endophyte**
- **agriculture**
- **biotic stress**
- **abiotic stress**
- **plant growth promotion**
- **sustainability**

REFERENCES

Abraham, S.; Basukriadi, A.; Pawiroharsono, S.; Sjamsuridzal W. Insecticidal Activity of Ethyl Acetate Extracts from Culture Filtrates of Mangrove Fungal Endophytes. *Mycobiology* **2015,** *43*, 137–149.

Aczel, M. What Is the Nitrogen Cycle and Why Is It Key to Life? *Front Young Minds* **2019,** *7*, 41.

Adhikari, P.; Pandey, A. Phosphate Solubilization Potential of Endophytic Fungi Isolated from *Taxus wallichiana* Zucc. Roots. *Rhizosphere* **2019,** *9*, 2–9.

Afridi, M. S.; Mahmood, T.; Salam, A.; et al. Induction of Tolerance to Salinity in Wheat Genotypes by Plant Growth Promoting Endophytes: Involvement of ACC Deaminase and Antioxidant Enzymes. *Plant Physiol. Biochem.* **2019,** *139*, 569–577.

Aktar, M. W.; Sengupta, D.; Chowdhury, A.; et al. Impact of Pesticides Use in Agriculture: Their Benefits and Hazards. *Interdiscip. Toxicol.* **2009,** *2*, 1–12.

Albermann, S.; Elter, T.; Teubner, A.; et al. Characterization of Novel Mutants with an Altered Gibberellin Spectrum in Comparison to Different Wild-Type Strains of *Fusarium fujikuroi*. *Appl. Microbiol. Biotechnol.* **2013,** *97*, 7779–7790.

Ali, A.; Abdelrahman, M.; Radwan, U.; et al. Effect of *Thermomyces* Fungal Endophyte Isolated from Extreme Hot Desert Adapted Plant on Heat Stress Tolerance of Cucumber. *Appl. Soil Ecol.* **2018,** *124*, 155–162.

Ali, S.; Charles, T. C.; Glick, B. R. Endophytic Phytohormones and Their Role in Plant Growth Promotion. In *Functional Importance of the Plant Microbiome*; Doty, S. L., Ed.; Springer: Berlin, 2017; pp 89–105.

Aly, A. H.; Debbab, A.; Proksch, P. Fungal Endophytes—Secret Producers of Bioactive Plant Metabolites. *Pharmazie* **2013,** *68*, 499–505.

Andrist-Rangel, Y.; Hillier, S.; Öborn, I.; et al. Assessing Potassium Reserves in Northern Temperate Grassland Soils: A Perspective Based on Quantitative Mineralogical Analysis and Aqua-Regia Extractable Potassium. *Geoderma* **2010,** *158*, 303–314.

Aon, M.; Khalid, M.; Hussain, S.; et al. Diazotrophic Inoculation Supplemented Nitrogen Demand of Flooded Rice Under Field Conditions. *Pak. J. Agric. Sci.* **2015,** *52*, 145–150.

Arora, J.; Ramawat K. G. An Introduction to Endophytes. In *Endophytes, Biology and Biotechnology*; *Maheshwari*, D. K., Ed.; Springer, 2017; Vol. 1, pp 1–23.

Babu, A. G.; Kim, J. D.; Oh, B. T. Enhancement of Heavy Metal Phytoremediation by *Alnus firma* with Endophytic *Bacillus thuringiensis* GDB-1. *J. Hazard. Mater.* **2013,** *250*, 477–483.

Bacon, C. W.; White, J. Physiological Adaptations in the Evolution of Endophytism in the Clavicipitaceae. In *Microbial Endophytes*; Bacon, C. W., White, J., Eds.; Marcel Dekker: New York, NY, 2000; pp 237–263.

Baghel, V.; Thakur, J. K.; Yadav, S. S.; et al. Phosphorus and Potassium Solubilization from Rock Minerals by Endophytic *Burkholderia* sp. strain FDN2-1 in Soil and Shift in Diversity of Bacterial Endophytes of Corn Root Tissue with Crop Growth Stage. *Geomicrobiol. J.* **2020,** *37*, 1–14.

Balk, J.; Schaedler, T. A. Iron Cofactor Assembly in Plants. *Annu. Rev. Plant Biol.* **2014,** *65*, 125–153.

Baltruschat, H.; Fodor, J.; Harrach, B. D.; et al. Salt Tolerance of Barley Induced by the Root Endophyte *Piriformospora indica* Is Associated with a Strong Increase in Antioxidants. *New Phytol.* **2008,** *180*, 501–510.

Barka, E. A.; Nowak, J.; Clément, C. Enhancement of Chilling Resistance of Inoculated Grapevine Plantlets with a Plant Growth-Promoting Rhizobacterium, *Burkholderia phytofirmans* Strain PsJN. *Appl. Environ. Microbiol.* **2006,** *72*, 7246–7252.

Bartholdy, B. A.; Berreck, M.; Haselwandter, K. Hydoxamate Siderophores Synthesis by *Phialocephala fortinii*, a Typical Dark Septate Fungal Root Endophyte. *BioMetals* **2001,** *14*, 33–42.

Bastián, F.; Cohen, A.; Piccoli, P.; et al. Production of Indole-3-Acetic Acid and Gibberellins A_1 and A_3 by *Acetobacter diazotrophicus* and *Herbaspirillum seropedicae* in Chemically-Defined Culture Media. *Plant Growth Regul.* **1998,** *24*, 7–11.

Bhore, S. J.; Ravichantar, N.; Loh, C. Y. Screening of Endophytic Bacteria Isolated from Leaves of Sambung Nyawa [*Gynura procumbens* (Lour.) Merr.] for Cytokinin-Like Compounds. *Bioinformation* **2010,** *5*, 191–197.

Bilal, L.; Asaf, S.; Hamayun, M.; et al. Plant Growth Promoting Endophytic Fungi *Aspergillus fumigatus* TS1 and *Fusarium proliferatum* BRL1 Produce Gibberellins and Regulates Plant Endogenous Hormones. *Symbiosis* **2018a,** *76*, 117–127.

Bilal, S.; Shahzad, R.; Khan, A. L.; et al. Endophytic Microbial Consortia of Phytohormones-Producing Fungus *Paecilomyces formosus* LHL10 and Bacteria *Sphingomonas* sp. LK11 to *Glycine max* L. Regulates Physio-Hormonal Changes to Attenuate Aluminum and Zinc Stresses. *Front. Plant Sci.* **2018b,** *9*, 1273.

Borah, M.; Das, P.; Pathak, S. S.; et al. Phosphate Solubilization by Endophytic Bacteria Isolated from *Oryza sativa*. *Int. J. Curr. Microbiol. Appl. Sci.* **2017,** *6*, 2713–2721.

Bou-Abdallah, F. The Iron Redox and Hydrolysis Chemistry of the Ferritins. *Biochim. Biophys. Acta* **2010,** *1800*, 719–731.

Chaves, M. M.; Oliveira, M. M. Mechanisms Underlying Plant Resilience to Water Deficits: Prospects for Water-Saving Agriculture. *J. Exp. Bot.* **2004,** *55*, 2365–2384.

Chen, L.; Luo, S.; Xiao, X.; et al. Application of Plant Growth-Promoting Endophytes (PGPE) Isolated from *Solanum nigrum* L. for Phytoextraction of Cd-Polluted Soils. *Appl. Soil Ecol.* **2010,** *46*, 383–389.

Chen, Y.; Wang, H. W.; Li, L.; Dai, C. C. The Potential Application of the Endophyte *Phomopsis liquidambari* to the Ecological Remediation of Long-Term Cropping Soil. *Appl. Soil Ecol.* **2013,** *67*, 20–26.

Chhabra, S.; Brazil, D.; Morrissey, J.; et al. Characterization of Mineral Phosphate Solubilization Traits from a Barley Rhizosphere Soil Functional Metagenome. *Microbiol. Open* **2013,** *2*, 717–724.

Chowdappa, S.; Jagannath, S.; Konappa, N.; et al. Detection and Characterization of Antibacterial Siderophores Secreted by Endophytic Fungi from *Cymbidium aloifolium*. *Biomolecules* **2020,** *10*, 1412.

Cohen, A.; Travaglia, C.; Bottini, R.; Piccoli, P. Participation of Abscisic Acid and Gibberellins Produced by Endophytic *Azospirillum* in the Alleviation of Drought Effects in Maize. *Botany* **2009,** *87*, 455–462.

Colla, G.; Rouphael, Y.; Bonini, P.; et al. Coating Seeds with Endophytic Fungi Enhances Growth, Nutrient Uptake, Yield and Grain Quality of Winter Wheat. *Int. J. Plant Prod.* **2015,** *9*, 171–189.

Coombs, J. T.; Michelson, P. P.; Franco, C. M. M. Evaluation of Endophytic Actinobacteria as Antagonists of *Gaeumannomyces graminis* var. *tritici* in Wheat. *Biol. Control* **2004,** *29*, 359–366.

Culliney, T. W. Crop Losses to Arthropods. In *Integrated Pest Management*; Pimentel, D., Peshin, R., Eds.; Springer, 2014; Vol. 3, pp 201–225.

Das, A.; Varma, A. Symbiosis: The Art of Living. In *Symbiotic Fungi, Principles and Practice*; Varma, A., Kharkwal, A. C., Eds.; Springer: Berlin, 2009; pp 1–28.

Dastogeer, K. M.; Li, H.; Sivasithamparam, K.; et al. Fungal Endophytes and a Virus Confer Drought Tolerance to *Nicotiana benthamiana* Plants Through Modulating Osmolytes, Antioxidant Enzymes and Expression of Host Drought Responsive Genes. *Environ. Exp. Bot.* **2018,** *149*, 95–108.

Davies, P. J. The Plant Hormone Concept: Concentration, Sensitivity and Transport. In *Plant Hormones*; Davies, P. J., Ed.; Springer, 1995; pp 13–38.

de Abreu, C. S.; Figueiredo, J. E.; Oliveira, C. A.; et al. Maize Endophytic Bactéria as Mineral Phosphate Solubilizers. *Genet. Mol. Res.* **2017,** *16*, 1.

Deng, Z.; Zhang, R.; Shi, Y.; et al. Enhancement of Phytoremediation of Cd- and Pb-Contaminated Soils by Self-Fusion of Protoplasts from Endophytic Fungus *Mucor* sp. CBRF59. *Chemosphere* **2013,** *91*, 41–47.

Dey, S.; Pal, S. Endophytes, Promising Option for Pest Management in Agriculture: Review. *J. Entomol. Zool. Stud.* **2020,** *8*, 157–162.

Dhungana, S. A.; Itoh, K. Effects of Co-Inoculation of Indole-3-Acetic Acid Producing and -Degrading Bacterial Endophytes on Plant Growth. *Horticulturae* **2019,** *5*, 2–10.

Dinckinson, N. M.; Baker, A. J. M.; Doronila, A.; et al. Phytoremediation of Inorganics: Realism and Synergies. *Int. J. Phytoremed.* **2009,** *11*, 97–114.

Dombrowski, J. E. Salt Stress Activation of Wound-Related Genes in Tomato Plants. *Plant Physiol.* **2003,** *132*, 2098–2107.

Doran, J. W.; Zeiss, M. R. Soil Health and Sustainability: Managing the Biotic Component of Soil Quality. *Appl. Soil Ecol.* **2000,** *15*, 3–11.

Duca, D.; Lorv, J.; Patten, C. L.; et al. Indole-3-Acetic Acid in Plant-Microbe Interactions. *Anton. Leeuw. Int. J. G.* **2014,** *106*, 85–125.

Eid, A. M.; Salim, S. S.; Hassan, S. E.-D.; et al. Role of Endophytes in Plant Health and Abiotic Stress Management. In *Microbiome in Plant Health and Disease, Challenges and Opportunities*; Kumar, V., *Prasad*, R., Kumar, M., Choudhary, D. K., Eds.; Springer: Singapore, 2019; pp 119–144.

Elbeltagy, A.; Nishioka, K.; Sato, T.; et al. Endophytic Colonization and in Planta Nitrogen Fixation by a *Herbaspirillum* sp Isolated from Wild Rice Species. *Appl. Environ. Microbiol.* **2001,** *67*, 5285–5293.

Estrada, G. A.; Baldani, V. L. D.; de Oliveira, D. M. Selection of Phosphate-Solubilizing Diazotrophic *Herbaspirillum* and *Burkholderia* Strains and Their Effect on Rice Crop Yield and Nutrient Uptake. *Plant Soil* **2013,** *369*, 115–129.

Etesami H.; Emami, S.; Alikhani, H. A. Potassium Solubilizing Bacteria (KSB): Mechanisms, Promotion of Plant Growth, and Future Prospects a Review. *J. Soil Sci. Plant Nutr.* **2017,** *17*, 897–911.

Fávaro, L. C.; Sebastianes, F. L.; Araújo, W. L. *Epicoccum nigrum* P16, a Sugarcane Endophyte, Produces Antifungal Compounds and Induces Root Growth. *PLoS One* **2012,** *7* (6), e36826.

Feng, J.; Barker, A. V. Ethylene Evolution and Ammonium Accumulation by Tomato Plants Under Water and Salinity Stresses. Part II. *J. Plant Nutr.* **1992,** *15*, 2471–2490.

Fernandez, O.; Vandesteene, L.; Feil, R.; et al. Trehalose Metabolism Is Activated upon Chilling in Grapevine and Might Participate in *Burkholderia phytofirmans* Induced Chilling Tolerance. *Planta* **2012,** *236*, 355–369.

Foley, J. A.; Defries, R.; Asner, G. P.; et al. Global Consequences of Land Use. *Science* **2005,** *309* (5734), 570–574.
Forchetti, G.; Masciarelli, O.; Alemano, S.; et al. Endophytic Bacteria in Sunflower (*Helianthus annuus*): Isolation, Characterization, and Production of Jasmonates and Abscisic Acid in Culture Medium. *Appl. Microbiol. Biotechnol.* **2007,** *76*, 1145–1152.
Fu, W.; Xu, M.; Sun, K.; et al. Biodegradation of Phenanthrene by Endophytic Fungus *Phomopsis liquidambari* In Vitro and In Vivo. *Chemosphere* **2018,** *203*, 160–169.
Fuentes-Ramirez, L. E.; Jiminez-Salgado, T.; Abarca-Ocampo, I. R.; et al. *Acetobacter diazotrophicus* an Indolacetic Acid-Producing Bacterium Isolated from Sugarcane Cultivars in Mexico. *Plant Soil* **1993,** *154*, 145–150.
Furhan, J. Adaptation, Production, and Biotechnological Potential of Cold-Adapted Proteases from Psychrophiles and Psychrotrophs: Recent Overview. *J. Genet. Eng. Biotechnol.* **2020,** *18*, 36.
Germaine, K. J.; Keogh, E.; Ryan, D.; Dowling, D. Bacterial Endophyte-Mediated Naphthalene Phytoprotection and Phytoremediation. *FEMS Microbiol. Lett.* **2009,** *296*, 226–234.
Glick, B. R.; Cheng, Z.; Czarny, J. Dual Promotion of Plant Growth by ACC Deaminase-Producing Soil Bacteria. *Eur. J. Plant Pathol.* **2007,** *119*, 329–339.
Govindarajan, M.; Balandreau, J.; Kwon, S. W.; et al. Effects of the Inoculation of *Burkholderia vietnamensis* and Related Endophytic Diazotrophic Bacteria on Grain Yield of Rice. *Microb. Ecol.* **2008,** *55*, 21–37.
Greene, E. M. Cytokinin Production by Microorganisms. *Bot. Rev.* **1980,** *46*, 25–74.
Guo, B.; Wang, Y.; Sun, X.; et al. Bioactive Natural Products from Endophytes: A Review. *Appl. Biochem. Microbiol.* **2008,** *44*, 136–142.
Gupta, N.; Gupta, A. K.; Gaur, V. S.; Kumar, A. Relationship of Nitrogen Use Efficiency with the Activities of Enzymes Involved in Nitrogen Uptake and Assimilation of Finger Millet Genotypes Grown Under Different Nitrogen Inputs. *Sci. World J.* **2012,** *2012*, 1–10.
Hallmann, J.; Quadt-Hallmann, A.; Mahaffee, W.; et al. Bacterial Endophytes in Agricultural Crops. *Can. J. Microbiol.* **1997,** *43*, 895–914.
Hallmann, J.; Sikora, R. A. Toxicity of Fungal Endophyte Secondary Metabolites to Plant Parasitic Nematodes and Soil Borne Plant Pathogenic Fungi. *Eur. J. Plant. Pathol.* **1996,** *102*, 155–162.
Hamayun, M.; Hussain, A.; Khan, S. A.; et al. Gibberellins Producing Endophytic Fungus *Porostereum spadiceum* AGH786 Rescues Growth of Salt Affected Soybean. *Front Microbiol.* **2017,** *8*, 686.
Han, X. F.; Zeng, H.; Bartocci, P.; et al. Phytohormones and Effects on Growth and Metabolites of Microalgae: A Review. *Fermentation* **2018,** *4*, 25.
Harman, G. E.; Doni, F.; Khadka, R. B.; Uphoff, N. Endophytic Strains of *Trichoderma* Increase Plants Photosynthetic Capability. *J. Appl. Microbiol.* **2019,** https://doi.org/10.1111/jam.14368.
Hashim, N. H.; Bharudin, I.; Nguong, D. L.; et al. Characterization of Afp1, an Antifreeze Protein from the Psychrophilic Yeast *Glaciozyma antarctica*. *Extremophiles* **2013,** *17*, 63–73.
He, W.; Megharaj, M.; Wu, C. Y.; et al. Endophyte-Assisted Phytoremediation: Mechanisms and Current Application Strategies for Soil Mixed Pollutants. *Crit. Rev. Biotechnol.* **2020,** *40*, 31–45.
Heck, M. Insect Transmission of Plant Pathogens: A Systems Biology Perspective. *mSystems* **2018,** *3* (2), e00168-17. https://doi.org/10.1128/mSystems.0016817.
Hendrickson, L.; Ball, M. C.; Wood, J. T.; et al. Low Temperature Effects on Photosynthesis and Growth of Grapevine. *Plant Cell Environ.* **2004,** *27*, 795–809.

Hu, H. J.; Chen, Y. L.; Wang, Y. F.; et al. Endophytic *Bacillus cereus* Effectively Controls *Meloidogyne incognita* on Tomato Plants Through Rapid Rhizosphere Occupation and Repellent Action. *Plant Dis.* **2017,** *101*, 448–455.

Hubbard, M.; Germida, J. J.; Vujanovic V. Fungal Endophytes Enhance Wheat Heat and Drought Tolerance in Terms of Grain Yield and Second-Generation Seed Viability. *J. Appl. Microbiol.* **2014,** *116*, 109–122.

Ihsan, M. Z.; Daur, I.; Alghabari, F.; et al. Heat Stress and Plant Development: Role of Sulphur Metabolites and Management Strategies. *Acta Agric. Scand. Sect. B Soil Plant Sci.* **2019,** *69*, 332–342.

Illmer, P.; Barbato, A.; Schinner, F. Solubilization of Hardly Soluble AlPO4 with P-Solubilizing Microorganisms. *Soil Biol. Biochem.* **1995,** *27*, 260–270.

Illmer, P. A.; Schinner, F. Solubilization of Inorganic Phosphates by Microorganisms Isolated from Forest Soil. *Soil Biol. Biochem.* **1999,** *24*, 389–395.

Ismail, M. H.; Hussain, A.; Iqbal, A.; et al. Endophytic Fungus *Aspergillus japonicas* Mediates Host Plant Growth Under Normal and Heat Stress Conditions. *Biol. Med. Res.* **2018,** *31*, 1–12.

Joe, M. M.; Devaraj, S.; Benson, A.; Sa, T. Isolation of Phosphate Solubilizing Endophytic Bacteria from *Phyllanthus amarus* Schum & Thonn: Evaluation of Plant Growth Promotion and Antioxidant Activity Under Salt Stress. *J. Appl. Res. Med. Aromat. Plants* **2016,** *3*, 71–77.

Kajula, M.; Tejesvi, M. V.; Kolehmainen, S.; et al. The Siderophore Ferricrocin Produced by Specific Foliar Endophytic Fungi *In Vitro*. *Fungal Biol.* **2010,** *114*, 248–254.

Kamran, S.; Shahid, I.; Baig, D. N.; et al. Contribution of Zinc Solubilizing Bacteria in Growth Promotion and Zinc Content of Wheat. *Front Microbiol.* **2017,** *8*, 2593.

Karthik M.; Oves R.; Thangabalu, R.; et al. *Cellulosimicrobium funkei*-Like Enhances the Growth of *Phaseolus vulgaris* by Modulating Oxidative Damage Under Chromium (VI) Toxicity. *J. Adv. Res.* **2016,** *7*, 839–850.

Khan, A. A.; Jilani, G.; Akhtar, M. S.; et al. Phosphorus Solubilizing Bacteria: Occurrence, Mechanisms and Their Role in Crop Production. *J. Agric. Biol. Sci.* **2009,** *1*, 48–58.

Khan, A. L.; Hamayun, M.; Ahmad, N.; et al. Salinity Stress Resistance Offered by Endophytic Fungal Interaction Between *Penicillium minioluteum* LHL09 and *Glycine max.* L. *J. Microbiol. Biotechnol.* **2011,** *21*, 893–902.

Khan, A. L.; Hamayun, M.; Kang, S. M.; et al. Endophytic Fungal Association Via Gibberellins and Indole Acetic Acid Can Improve Plant Growth Under Abiotic Stress: An Example of *Paecilomyces formosus* LHL10. *BMC Microbiol.* **2012,** *12*, 3.

Khan, A. L.; Waqas, M.; Kang, S. M. Bacterial Endophytes *Sphingomonas* sp LK11 Produces Gibberellins and IAA and Promotes Tomato Plant Growth. *J. Microbiol.* **2014,** *52*, 689–695.

Khan, A. L.; Halo, B. A.; Elyassi, A.; et al. Indole Acetic Acid and ACC Deaminase from Endophytic Bacteria Improves the Growth of *Solanum lycopersicum*. *Electron J. Biotechnol.* **2016,** *21*, 58–64.

Khan, B.; Yan, W.; Wei, S.; et al. Nematicidal Metabolites from Endophytic Fungus *Chaetomium globosum* YSC5. *FEMS Microbiol. Lett.* **2019,** *366*, fnz169.

Khan, M. A.; Asaf, S.; Khan, A. L.; et al. Thermotolerance Effect of Plant Growth-Promoting *Bacillus cereus* SA1 on Soyabean During Heat Stress. *BMC Microbiol.* **2020,** *20*, 175. https://doi.org/10.1186/s12866-020-01822-7.

Kieber, J. J. Cytokinins. In *The Arabidopsis Book*, Somerville, C. R., Meyerowitz, E. M., Eds.; American Society of Plant Biologists: Rockville, MD, 2002; http://www.aspb.org/publications/arabidopsis/toc.cfm.

Kumar, J.; Singh, D.; Ghosh, P.; et al. Endophytic and Epiphytic Modes of Microbial Interactions and Benefits. In *Plant-Microbe Interactions in Agro-Ecological Perspectives*; *Singh*, D. P., *Singh*, H. B., Prabha, R., Eds.; Springer Nature: Singapore, 2017b; pp 227–253.
Kumar, M.; Saxena, R.; Tomar, R. S. Endophytic Microorganisms: Promising Candidate as Biofertilizer. In *Microorganisms for Green Revolution*, Panpatte, D., Jhala, Y. K., Vyas, R. V., *Shelat*, H., Eds.; Springer International Publishing Science: Singapore, 2017a; Vol. 1., pp 77–85.
Kumar, R.; Chaurasiya, P. C.; Singh, R. N.; et al. A Review Report: Low Temperature Stress for Crop Production. *Int. J. Pure Appl. Biosci.* **2018,** *6*, 575–598.
Kusari, S.; Lamshöft, M.; Zuhlke, S.; *Spiteller*, M. An Endophytic Fungus from *Hypericum perforatum* that Produces Hypericin. *J. Nat. Prod.* **2008,** *71*, 159–162.
Kusari, S.; Spiteller, M. Camptothecin: Recent Advances in Plant Endophyte Research. In *Natural Resources Conservation and Management*; Patro, L. R., Ed.; Manglam Publications: New Delhi, 2012; pp 1–32.
Kushwaha, P.; Kashyap, P. L.; Srivastava, A. K.; Tiwari, R. K. Plant Growth Promoting and Antifungal Activity in Endophytic Bacillus Strains from Pearl Millet (*Pennisetum glaucum*). *Braz. J. Microbiol.* **2020,** *51*, 229–241.
Laabs, V.; Amelung, W.; Pinto, A.; et al. Leaching and Degradation of Corn and Soybean Pesticides in an Oxisol of Brazilian Cerrados. *Chemosphere* **2000,** *41*, 1441–1449.
Lacava, P. T.; Azevedo, J. L. Endophytic Bacteria: A Biotechnological Potential in Agrobiology System. In *Bacteria in Agrobiology: Crop Productivity*; *Maheshwari*, D. K., Saraf, M., Aeron, A., Eds.; Springer: Berlin Heidelberg, 2013; pp 1–44.
Lacava, P. T.; Silva-Stenico, M. E.; Araújo, W. L.; et al. Detection of Siderophores in Endophytic Bacteria *Methylobacterium* spp. Associated with *Xylella fastidiosa* subsp. *Pauca. Pesq. Agropec. Bras.* **2008,** *43*, 521–528.
Laghari, S. J.; Wahocho, N. A.; Laghari, G. M.; et al. Role of Nitrogen for Plant Growth and Development: A Review. *Adv. Environ. Biol.* **2016,** *10*, 209–218.
Lata, R.; Chowdhury, S.; Gond, S. K.; White, J. F. Jr. Induction of Abiotic Stress Tolerance in Plants by Endophytic Microbes. *Lett. Appl. Microbiol.* **2018,** *66*, 268–276.
Loper, J. E.; Henkels, M. D. Utilization of Heterologous Siderophores Enhances Levels of Iron Available to *Pseudomonas putida* in the Rhizosphere. *Appl. Environ. Microbiol.* **1999,** *65*, 5357–5363.
Madhaiyan, M.; Poonguzhali, S.; Sa, T. Metal Tolerating Methylotrophic Bacteria Reduces Nickel and Cadmium Toxicity and Promotes Plant Growth of Tomato (*Lycopersicon esculentum* L.). *Chemosphere* **2007,** *69*, 220–228.
Maheshwari, R.; Bhutani, Suneja, R. Screening and Characterization of Siderophore Producing Endophytic Bacteria from *Cicer arietinum* and *Pisum sativum* plants. *J. Appl. Biol. Biotechnol.* **2019,** *7*, 7–14.
Mahmood, I.; Imadi, S. R.; Shazadi, K.; et al. Effects of Pesticides on Environment. In *Plant, Soil and Microbes*; *Implications in Plant Science*; Hakeem, K. R., Akhtar, M. S., Abdullah, S. N. A., Eds.; Springer International Publishing, 2016; Vol. 1, pp 253–269.
Majeed, A. Application of Agrochemicals in Agriculture: Benefits, Risks and Responsibility of Stakeholders. *J. Food Sci. Toxicol.* **2018,** *2*, 1.
Makino, A.; Mae, T. Photosynthesis and Plant Growth at Elevated Levels of CO_2. *Plant Cell Physiol.* **1999,** *40*, 999–1006.
Marschner, H. *Mineral Nutrition of Higher Plants*; Academic Press: London, 1986.

Mehnaz, S.; Bauer, J. S.; Gross, H. Complete Genome Sequence of the Sugar Cane Endophyte *Pseudomonas aurantiaca* PB-St2, a Disease-Suppressive Bacterium with Antifungal Activity Toward the Plant Pathogen *Colletotrichum falcatum*. *Genome Announc.* **2014,** *2*, e01108–e01113.

Morsy, M.; Cleckler, B.; Armuelles-Millican, H. Fungal Endophytes Promote Tomato Growth and Enhance Drought and Salt Tolerance. *Plants* **2020,** *9* (7), 877.

Mousa, W. K.; Raizada, M. N. The Diversity of Anti-Microbial Secondary Metabolites Produced by Fungal Endophytes: An Interdisciplinary Perspective. *Front Microbiol.* **2013,** *4*, 65.

Mukherjee, G.; Saha, C.; Naskar, N.; et al. An Endophytic Bacterial Consortium Modulates Multiple Strategies to Improve Arsenic Phytoremediation Efficacy in *Solanum nigrum*. *Sci. Rep.* **2018,** *8*, 6979.

Nagendran, K.; Karthikeyan, G.; Peeran, M. F.; et al. Management of Bacterial Leaf Blight Disease in Rice with Endophytic Bacteria. *World Appl. Sci. J.* **2013,** *28*, 2229–2241.

Nath, R.; Sharma, G.; Barooah, M. Plant Growth Promoting Endophytic Fungi Isolated from Tea (*Camellia sinensis*) Shrubs of Assam, India. *Appl. Ecol. Environ. Res.* **2015,** *13*, 877–891.

Nettles, R.; Watkins, J.; Ricks, K.; et al. Influence of Pesticide Seed Treatments on Rhizosphere Fungal and Bacterial Communities and Leaf Fungal Endophyte Communities in Maize and Soybean. *Appl. Soil Ecol.* **2016,** *102*, 61–69.

Newman, L. A.; Reynolds, C. M. Bacteria and Phytoremediation: New Uses for Endophytic Bacteria in Plants. *Trends Biotechnol.* **2005,** *23*, 6–8.

Ngwene, B.; Boukail, S.; Söllner, L.; et al. Phosphate Utilization by the Fungal Root Endophyte *Piriformospora indica*. *Plant Soil* **2016,** *405*, 1–11.

Nieves-Cordones, M.; García-Sánchez, F.; Pérez-Pérez, J. G.; et al. Coping with Water Shortage: An Update on the Role of K+, Cl^-, and Water Transport Mechanisms on Drought Resistance. *Front Plant Sci.* **2019,** *10*, 1619.

Numponsak, T.; Kumla, J.; Suwannarach, N.; et al. Biosynthetic Pathway and Optimal Conditions for the Production of Indole-3-Acetic Acid by an Endophytic Fungus, *Colletotrichum fructicola* CMU-A109. *PLoS One* **2018,** *13* (10), e0205070.

Ohra, J.; Morita, K.; Tsujino, Y.; et al. Production of the Phytotoxic Metabolite, Ferricrocin, by the Fungus *Colletotrichum gloeosporioides*. *Biosci. Biotechnol. Biochem.* **1995,** *59*, 113–114.

Patil, R. H.; Patil, M. P.; Maheshwari, V. L. Bioactive Secondary Metabolites from Endophytic Fungi: A Review of Biotechnological Production and Their Potential Applications. In *Studies in Natural Products Chemistry*; Rahman, A. U., Ed.; Elsevier: The Netherlands, 2016; pp 89–205.

Pearce, R. S. Plant Freezing and Damage. *Ann. Bot. (Lond.)* **2001,** *87*, 417–424.

Phetcharat, P.; *Duangpaeng*, A. Screening of Endophytic Bacteria from Organic Rice Tissue for Indole Acetic Acid Production. *Procedia Eng.* **2012,** *32*, 177–183.

Pirhadi, M.; Enayatizamir, N.; Motamedi, H.; Sorkheh K. Screening of Salt Tolerant Sugarcane Endophytic Bacteria with Potassium and Zinc for Their Solubilizing and Antifungal Activity. *Biosci. Biotech. Res. Commun.* **2016,** *9*, 530–538.

Ponpandian, L. N.; Rim, S. O.; Shanmugam, G.; et al. Phylogenetic Characterization of Bacterial Endophytes from Four *Pinus* Species and Their Nematicidal Activity Against the Pine Wood Nematode. *Sci. Rep.* **2019,** *9*, 12457.

Prasad, M. N. V. Phytoremediation of Metals and Radionuclides in the Environment: The Case for Natural Hyperaccumulators, Metal Transporters, Soil-Amending Chelators and

Transgenic Plants. In *Heavy Metal Stress in Plants, From Molecules to Ecosystems,* 2nd ed.; Prasad, M. N. V., Ed.; Berlin: Springer, 2004; pp 345–391.

Puente, M. E.; Li, C. Y.; Bashan, Y. Rock-Degrading Endophytic Bacteria in Cacti. *Environ. Exp. Bot.* **2009,** *66*, 389–401.

Puri, A.; Padda, K. P.; Chanway, C. P. Seedling Growth Promotion and Nitrogen Fixation by a Bacterial Endophyte *Paenibacillus polymyxa* P2b-2R and Its GFP Derivative in Corn in a Long-Term Trial. *Symbiosis* **2016,** *69*, 123–129.

Radhakrishnan, R.; Khan, A. L.; Kang, S. M.; Lee, I. J. A Comparative Study of Phosphate Solubilization and the Host Plant Growth Promotion Ability of *Fusarium verticillioides* RK01 and *Humicola* sp. KNU01 Under Salt Stress. *Ann. Microbiol.* **2015,** *65*, 585–593.

Rai, M.; Rathod, D.; Agarkar, G.; et al. Fungal Growth Promotor Endophytes: A Pragmatic Approach Towards Sustainable Food and Agriculture. *Symbiosis* **2014,** *62*, 63–79.

Ramanuj, K. B.; Shelat, H. N. Plant Growth Promoting Potential of Bacterial Endophytes from Medicinal Plants. *Adv. Res.* **2018,** *13*, 1–15.

Redman, R. S.; Freeman, S.; Clifton, D. R.; et al. Biochemical Analysis of Plant Protection Afforded by a Non-Pathogenic Endophytic Mutant of *Colletotrichum magna*. *Plant Physiol.* **1999,** *119*, 795–804.

Redman, R. S.; Kim, Y. O., Woodward, C. J.; et al. Increased Fitness of Rice Plants to Abiotic Stress via Habitat Adapted Symbiosis: A Strategy for Mitigating Impacts of Climate Change. *PLoS One* **2011,** *6* (7), e14823.

Rehm, G.; Schmitt, M. *Potassium for Crop Production*; 2002. Retrieved February 2, 2011, from Regents of the University of Minnesota Website: http://www.extension.umn.edu/distribution/cropsystems/dc6794.html.

Rho, H.; Doty, S. L.; Kim, S. H. Endophytes Alleviate the Elevated CO_2-Dependent Decrease in Photosynthesis in Rice, Particularly Under Nitrogen Limitation. *J. Exp. Bot.* **2019,** *71*, 707–718.

Rho, H.; Kim, S. H. Endophyte Effects on Photosynthesis and Water Use of Plant Hosts: A Meta-Analysis. In *Functional Importance of the Plant Microbiome*; Doty, S., Ed.; Springer Nature: Cham, 2017; pp 43–70.

Ribeiro, M.; *Simões*, M. Advances in the Antimicrobial and Therapeutic Potential of Siderophores. *Environ. Chem. Lett.* **2019,** *17*, 1485–1494.

Rosa, L. H.; Vieira, M. L. A.; Santiago, I. F.; et al. Endophytic Fungi Community Associated with the Dicotyledonous Plant *Colobanthus quitensis* (Kunth) Bartl. (Caryophyllaceae) in Antarctica. *FEMS Microbiol. Ecol.* **2010,** *73*, 178–189.

Rout, G. R.; Gyana, S. Role of Iron in Plant Growth and Metabolism. *Rev. Agric. Sci.* **2015,** *3*, 1–2.

Rozpądek, P.; Wężowicz, K.; Nosek, M.; et al. The Fungal Endophyte *Epichloë typhina* Improves Photosynthesis Efficiency of Its Host Orchard Grass (*Dactylis glomerata*). *Planta* **2015,** *242*, 1025–1035.

Rungin, S.; Indanand, C.; Suttiviriya, P.; et al. Plant Growth Enhancing Effects by a Siderophore Producing Endophytic Streptomycete Isolated from a Thai Jasmine Rice Plant (*Oryza sativa* L. cv. KDML105). *Ant. van. Leeuw.* **2012,** *102*, 463–472.

Sadrati, N.; Daoud, H.; Zerrog, A.; et al. Screening of Antimicrobial and Antioxidant Secondary Metabolites from Endophytic Fungi Isolated from Wheat (*Triticum durum*). *J. Plant Prot. Res.* **2013,** *53*, 128–136.

Saha, M.; Sarkar, S.; Sarkar, B.; et al. Microbial Siderophores and Their Potential Applications: A Review. *Environ. Sci. Pollut. Res.* **2016,** *23*, 3984–3999.

Sahoo, S.; Sarangi, S.; Kerry, R. G. Bioprospecting of Endophytes for Agricultural and Environmental Sustainability. In *Microbial Biotechnology*; Patra, J. K., Vishnuprasad, C. N., Das, G., Eds.; Springer: Berlin Heidelberg, 2018; Vol. 1. pp 429–458.

Saikkonen, K.; Faeth, S. H.; Helander, M.; et al. Fungal Endophytes: A Continuum of Interactions with Host Plants. *Annu. Rev. Ecol. Syst.* **1998,** *29*, 319–343.

Sampangi-Ramaiah, M. H.; Dey, P.; Jambagi, S.; et al. An Endophyte from Salt-Adapted Pokkali Rice Confers Salt-Tolerance to a Salt-Sensitive Rice Variety and Targets a Unique Pattern of Genes in Its New Host. *Sci. Rep.* **2020,** *10*, 1–14.

Sandhya, V.; Shrivastava, M.; Ali, S. Z.; Prasad, V. S. S. K. Endophytes from Maize with Plant Growth Promotion and Biocontrol Activity Under Drought Stress. *Russ. Agric. Sci.* **2017,** *43*, 22–34.

Sangakkara, U. R.; Frehner, M.; Nosberger, J. Effect of Soil Moisture and Potassium Fertilizer on Shoot Water Potential, Photosynthesis and Partitioning of Carbon in Mungbean and Cowpea. *J. Agron. Crop Sci.* **2000,** *185*, 201–207.

Saravanakumar, D.; Samiyappan, R. ACC Deaminase from *Pseudomonas fluorescens* Mediated Saline Resistance in Groundnut (*Arachis hypogea*) Plants. *J. Appl. Microbiol.* **2007,** *102*, 1283–1292.

Schulz, B.; Boyle, C. The Endophytic Continuum. *Mycol. Res.* **2005,** *109*, 661–686.

Schulz, B.; Boyle, C.; Draeger, S.; et al. Endophytic Fungi: A Source of Novel Biologically Active Secondary Metabolites. *Mycol. Res.* **2002,** *109*, 996–1004.

Selim, K. A.; El-Beih, A. A.; Abdel-Rahman, T. M.; et al. Biodiversity and Antimicrobial Activity of Endophytes Associated with Egyptian Medicinal Plants. *Mycosphere* **2011,** *2*, 669–678.

Shahzad, R.; Khan, A. L.; Bilal, S.; et al. Plant Growth-Promoting Endophytic Bacteria Versus Pathogenic Infections: An Example of *Bacillus amyloliquefaciens* RWL-1 and *Fusarium oxysporum* f. sp. *lycopersici* in Tomato. *Peer J.* **2017,** *5*, e3107.

Sharma, P.; Kunawat, K. C.; Kaur, S.; Kaur, N. Assessment of Zinc Solubilization by Endophytic Bacteria in Legume Rhizosphere. *Ind. J. Appl. Res.* **2014,** *4*, 439–441.

Sharma, S. B.; Sayyed, R. Z.; Trivedi, M. H.; Gobi, T. A. Phosphate Solubilizing Microbes: Sustainable Approach for Managing Phosphorus Deficiency in Agricultural Soils. *Springer Plus* **2013,** *2*, 1.

Sheldon, A.; Menzies, N. W.; So, H. B.; Dalal, R. The Effect of Salinity on Plant Available Water. In *3rd Australian New Zealand Soils Conference (December): 2004, 5*. http://www.regional.org.au/au/asss.

Shi, Y.; Xie, H.; Cao, L.; et al. Effects of Cd- and Pb-Resistant Endophytic Fungi on Growth and Phytoextraction of *Brassica napus* in Metal-Contaminated Soils. *Environ. Sci. Pollut. Res.* **2017,** *24*, 417–426.

Shi, Y. W.; Lou, K.; Li, C. Growth and Photosynthetic Efficiency Promotion of Sugar Beet (*Beta vulgaris* L.) by Endophytic Bacteria. *Photosynth. Res.* **2010,** *105*, 5–13.

Shi, Y. W.; Zhang, X.; Lou, K. Isolation, Characterization, and Insecticidal Activity of an Endophyte of Drunken Horse Grass, *Achnatherum inebrians*. *J. Insect Sci.* **2013,** *13*, 151.

Shrivastava, P.; Kumar, R. Soil Salinity: A Serious Environmental Issue and Plant Growth Promoting Bacteria as One of the Tools for Its Alleviation. *Saudi J. Biol. Sci.* **2015,** *22*, 123–131.

Singh, D.; Geat, N.; Rajawat, M. V. S.; et al. Prospecting Endophytes from Different Fe or Zn Accumulating Wheat Genotypes for Their Influence As Inoculants on Plant Growth, Yield and Micronutrient Content. *Ann. Microbiol.* **2018,** *68*, 815–833.

Soleimani, M.; Afyuni, M.; Hajabbasi, M. A.; et al. Phytoremediation of an Aged Petroleum Contaminated Soil Using Endophyte Infected and Non-Infected Grasses. *Chemosphere* **2010,** *81*, 1084–1090.

Spagnoletti, F.; Tobar, N.; Di Pardo, A. F.; et al. Dark Septate Endophytes Present Different Potential to Solubilize Calcium, Iron and Aluminium Phosphates. *Appl. Soil Ecol.* **2017,** *111*, 25–32.

Sparks, D. L. Potassium Dynamics in Soils. *Adv. Soil Sci.* **1987,** *6*, 1–63.

Stella, M.; Halimi, M. Gluconic acid Production by Bacteria to Liberate Phosphorus from Insoluble Phosphate Complexes. *J. Trop. Agric. Food Sci.* **2015,** *43*, 41–53.

Stuart, A. K.; Stuart, R. M.; Pimentel, I. C.; et al. Effect of Agrochemicals on Endophytic Fungi Community Associated with Crops of Organic and Conventional Soybean (*Glycine max* L. Merril). *Agric. Nat. Resour.* **2018,** *52*, 388–392.

Stuart, R. M.; Romão, A. S.; Pizzirani-Kleiner, A. A.; et al. Culturable Endophytic Filamentous Fungi from Leaves of Transgenic Imidazolinone-Tolerant Sugarcane and Its Non-Transgenic Isolines. *Arch. Microbiol.* **2010,** *192*, 307–313.

Sudha, V.; Govindaraj, R.; Baskar, K.; et al. Biological Properties of Endophytic Fungi. *Braz. Arch. Biol. Technol.* **2016,** *59*, e16150436.

Suman, A.; Gaur, A.; Shrivastava, A.; et al. Improving Sugarcane Growth and Nutrient Uptake by Inoculating *Gluconacetobacter diazotrophicus*. *Plant Growth Regul.* **2005,** *47*, 155–162.

Suman, A.; Shrivastava, A. K.; Gaur, A.; et al. Nitrogen Use Efficiency of Sugarcane in Relation to Its BNF Potential and Population of Endophytic Diazotrophs at Different n Levels. *Plant Growth Regul.* **2008,** *54*, 1–11.

Suman, A.; Yadav, A. N.; Verma, P. Endophytic Microbes in Crops: Diversity and Beneficial Impact for Sustainable Agriculture. In *Microbial Inoculants in Sustainable Agricultural Productivity*; *Research Perspectives*; Singh, D. P., Singh, H. B., Prabha, R., Eds.; Springer India: New Delhi, 2016; Vol. 1, pp 117–143.

Sun, C.; Johnson, J. M.; Cai, D.; et al. *Piriformospora indica* Confers Drought Tolerance in Chinese Cabbage Leaves by Stimulating Antioxidant Enzymes, the Expression of Drought-Related Genes and the Plastid-Localized CAS Protein. *J. Plant Physiol.* **2010,** *167*, 1009–1017.

Syamsia, S.; Idhan, A.; Noerfitryani, N.; et al. Paddy Chlorophyll Concentrations in Drought Stress Condition and Endophytic Fungi Application. *IOP Conf. Ser. Earth Environ. Sci.* **2018,** *156* (1), 012040.

Syamsia, S.; Kuswinanti, T.; Syamun, E.; et al. The Potency of Endophytic Fungal Isolates Collected from Local Aromatic Rice as Indole Acetic Acid (IAA) Producer. *Procedia Food Sci.* **2015,** *3*, 96–103.

Thakur, A.; Singh, V.; Kaur, A.; et al. Insecticidal Potential of an Endophytic Fungus, *Cladosporium uredinicola*, Against *Spodoptera litura*. *Phytoparasitica* **2013,** *41*, 373–382.

Tivendale, N. D.; Ross, J. J.; Cohen, J. D. The Shifting Paradigms of Auxin Biosynthesis. *Trends Plant Sci.* **2014,** *19*, 44–51.

Ullah, A.; Farooq, M.; Nadeem, F.; et al. Zinc Application in Combination with Zinc Solubilizing *Enterobacter* sp. MN17 Improved Productivity, Profitability, Zinc Efficiency and Quality of Desi Chickpea. *J. Soil Sci. Plant Nutr.* **2020,** *20*, 2133–2144.

Ullah, A.; Nisar, M.; Ali, H.; et al. Drought Tolerance Improvement in Plants: An Endophytic Bacterial Approach. *Appl. Microbiol. Biotechnol.* **2019,** *103*, 7385–7397.

UmaMaheswari, T.; Anbukkarasi, K.; Hemalatha, T.; Chendrayan, K. Studies on Phytohormone Producing Ability of Indigenous Endophytic Bacteria Isolated from Tropical Legume Crops. *Int. J. Curr. Microbiol. Appl. Sci.* **2013,** *2*, 127–136.

Vergara, C.; Araujo, K. E. C.; Urquiaga, S.; et al. Dark Septate Endophytic Fungi Help Tomato to Acquire Nutrients from Ground Plant Material. *Front Microbiol.* **2017,** *8*, 2437.

Verma, P.; Yadav, A. N.; Kazy, S. K.; et al. Evaluating the Diversity and Phylogeny of Plant Growth Promoting Bacteria Associated with Wheat (*Triticum aestivum*) Growing in Central Zone of India. *Int. J. Curr. Microbiol. Appl. Sci.* **2014,** *3*, 432–447.

Verma, P.; Yadav, A. N.; Khannam, K. S.; et al. Assessment of Genetic Diversity and Plant Growth Promoting Attributes of Psychrotolerant Bacteria Allied with Wheat (*Triticum aestivum*) from the Northern Hills Zone of India. *Ann. Microbiol.* **2015,** *65*, 1885–1899.

Vigani, G.; Rolli, E.; Marasco, R.; et al. Root Bacterial Endophytes Confer Drought Resistance and Enhance Expression and Activity of a Vacuolar H+-Pumping Pyrophosphatase in Pepper Plants. *Environ. Microbiol.* **2018,** *21*, 3212–3228.

Vyas, R. V.; Panpatte, D.; Jhala, Y. K.; et al. Wonders of Microbes in Agriculture for Productivity and Sustainability. In *Microorganisms for Green Revolution*; Panpatte, D., Jhala, Y. K., Vyas, R. V., Shelat, H., Eds.; Springer International Publishing Science: Singapore, 2017; Vol. 1, pp 1–23.

Walitang, D. I.; Kim, K.; Madhaiyan, M.; et al. Characterizing Endophytic Competence and Plant Growth Promotion of Bacterial Endophytes Inhabiting the Seed Endosphere of Rice. *BMC Microbiol.* **2017,** *17*, 209.

Wang, J.; Hou, W.; Christensen, M. J.; et al. Role of *Epichloë* Endophytes in Improving Host Grass Resistance Ability and Soil Properties. *J. Agric. Food Chem.* **2020a,** *68*, 6944–6955.

Wang, Q.; Ge, C.; Xu, S.; et al. The Endophytic Bacterium *Sphingomonas* SaMR12 Alleviates Cd Stress in Oilseed Rape Through Regulation of the GSH-AsA Cycle and Antioxidative Enzymes. *BMC Plant Biol.* **2020b,** *20*, 63. https://doi.org/10.1186/s12870-020-2273-1.

Wani, Z. A.; Ashraf, N.; Mohiuddin, T.; et al. Plant Endophyte Symbiosis: An Ecological Perspective. *Appl. Microbiol. Biotechnol.* **2015,** *99*, 2955–2965.

Waqas, M.; Khan, A. L.; Kamran, M.; et al. Endophytic Fungi Produce Gibberellins and Indole-Acetic Acid and Promotes Host-Plant Growth During Stress. *Molecules* **2012,** *17*, 10754–10773.

Waqas, M.; Khan, A. L.; Shahzad, R.; et al. Mutualistic Fungal Endophytes Produce Phytohormones and Organic Acids that Promote Japonica Rice Plant Growth Under Prolonged Heat Stress. *J. Zhejiang. Univ.-Sci. B (Biomed. Biotechnol.)* **2015,** *16*, 1011–1018.

Waqas, M. A.; Khan, I.; Akhter, M. J.; et al. Exogenous Application of Plant Growth Regulators (PGRs) Induces Chilling Tolerance in Short-Duration Hybrid Maize. *Environ. Sci. Pollut. Res.* **2017,** *24*, 11459–11471.

Weatherby, K.; Carter, D. Chromera velia: The Missing Link in the Evolution of Parasitism. *Adv. Appl. Microbiol.* **2013,** *85*, 119–144.

Wei, C.-Y.; Lin, L.; Luo, L.-J.; et al. Endophytic Nitrogen-Fixing *Klebsiella variicola* Strain DX120E Promotes Sugarcane Growth. *Biol. Fertil. Soils* **2014,** *50*, 657–666.

Weyens, N.; Croes, S.; Dupae, J.; et al. Endophytic Bacteria Improve Phytoremediation of Ni and TCE Co-Contamination. *Environ. Pollut.* **2010,** *158*, 2422–2427.

Whipps, J. Microbial Interactions and Biocontrol in the Rhizosphere. *J. Exp. Bot.* **2001,** *b52*, 487–511.

White, J. F., Jr; Torres, M. S.; Johnson, H.; et al. A Functional View of Plant Microbiomes: Endosymbiotic Systems that Enhance Plant Growth and Survival. In *Advances in Endophytic Research*; Verma, V. C., Gange, A. C., Eds.; Springer India: New Delhi, 2014; pp 425–439.

Xia, C.; Li, N.; Zhang, X.; et al. An *Epichloë* Endophyte Improves Photosynthetic Ability and Dry Matter Production of Its Host *Achnatherum inebrians* Infected by *Blumeria graminis* Under Various Soil Water Conditions. *Fungal Ecol.* **2016,** *22*, 26–34.

Xie, X. G.; Huang, C. Y.; Fu, W. Q.; Dai, C. C. Potential of Endophytic Fungus *Phomopsis liquidambari* for Transformation and Degradation of Recalcitrant Pollutant Sinapic Acid. *Fungal Biol.* **2015,** *120*, 402–413.

Xie, X. G.; Zhang, F. M.; Yang, T.; et al. Endophytic Fungus Drives Nodulation and N2 Fixation Attributable to Specific Root Exudates. *mBio* **2019,** *10*, e00728-e00719.

Xu, M.; Sheng, J.; Chen, L.; et al. Bacterial Community Compositions of Tomato (*Lycopersicum esculentum* Mill.) Seeds and Plant Growth Promoting Activity of ACC Deaminase Producing *Bacillus subtilis* (HYT-12-1) on Tomato Seedlings. *World J. Microbiol. Biotechnol.* **2014,** *30*, 835–845.

Yadav, A. N. Biodiversity and Biotechnological Applications of Host-Specific Endophytic Fungi for Sustainable Agriculture and Allied Sectors. *Acta Sci. Microbiol.* **2018,** *1*, 01–05.

Yadav, A. N.; Sharma, D.; Gulati, S.; et al. Haloarchaea Endowed with Phosphorus Solubilization Attribute Implicated in Phosphorus Cycle. *Sci. Rep.* **2015,** *5*, 12293.

Yadav, A. N.; Verma, P.; Kumar, V.; et al. Extreme Cold Environments: A Suitable Niche for Selection of Novel Psychrotrophic Microbes for Biotechnological Applications. *Adv. Biotechnol. Microbiol.* **2017b,** *2*, 1–4.

Yadav, A. N.; Verma, P.; Singh, B.; et al. Plant Growth Promoting Bacteria: Biodiversity and Multifunctional Attributes for Sustainable Agriculture. *Adv. Biotechnol. Microbiol.* **2017a,** 5, 1–16.

Yadav, S.; Sandhu, N.; Majumder, R. R.; et al. Epistatic Interactions of Major Effect Drought QTLs with Genetic Background Loci Determine Grain Yield of Rice Under Drought Stress. *Sci. Rep.* **2019,** *22*, 1–3.

Yaish, M. W.; Antony, I.; Glick, B. R. Isolation and Characterization of Endophytic Plant Growth-Promoting Bacteria from Date Palm Tree (*Phoenix dactylifera* L.) and Their Potential Role in Salinity Tolerance. *Antonie Van Leeuwenhoek* **2015,** *107*, 1519–1532.

Yan, N.; Wang, X. Q.; Xu, X. F.; et al. Plant Growth and Photosynthetic Performance of *Zizania latifolia* Are Altered by Endophytic *Ustilago esculenta* Infection. *Physiol. Mol. Plant Pathol.* **2013,** *83*, 75–83.

Yang, B.; Wang, X. M.; Ma, H. Y.; et al. Fungal Endophyte *Phomopsis liquidambari* Affects Nitrogen Transformation Processes and Related Microorganisms in the Rice Rhizosphere. *Front Microbiol.* **2015,** *6*, 982.

Yousaf, S.; Andria, V.; Reichenauer, T. G.; et al. Phylogenetic and Functional Diversity of Alkane Degrading Bacteria Associate with Italian Ryegrass (*Lolium multiflorum*) and Birds Foot Trefoil (*Lotus corniculatus*) in a Petroleum Oil-Contaminated Environment. *J. Hazard Mater.* **2010,** *184*, 523–532.

Yu, J.; Yu, Z.; Fan, G. Q.; et al. Isolation and Characterization of Indole Acetic Acid Producing Root Endophytic Bacteria and Their Potential for Promoting Crop Growth. *J. Agric. Sci. Technol.* **2016,** *18*, 1381–1391.

Yuan, Z. S.; Liu, F.; Zhang, G. F. Characteristics and Biodiversity of Endophytic Phosphorus- and Potassium-Solubilizing Bacteria in Moso Bamboo (*Phyllostachys edulis*). *Acta Biol. Hung.* **2015,** *66*, 449–459.

Zafar, S. A.; Hameed, A.; Nawaz, M. A.; et al. Mechanisms and Molecular Approaches for Heat Tolerance in Rice (*Oryza sativa* L.) Under Climate Change Scenario. *J. Integ. Agric.* **2018,** *17*, 726–738.

Zahoor, M.; Irshad, M.; Rahman, H.; et al. Alleviation of Heavy Metal Toxicity and Phytostimulation of *Brassica campestris* L. by Endophytic *Mucor* sp. MHR-7. *Ecotoxicol. Environ. Saf.* **2017,** *142*, 139–149.

Zhang, T.; Yao, Y. F. Endophytic Fungal Communities Associated with Vascular Plants in the High Arctic Zone Are Highly Diverse and Host Plant Specific. *PLoS One* **2015,** *10*, e0130051. https://doi.org/10.1371/journal.pone.0130051.

Zhang, Y.; He, L.; Chen, Z.; et al. Characterization of Lead-Resistant and ACC Deaminase-Producing Endophytic Bacteria and Their Potential in Promoting Lead Accumulation of Rape. *J. Hazard Mater.* **2011,** *186*, 1720–1725.

Zhao, J.; Zhou, L.; Wang, J.; et al. Endophytic Fungi for Producing Bioactive Compounds Originally from Their Host Plants. In *Current Research, Technology and Education Topics in Applied Microbiology and Microbial Biotechnology*; Vilas, M. A., Ed.; Formatex: Spain, 2010; Vol. 1, pp 567–576.

Zheng, L. J.; Li, G. H.; Wang, X. B.; et al. Nematicidal Endophytic Bacteria Obtained from Plants. *Ann. Microbiol.* **2008,** *58*, 569–572.

Zörb, C.; Geilfus, C. M.; Dietz, K. J.; et al. Salinity and Crop Yield. *Plant Biol.* **2018,** *21*, 31–38.

CHAPTER 13

Scaling Up Strategies for Endophytic Biomolecules

TANVI GUPTA[1,2] and VIDUSHI ABROL[1,3*]

[1]*CSIR-Indian Institute of Integrative Medicine, Jammu 180001, India*

[2]*Institute of Clinical Medicine, National Cheng Kung University, Tainan, Taiwan*

[3]*Shri Mata Vaishno Devi University, Katra 182320, India*

[*]*Corresponding author. E-mail: vidushiabrol.1285@gmail.com*

ABSTRACT

Solid-state fermentation (SSF) has made credible advances in biotechnological industries for their potential applications especially for bioactive metabolites and pharmaceutical products. After the detailed understanding of biochemical manipulations particularly in the field of fermenter modeling and designs, SSF have been successfully continued for the scale-up and commercialization of desired products. In the present scenario, the large-scale production of bio molecules produced by natural products is the need of the hour. Henceforth, SSF showed intensive application in the field of biotechnology for the large scale production of pharmacologically active molecules along with extensive stability of the desired product at economical level. In this chapter, we highlighted various technological interventions of the solid-state fermentation (SSF) for the enhanced production of endophytic bioactive secondary metabolites and various optimizing strategies that are mandatory in the coming times for their growth and development.

13.1 INTRODUCTION

Endophytes are those microorganisms that are in mutual association with developing plant tissues in a synergistic fashion. This synergistic fashion defends the plant against foreign agents and helps the endophytes to uptake essential nutrients and provides them with living environment. The endophyte term was coined by Anton de Barry in the year 1866 to differentiate the aerophytes persisting on the plants surface. Endophyte comprises rich a variety of taxon, which includes fungi, bacteria, archae, and protists, and are considered as symbiotic organisms. The prehistoric remains of the plants prove that terrestrial plants were in association with endophytes more than 400 million years ago. Plants based on their evolutionary properties changed their habitat, which led to nutrient deficiency, alteration in temperature, and water accessibility (Fig. 13.1). Due to these properties, microbes present in the plants become resistant against harsh conditions (Arora and Ramawat, 2017).

FIGURE 13.1 Evolution of endophytes.

The endophytes reside inside the growing plant tissues for a short time or throughout their cycle (Greenfield et al., 2015). The isolation of endophytes is performed by the surface sterilization of living tissue of the plant by causing no damage to the plant (Rashid et al., 2012). The surface sterilization of the explant is an important step, as it is necessary to remove any epiphytes/aerophytes or any microorganisms present on the plant surface (Kjer et al., 2010). Most commonly used sterilants are sodium hypochlorite and ethanol. The duration of the exposure of explants to sterilant varies according to the living plant tissue, responsiveness, life span, and density, as it should be able to sterilize the surface and not damage the plant tissue (Shah et al., 2019). There is huge diversity of endophytes in nature, and it is reported that more than 1 million fungal endophytes and over 200 bacterial groups from 16 phyla exist in nature. The vital endophytes belong to three phyla that include Proteobacteria, Actinobacteria, Firmicutes, and involves association with *Gluconobacter*, *Enterobacter*, *Bacillus*, *Pseudomonas*, and *Serratia* (Rajamanikyam et al., 2017).

There is a huge demand for novel compounds in market to impart support and reassurance against various needs arisen by humans in various sectors including health, agriculture, medicine, food, etc. Various products extracted from different sources like bacteria, fungi, yeast, etc. showed bioactive potential and are known as secondary metabolites that can be used as potential drugs or antibiotics to treat against various infections or diseases in humans. For example, Cryptocin, produced by *Cryptosporiopsis quercina*, shows activity against various pathogenic fungi. Various volatile antibiotics produced by *Cinnamomum zeylanicum* (Pena et al., 2019), anticancer agents such as Paclitaxel produced by *Taxomyces andreanae* (Naik, 2019), torreyanic acid produced by *Pestalotiopsis microspora* allied with *Taxus taxifolia* tree (Rana et al., 2019), anti-insecticide like nodulisporic acid produced by an endophyte isolated from *Bontia daphnoides* (Chhipa and Deshmukh, 2019), anti-diabetic such as L-783,281 compound produced by *Pseudomassaria* sp. (Khan et al., 2019), immunosuppressive agents such as subglutinol A and B produced by *Fusarium subglutinans* (Gautam and Avasthi, 2019). There are many endophytes isolated from plants of different regions of the world that possesses variety of bioactive compounds and are used for treatment of multiple ailments. *Artocarpus* used as medication against malarial fever, diarrhea, infection of tapeworm, healing of wounds, syphilis. *Vateria indica* used as medication against piles, throat infection, diarrhea, boils, bronchitis, inflammation, and pain in joints (Ruma et al., 2011). Cellulolytic enzymes are extracted from *Trichoderma reesei*, *Aspergillus niger*, and *Penicillium* sp. (Syed et al., 2013). The production of extracellular lipase enzyme by *Colletotrichum gloeosporioides* (Kumar et al., 2011), the production of glutaminase free L-asparaginase by *Alternaria* sp. (Moharram et al., 2016), production of piperine by *C. gloeosporioides* (Chithra et al., 2014), production of amylase enzyme by *Preussia minima* (Mishra et al., 2019). In this chapter, we will discuss the large-scale production of bioactive compounds by endophytes by various fermentation techniques and parameters related to the operation and regulation of solid-state fermentation (SSF).

13.2 MICROBIAL FERMENTATION

Fermentation processes is a technique that involves solid and liquid material transformed into useful product with the use of various microorganisms (Ashok et al., 2017). There are two types of fermentation processes involved for large-scale production in industries such as submerged fermentation (SmF) and SSF. SmF is a process that involves the multiplication of microorganisms in broth culture consisting essential nutrients for their growth. SSF is a process that involves the growth of microorganisms and development of product without water but involves a moist surface on which microorganism grows. SSF is better as compared to SmF as it leads higher production of products, has low capital costs, low cost of horticultural and industrial residual matter treated as substrates, less water requirement, and less amount used in sterilization. The major issue of SSF lies in the uniformity of the culture and conduction of heat, which could be avoided with the help of certain modifications in the bioreactors (Martins et al., 2011).

The SSF involves microbial fermentation in the absence or near absence of free water providing a natural environment to the selected microorganisms, especially fungi. The production of biomolecules using culture substrate in solid form of wheat bran, rice husk is SSF. In SSF, the microorganisms are inoculated on flat bed solid substrates with controlled temperatures. The water supply in solid fermentation is controlled unlike in liquid state where continuous stirring is required for aeration. The SSF supports mycelial growth in fungi by providing natural habitat like conditions. During the initiation of growth, fungus utilizes nonsoluble compounds from the substrate and solid culture to meet its essential nutrient requirements. Later during growth fungus produces enzymes to meet its growth requirements and in turn giving us the desired metabolites as by product. The process of SSF can be regulated in batch, fed-batch, and continuous mechanisms. The necessary condition needed to be taken care during the planning of the bioreactor is the response of the material or the microorganisms to the different forces caused by stirring. For example, if the agitation speed is increased from 10 to 50 rpm, the mycelium of the fungi can be damaged due to the shear forces inside the bioreactor. The periodic stirring for long interval of time of fixed action can also damage the mycelium of fungi. The other aspect is the microbial degradation of the material, leading to the deformation of the substrate bed dragging aside from the surface of the bioreactor, which emerges out in the inescapable air in between the surface and bed. This could be dodged by the high surface area of the solid material not invaded by any microorganism (Raghavarao et al., 2003).

Fermentation strategies are required for the production of biomolecules that are commonly known metabolites extracted from the microorganisms that help in the scaling up of the large biomolecules including antibiotics, detergents, enzymes, and many other useful products (Perez et al., 2019). The design of bioreactor/fermenters needs to overcome certain parameters such as mass and heat transfer, ease of diffusion, and extraction of secondary metabolite. There are various types of bioreactors considered for the solid and liquid state fermentation processes like: tray bioreactor, rotating-drum bioreactor, packed-bed bioreactor, trickle-bed bioreactor, bubble column bioreactor, airlift bioreactor, stirred aerated bed bioreactor, rocking drum bioreactor, and gas–solid fluidized-bed bioreactor (Mitchell and Krieger, 2019; Huerta-Ochoa et al., 2019) (Fig. 13.2). Earlier reports have laid impact on tray and drum bioreactors, now packed-bed bioreactor is also considered as it is superior to others and comfortable in use. There are some characteristics of solid-state fermenter, which involve that the substrate should be economical, static, noncorrosive. There should be no occurrence of contamination to avoid coincidental threats due to deterioration of the biological material. The management for the supervision of the practical framework needs to be productive. The consistency of the organic material needs to be maintained. The design of the bioreactor has to be clarified for the purification and recovery of product (Krishania et al., 2018). There are various parameters needed to increase the productivity of SSF, which include the selection and optimization of operation, agitation, aeration, oxygen transfer, and density of inoculum, size of particle, composition of media, pH, temperature, moisture content, humidity, sterilization, extraction process, and downstream processing (Rudakiya, 2019). It is necessary to check all the parameters that provide satisfying results for the large-scale production.

13.3 CLASSIFICATION OF BIOREACTOR DESIGNS USED IN MICROBIAL FERMENTATION

The main objectives of designing bioreactors are adequate heat removal, maintenance of sufficient water activities, and high oxygen concentration within the interparticle voids (Manan and Webb, 2018). Bioreactors can be categorized into four main groups based on the type of aeration and mixing/agitation (Mitchell et al., 2006; Robinson and Nigam, 2003):

- Group I: No forced aeration, without mixing
- Group II: Forced aeration, without mixing
- Group III: No forced aeration, with continuous or intermittent mixing
- Group IV: Forced aeration, with continuous or intermittent mixing

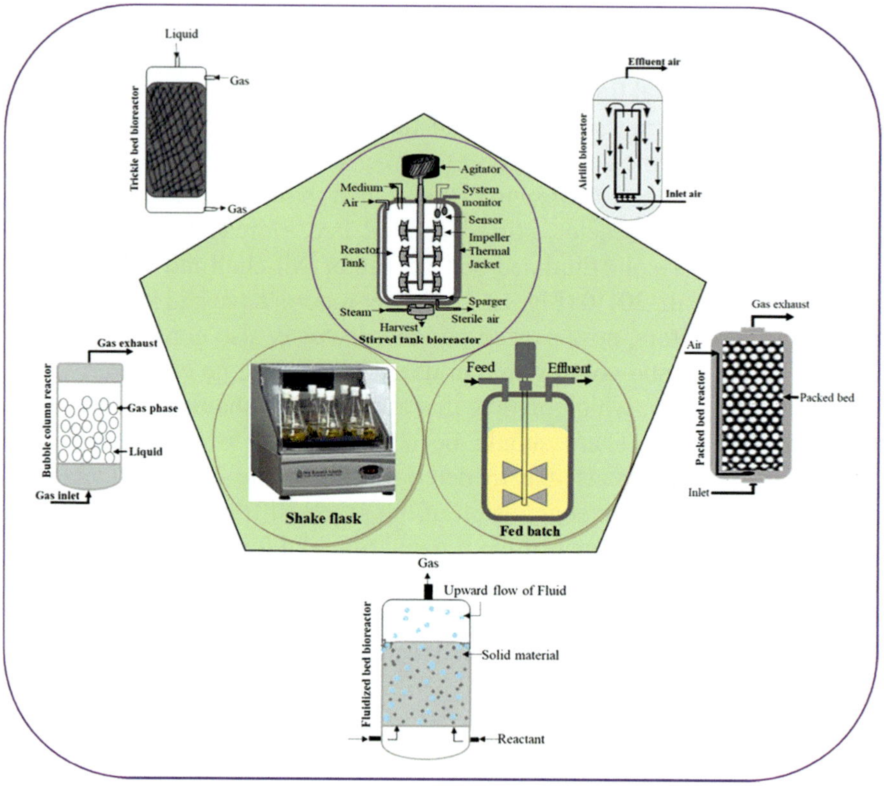

FIGURE 13.2 Bioreactor types for large-scale production.

Group I bioreactors

Tray bioreactors: These are the simplest of all types of bioreactors used in SSF process. The construction material could be wooden or stainless steel. The trays are stacked over each other with a suitable gap between them. The fermentation is carried in a chamber with controlled environment, where trays are placed, to maintain the temperature by circulating cold or warm air and relative humidity by subjecting saturated air inside the chamber. The thickness of the substrate spread on the trays varies from 5 to 15 cm. It is very easy to scale up the tray fermentation process by increasing the number of trays and not the thickness of the substrate. There are some limitations to the use of tray bioreactors as they require a large operation area and are labor intensive. Separate sterilization is also required for the substrate in the tray fermenters.

Group II bioreactors

Packed-bed bioreactors: A typical example of packed-bed reactor is column bioreactor. The column bioreactor is made up of cylindrical glass or plastic columns. The diameters are variable and the length of the bioreactor depends on the scale of the operation. The water bath, used for humidifying the air, generates steam that can be used for the in situ sterilization of the reactor. The sterile air is passed through the bottom of the column through a sieve that supports the substrate (Durand, 2003). A few kilograms of dry solid medium can be processed by the packed-bed bioreactors. The water bath can be used to control the temperature by placing the column inside it or by circulating the water in a double-walled or jacketed column. The application of column bioreactors is mostly at laboratory scale for producing enzymes, organic acids, biologically active secondary metabolites, spores, etc.

The packed-bed bioreactors should be designed in such a way that they allow mass and heat transfer, and there should be a provision to regulate the pressure drop by using larger catalyst particles. Most of the heat is removed by convection and water evaporation as a result of which the bed dries out, and there is a need to add water. The agitation is required to distribute the added water evenly (Ranjbar and Hejazi, 2019; Sangsurasak and Mitchell, 1998). Another concept was introduced to reduce the need for strong aeration, which included the use of perforated trays with heat exchangers introduced directly beneath them. A similar strategy was demonstrated for Zymotis bioreactor used on a lab scale but with vertical heat exchangers (Roussos et al., 1993). The first bioreactor was patented and used by a German company (Prophyta) for the production of biopesticides in sterile condition, which was based on this principle (nee'Nigam and Pandey, 2009). The metabolic heat is eliminated by conduction in this bioreactor. An Indian company, Biocon, patented a similar bioreactor named as Plafractor™ (Durand, 2003).

Group III bioreactors

Rotating drum bioreactors: No forced aeration is used and mixing is done intermittently, which may be in continuous or semicontinuous mode that prevents the overheating issue generated by microbial activity. It is a horizontal drum, semifilled with a bed of substrate (Nava et al., 2011). The intermittent mixing results in uniform growth and causes less damage to fungal mycelium, but continuous mixing increases the damage to the fungal mycelium due to the generation of shear stress. To limit the height of the substrate layer, the intermittently rotating drum, between two agitations/rotations, operates like a tray reactor (Manan and Webb, 2017; Prabhakar et al., 2005).

Group IV bioreactors

Fluidized-bed bioreactors: They are typically constructed from a vertical chamber with a perforated plate at the bottom where forced aeration can be applied from the bottom at a sufficient speed to fluidize the solid substrate particles and cause mixing. The fluidized-bed bioreactor consists of an agitator (clump breaker) that breaks the agglomerates/clumps formed and settle at the bottom of reactor (Sindhu et al., 2015; Manan and Webb, 2017). This type of bioreactor shows good mixing behavior for every state of matter, that is, solid, liquid, and gas (Mitchell et al., 2010). The properties of substrate used in gas–solid fluidized-bed bioreactors have a greater impact on the effectiveness of the bioreactor. For example, the large clumps formed by the sticky substrate cannot be fluidized (Ali and Zulkali, 2011; Foong et al., 2009a, 2009b). The difference in the size of the particles of the substrate also affects its fluidization (Jang and Yang, 2008). The temperature control is not an issue with these bioreactors.

Spouted-bed bioreactors: It is a variant of fluidized-bed bioreactor and the air is blown upward through the central axis of solid bed, as a result of which only part of the bed is fluidized at a particular time. The designing of this type of bioreactor allows the process to work in continuous cycles (Manan and Webb, 2018). Unlike fluidized-bed bioreactors, the substrates with sticky nature, irregular sizes, and texture can be treated in spouted-bed bioreactors (Ali et al., 2019). The problems face by tray and packed-bed bioreactors can be overcome by spouted-bed bioreactors.

Rocking drum bioreactors: Rocking drum bioreactors consist of three concentric drums: an inner, middle, and an outer drum. The outer drum is solid cylinder, inner and middle drums are perforated, and in between these two drums, there is a substrate bed that is packed loosely. The mixing of the substrate bed is done by rotation of the two outer cylinders in relation with the inner cylinder. The air enters through the central cylinder, passes through the substrate bed, and then passes between the outer perforated cylinder and the outer solid cylinder (Arora et al., 2018).

According to the mode of action, there are two major types of bioreactors—small-scale laboratory and pilot-scale bioreactor.

- **Small-scale laboratory:** It includes Petri dishes, beakers, flasks, bottles, roller bottles. In these reactors, the addition of material should be weighed in grams. It is an easy mode of action as it doesn't requires involuntary agitation and aeration. In this system, only temperature of the surroundings of bioreactor needs to be maintained. It involves various aspects like pH, temperature, size of inoculum, and incubation time.
- **Pilot-scale bioreactor:** It is a complex method as it involves huge amount of substrate. The regulation should be adequate enough to meet the desired conditions that will enhance growth and secondary metabolite production. The two important limiting factors are moisture and temperature of the substrate bed. For the enhanced production, these factors need to be maintained in the large-scale industrial process. The challenging issue in this bioreactor is to adjust its limiting factors manually, as it requires analysis and administration to be done at the same time. If the regulation takes place without control, there would alterations in temperature in the substrate bed that could lead to low production, contamination, and loss of experiment (Krishania et al., 2018).

13.4 OPERATION AND REGULATION OF BIOREACTORS

The main goal of regulation of the bioreactor is maintaining temperature and water content of the solid bed for the optimum growth and production. The factors responsible for the designing of the bioreactor are temperature, flow rate, humidity, and agitation. The change in temperature and flow rate of air/water is feasible through the heat exchangers and jackets around the walls. For the addition of water, nutrients and buffer appropriate amount of measurement need to be done timely (Raghavarao et al., 2003). In the scale-up process, heat, mass transfer, and the energy generated in the working of the SSF is planned accordingly. The essential criteria's on which the solid-state bioreactor works for the scaling up are listed below.

- Agitation
- Transfer of oxygen and aeration
- Temperature of the solid bed
- Moisture content of solid bed
- Humidity of the fermenters

The easiest way to associate the aeration and moisture content in solid bed is by moving the humidified air. This process as far as maintaining the temperature and moisture content of solid bed benefits the microorganisms for aerobic respiration (Raghavarao et al., 2003).

13.5 FACTORS INFLUENCING THE EFFICIENCY OF SOLID-STATE FERMENTERS

There are various factors that influence the efficiency of solid-state fermenters, which can be broadly divided into three major categories:

A. Biological factors
B. Physiochemical factors
C. Mechanical factors

A. Biological factors

The selection of microorganism for SSF plays crucial role in fermentation process, which is required for the fulfillment of the growth requirements and for the production of desired products (Krishna, 2005). The filamentous fungi have been proved as important microorganism in SSF because of their mycelial growth and neutral physiological capabilities (Manan and Webb, 2018). In most of the fermentation processes, the inoculum age, cultivation medium, and the physiological state of the microorganisms have much importance. A significant loss in the yield of secondary metabolites occurs because of the incorrect physiological state of the inoculum (Crafack et al., 2014). It is suggested that an increase in the inoculum quantity can decrease the substrate utilization time (Nigam and Singh, 1994; Adi et al., 2019).

B. Physiochemical factors

The solid substrate is another major component in SSF. Along with supplying nutrients such as carbon and nitrogen, the solid substrate also provides physical support to the growing microorganisms (Rodríguez, 2008; Marzo et al., 2019). The selection of substrate also depends on the water holding capacity of the fermented substrate to maintain the moisture content (Lüth and Eiben, 2003). Furthermore, to replicate the typical SSF conditions, defined media and various inert carriers such as vermiculite, perlite, polyurethane foam, polystyrene, etc. can be employed. A few advantages of using inert carriers include enhancement in the homogenous aerobic conditions, better water activity control, improved mass and heat transfer control, better temperature control, and easy product recovery (Xu et al., 2011; Ganaie et al., 2017; Subbalaxmi and Murty, 2016; Rudakiya, 2019).

The moisture content is one of the major attributes in SSF. The water activity (a_w)—water requirement of microbes for their microbial activity—determines the amount of free water available in the substrate and the type of microorganism suitable for an SSF process. The a_w value for bacteria, yeast, and filamentous fungi is 0.9, 0.8, and 0.6–0.7, respectively (Ahmad and Munaim, 2019). Microorganisms, capable of growing at lower values of a_w, are more suitable for SSF process (Valta et al., 2019). The moisture content of the microorganisms plays a crucial role whether its bacteria or fungi, both requires different moisture content. In case of fungi, it requires low moisture content of about 40–60%, whereas in bacteria, it requires high moisture content of 60–85%. The ideal moisture content in an SSF requires appropriate nutrient level, oxygen and carbon dioxide dispersion during the fermentation process. If there is high moisture content, it will reduce the permeability and deformity in particles structure and create obstruction in oxygen dispersal. On the other hand, low moisture content could restrain the solubility of nutrient affecting the growth of microorganism (Darabzadeh et al., 2019). If there is increase in temperature, the growth of microorganism also increases but sometimes increase in temperature will have negative impact on the growth of microorganism. In certain cases, rise in temperature leads to increase in yield production of an enzyme (Sala et al., 2019). It is very difficult to measure and maintain pH in SSF because of very low water content and absence of other methods for pH measurement (Behera and Ray, 2016). Usually, microbes growing over a wide range of pH are recommended in SSF.

The microbial activity generates a lot of heat that gets accumulated in the system leading to the problem of increase in temperature, and it becomes difficult to handle the process (Lüth and Eiben, 2003). The heat generated during the process should be removed to avoid overheating that hinders the growth of microorganisms and affects the formation of products (Pandey et al., 2001). In large-scale systems, overheating processes leads to great loss in the moisture content that can be neutralized by blowing air inside the system and to release the excess heat through gas outlets (Castro et al., 2015). Adequate supply of oxygen and lower carbon dioxide levels are required for efficient SSF process. This can be done by proper aeration of the system. The quality of air and its flow rate are important points to be considered. The temperature and moisture gradients of the solid substrate are controlled by the rate of aeration (Umsza-Guez et al., 2011; Garcia-Galindo et al., 2019). The surface-area-to-volume ratio of the solid substrate is affected by the particle size (Chakraborty et al., 2019). Suitable particle size is observed as important parameter for mycelial growth, porosity, size of voids between particles and the oxygen and nutrient requirements (Reisman, 2019; Hölker et al., 2004; Pessoa et al., 2019).

C. Mechanical factors

The agitation has the same function as that of aeration. It might also improve the homogeneity of the process (Nava et al., 2011; von Meien et al., 2004; Huanes, 2019). Another positive aspect of agitation is the even distribution of airflow. However, it has some negative aspects as well. The shear forces due to continuous agitation can destroy the cells as in the case of filamentous fungi (Manan and Webb, 2018; Suryanarayan, 2003). Therefore, slower or intermittent agitation is employed to avoid any serious damage.

13.6 PARAMETERS INVOLVED IN THE FERMENTATION PROCESS

There are certain parameters required for the optimization of the SSF process at a large-scale level listed below:

1. **pH:** The range of pH varies from 5 to 7 in the broth culture for the study of various pH levels to optimize the effect of pH. The pH in the media was maintained by adding 0.1 N HCl or 0.1 N NaOH. Sterilization was done at 121°C at 15 psi for 20 minutes. The inoculation of the culture was done with the 5–7 days old strain of culture. The biomass production of the fungal culture was estimated as the dry weight.
2. **Basal media:** For the large-scale production, Potato Dextrose Broth is used as basal media. The medium is sterilized and inoculated with 5–7 days old culture and kept at room temperature. The fungal

production treated as a biomass and the by-product is taken out after 2 days by separating it from the media with filter paper, pretreated and washed out with water. It is further dried in an oven at 550°C, and the dry weight is calculated out for biomass. The liquid–liquid extraction is done by ethyl acetate and the separation of the extract is done by separating funnel and dried. The metabolite production is calculated out from the extract.

3. **Culture media:** For the applicable growth of media, the strain culture was added in various types of media like Czapek Dox Broth, Potato Dextrose Broth, Nutrient Broth, Tryptic Soya Broth, and Sabouraud Broth. The inoculation with the strain culture was done with 5–7 days old culture and kept at room temperature. The biomass production and the secondary metabolite production were estimated by its dry weight. The bioactive compound was measured by densitometry thin layer chromatography.
4. **Carbon source:** To analyze the carbon source of endophytes, addition of 1% of starch, glucose, fructose, maltose, sucrose in the media. Every flask has different carbon source with an inoculation of 5–7 days old culture. The biomass production and secondary metabolite production were estimated as the dry weight. The bioactive compound was measured by densitometry thin layer chromatography.
5. **Nitrogen source:** To analyze the nitrogen source of endophytes, 1% peptone, ammonium chloride, sodium nitrate, and beef and yeast extract are added in the media. Every flask has different nitrogen source with an inoculation of 5–7 days old culture. The biomass production and secondary metabolite production were estimated as the dry weight. The bioactive compound was measured by densitometry thin layer chromatography.
6. **Sodium chloride concentration:** To analyze the outcome of salinity on the growth of secondary metabolite and production of culture strain with different concentrations of about 3–7 g/L with respect to other parameters. The fungal and secondary metabolite production is calculated by its dry weight, and the bioactive compound was measured by densitometry thin layer chromatography.
7. **Incubation temperature:** To analyze the appropriate temperature for the growth and production of secondary metabolite, the temperature should lie between 25°C and 30°C and room temperature is chosen for the basal media. The sterilization is done at 121°C at 15 psi for 20 minutes, and inoculation is done with culture strain within the

required temperature. The fungal and secondary metabolite production was estimated by its dry weight. The bioactive compound was measured by densitometry thin layer chromatography.

13.7 APPLICATIONS OF SSF

The isolation and screening of microorganisms are the important steps for the scaling up of the secondary metabolites. The isolation of the bioactive metabolites is processed by chemical processes that involve use of organic solvents such as ethyl acetate, methanol, etc. The metabolites are extracted out from the liquid growth culture of microorganisms. There are certain parameters that enhance the growth of microorganisms that include nitrogen source, carbon source, nutrients, and pH. SSF is more favored as compared to the SmF due to high production rate, high yield, quality of the product, operation rate is lower as cheap materials can be utilized as substrates. It is a cost-effective process as it involves less amount of water usage, which helps in the reduction of size of bioreactor, downstream processing, less involvement of stirring, and decreased cost for sterilization (Chakravarty and Gaur, 2018).

There are different types of microorganisms used in the SSF, which include bacteria, fungi, and yeasts. The common genera of yeasts that are found in the SSF process are *Candida* and *Saccharomyces*, whereas for the genera of fungi are *Penicillium*, *Rhizopus*, and *Penicillium*, which can be used to extract out bioactive compounds (Table 13.1). The familiar genera of bacteria used are *Streptomyces* and *Bacillus* (Table 13.2). There are many reports on filamentous fungi used in the SSF for the production of heat sensitive enzymes at an industrial-scale level. The selection of the plant material from which the microorganisms are isolated and used as a substrate plays an important role in the production of the desired product (Abu Yazid et al., 2017). The various applications of the SSF process are as follows.

- **Production of enzymes:** Bacteria, yeasts, and fungi are most commonly used for the enzyme production, and the environment is also suitable for their production in the SSF. Cellulase production can be done from *T. reesei*, *Thermoascus aurantiacus*, *A. niger*, *Penicillium* sp., *Trichoderma viride*, *Candida tropicalis*, *Trametes hirsute*, *Aspergillus fumigatus*, *Trichoderma harzianum*, *Trichoderma asperellum*, *Phanerochaete chrysosporium*, *Aspergillus nidulans.* Amylase production can be done from *Bacillus* sp., *Thermomyces*

TABLE 13.1 SSF of Fungal Endophytes Isolated from Different Plant Sources.

Name of plant	Part of plant	Biological source	Parameters							References
			Media	Carbon source	Nitrogen source	Temperature (°C)	pH	Incubation period	Agitation speed (rpm)	
Red Betel	Stems	*Athelia rolfsii*	Sabouraud Dextrose Broth			29	5.0			Yuniati et al. (2018)
Taxus chinensis	Stems	*Acremonium alternatum*	Potato Dextrose Broth							Liu et al. (2009)
Eremophila longifolia		*Preussia minima*		Starch	L-Asparagine	25	9.0			Zaferanloo et al. (2014)
Chlorophytum comosum	Leaf	*Penicillium* sp.		Cellulose	Peptone	28	5.5			Syed et al. (2013)
Withania somnifera	Leaf	*Alternaria* sp.		Glucose	L-Asparagine	30				Nagarajan et al. (2014)
Piper nigrum	Stem	*Colletotrichum gloeosporioides*	Potato Dextrose Broth			30				Chithra et al. (2013)
Tomato	Segments	*Cladosporium sphaerospermum*				50–70	7.0			Hafez et al. (2016)
Taxodium distichum	Bark	*Aspergillus fumigatus*	Potato Dextrose Broth			25	6.0			Ismaiel et al. (2017)
Taxodium arjuna		*Alternaria tenuissima*								
Curcuma amada	Rhizomes	*Talaromyces pinophilus*	Modified Czapek Dox			28	6.4			Prajna Rao et al. (2016)
Digitalis lanata	Leaves	*Alternaria* sp. *Penicillium* sp. *Aspergillus* sp.	Potato Dextrose Broth			27				Kaul et al. (2013)

TABLE 13.1 *(Continued)*

Name of plant	Part of plant	Biological source	Parameters							References
			Media	Carbon source	Nitrogen source	Temperature (°C)	pH	Incubation period	Agitation speed (rpm)	
Bruguiera gymnorrhiza	Root	*Penicillium thomi*		Sucrose	Sodium nitrate	25	7.0			Chen et al. (2007)
Nothapodytes foetida	Stems	*Entrophospora infrequens*	Sabouraud Dextrose Broth			28				Puri et al. (2005)
Podophyllum hexandrum	Rhizomes	*Fusarium solani*	Potato Dextrose Broth			28	5.6			Nadeem et al. (2012)
Ocimum sanctum	Leaf	*Streptomyces coelicolor*	Potato Dextrose Broth			30	6.8			EL-Moslamy (2018)
E. longifolia	Leaf	*P. minima*	Potato Dextrose Broth			25	5.5			Zaferanloo et al. (2012)
E. longifolia	Leaf	*Alternaria alternata*	Czapek Dox Broth			37	7.0			Zaferanloo et al. (2014)
Nothapodytes nimmoniana	Leaf	*Colletotrichum fructicola*, *Corynespora cassiicola*	Whey Complex Media			30	6.0	7 days	110	Bhalkar et al. (2016)
Taxus baccata	Leaf	*Paraconiothyrium variabile*, *Epicoccum nigrum*	Milled Yew Broth	Salicyclic acid	Serine	25	5.5			Somjaipeng et al. (2016)
Brucea mollis	Leaf, roots and bark	*Geosmithia pallida*	Potato Dextrose Broth			25	6.5			Deka and Jha (2018)

TABLE 13.1 *(Continued)*

Name of plant	Part of plant	Biological source	Parameters							References
			Media	Carbon source	Nitrogen source	Temperature (°C)	pH	Incubation period	Agitation speed (rpm)	
Schinus terebinthifolius	Leaf	*A. alternata*	Czapek Dox Broth			35	5.5			Tonial et al. (2015)
Sophora flavescens	Seed	*Aspergillus terreus*	Potato Dextrose Broth			25			150	Zhang et al. (2017)
Psidium guajava	Leaf	*Bionectria ochroleuca*	Potato Dextose Broth			25	6.5		150	Li et al. (2016)
Bacopa monnieri		*Trichoderma lixii*	Rose Bengal Broth			25		15 days		Katoch et al. (2019)
T. baccata	Twig segments	*P. variabile*		Sucrose	Ammonium tartrate	25	5.5	21 days		Somjaipeng et al. (2015)
Senecio stapeliiformis	Aerial part	*Chaetomium* sp.		Yeast and malt extract	Peptone		7.4	30 days		Tawfik et al. (2016)
Narcissus tazetta	Bulb tissue	*F. solani*	Potato Dextrose Broth			25		16 days		Wang et al. (2015)
Adiantum capillus-veneris	Leaf	*Chaetomium globosum*	Potato Dextrose Broth			30		17 days		Selim et al. (2015)
Rhodiola angusta		*Phialocephala fortinii*		Sucrose	Sodium nitrate	25	6.5	10 days	150	Cui et al. (2016)
Mikania glomerata	Leaf	*Diaporthe citri*	Potato Dextrose Broth			28	7.0	22 days		Polonio et al. (2016)
Angelica sinensis	Roots	*A. tenuissima*		Mannose	Yeast extract	30	7.0		180	Wang et al. (2019)

TABLE 13.1 *(Continued)*

Name of plant	Part of plant	Biological source	Parameters							References
			Media	Carbon source	Nitrogen source	Temperature (°C)	pH	Incubation period	Agitation speed (rpm)	
Tinospora cordifolia	Stem, leaf, roots	*F. solani*	Modified Czapek Dox Broth			30	7.0		120	Uzma et al. (2016)
Achyranthus aspera	Stem, root, leaf	*A. terreus*	Potato Dextrose Broth			26		21 days		Goutam et al. (2016)
Andrographis paniculata		*Pestalotiopsis* sp.	Potato Dextrose Broth			28	6.0	7 days	120	Mahapatra and Banerjee (2015)
Cajanus cajan	Leaf	*Dichotomopilus funicola*	Potato Dextrose Broth			30		6 days	130	Gu et al. (2018)
Alpinia calcarata	Leaf	*Cylindrocephalum* sp.		Starch	Sodium nitrate	30	5.0			Sunitha et al. (2012)
Juniperus recurva	Stem, root, leaf	*Fusarium oxysporum*	Sabouraud Dextrose Broth			28	5.6			Kour et al. (2008)
Sueada maritime	Leaf	*A. terreus*	Modified Czapek Dox Broth			30	7.0			Kalyana-sundaram et al. (2015)
Solanum nigrum	Leaf, stem, fruits	*Aspergillus flavus*	Malt Extract Broth			20		4 weeks		El-Hawary et al. (2016)
Sabina chinensis	Stem	*Microsphaeropsis* sp.	Potato Dextrose Broth with wheat straw and bran			27			200	Peng and Chen (2007)
T. chinensis		*Sclerocystis* sp.								
Keteleeria evelyniana		*Nigrospora* sp.								

TABLE 13.1 *(Continued)*

Name of plant	Part of plant	Biological source	Parameters							References
			Media	Carbon source	Nitrogen source	Temperature (°C)	pH	Incubation period	Agitation speed (rpm)	
Pinus massoniana		*Phomopsis* sp.								
Keteleeria davidiana		*Phomopsis* sp.								
Cupressus torulosa		*Cephalosporium* sp.								
N. foetida	Stem	*E. infrequens*	Sabouraud Dextrose Broth			28				Puri et al. (2005)
N. foetida	Seeds, leaf, stem	*Neurospora crassa*	Sabouraud Dextrose Broth			28			220	Rehman et al. (2006)
C. torulosa	Leaf	*A. alternata*		Glucose	Yeast extract	27	7.0			Rajput et al. (2016)
Cedrus sp.	Leaf	*Monodictys castaneae*	Czapek Yeast Extract Broth			24	5.0			Visalakchi and Muthumary (2009)
Cinnamomum zeylanicum	Leaf and stem	*Nigrospora aurantiaca*	Czapek Dox Broth			27	5.0	5 days		Suwannarah et al. (2019)
Phoenix dactylifera	Roots	*Penicillium bilaiae*	Czapek Dox Broth			24	6.2		150 rpm	Mefteh et al. (2019)
Fritillaria unibracteata	Bulb	*Fusarium redolens*	Potato Dextrose Broth			28		7 days	130 rpm	Pan et al. (2015)
Ginkgo biloba	Roots and stems	*Pestalotiopsis uvicola*	Potato Dextrose Broth			28		10 days		Qian et al. (2016)

TABLE 13.2 SSF of Bacterial Endophytes Isolated from Different Plant Sources.

Name of plant	Part of plant	Biological source	Parameters							References
			Media	Carbon source	Nitrogen source	Temperature (°C)	pH	Incubation period	Agitation speed (rpm)	
Citrus maxima	Tissues of seed and peels	*Bacillus aryabhattai*, *Bacillus velezensis*, *Bacillus wiedmannii*, *Bacillus cereus*, *Bacillus aerius*, *Bacillus stratosphericus*	Inorganic salt medium	Glucose	–	37		14 hours	170	Zhang et al. (2018)
Platycodon grandiflorum		*Lucteibacter* sp.		Yeast extract	Ammonium chloride	30	7.0	7 days		Cui et al. (2015)
Catharanthus roseus	Leaf	*Microbacterium* sp.	Nutrient Broth			37	7.0		160	Anjum and Chandra (2018)
Salsola imbricata	Roots	*Streptomyces enissocaesilis*	R2A Broth			28		48 hours	140	Bibi et al. (2017)
Avicennia marina	Aerial roots	*Nocardioides aromaticivorans*, *Streptomyces spectabilis*, *Nocardioides albus*, *Inquilinus limosus*	R2A Broth			28	7.5	48 hours	140	Bibi et al. (2017)
Haplopeplis perfoliata	Roots	*Inquilinus alexandrii*	R2A Broth			28	7.5	48 hours	140	Bibi et al. (2017)
C. roseus	Root, stem, leaf	*Micrococcus yunnanensis*	Tryptone Soya Broth			30	7.0	9–12 days		Ranjana and Jadeja (2017)

TABLE 13.2 *(Continued)*

Name of plant	Part of plant	Biological source	Parameters							References
			Media	Carbon source	Nitrogen source	Temperature (°C)	pH	Incubation period	Agitation speed (rpm)	
Narcissus tazetta	Bulb tissue	*Achromobac ter xylosoxidans*	Nutrient Broth			30	7.0	9–12 days		Wang et al. (2015)
W. somnifera	Stem	*Bacillus licheniformis*		Glucose, lactose		40	8.0	48–72 hours		Joshi and Kulkarni (2016)
O. sanctum	Leaves	*Bacillus pseudomycoides*		Glucose, lactose		40	8.0	48–72 hours		
C. roseus	Fruits, roots	*Paenibacillus*		Glucose, lactose		40	8.0	48–72 hours		
Oryza sativa	Seeds	*Bacillus amyloliquefaciens*	Luria Bertani			27		5 days		Shahzad et al. (2016)
Hibiscus rosa-sinensis	Leaf and stem	*Pseudomonas oryzihabitans*		Glucose	Asparagine	30	7.0	18 hours		Bhagat et al. (2016)
Camptotheca acuminata		*Paenibacillus polymyxa*	Sabouraud Dextrose Broth			27	7.0		100	Pu et al. (2015)
Helianthus annuus	Leaf	*B. cereus*		Sucrose	Yeast extract	32	7.6		120	Das et al. (2016)
Lycoris aurea	Leaf, stem, flowers	*Burkholderia gladioli*		Glucose	Peptone	30	7.2	48 hours	200	Li et al. (2019)
Ricinus communis	Leaf	*B. cereus*		Glucose	Yeast extract	32	6.2	96 hours	120	Das et al. (2017)

TABLE 13.2 *(Continued)*

Name of plant	Part of plant	Biological source	Parameters							References
			Media	Carbon source	Nitrogen source	Temperature (°C)	pH	Incubation period	Agitation speed (rpm)	
Eucalyptus spp.	Wood	*Pseudomonas aeruginosa*	Luria Bertani			33	4.5			Ramnath et al. (2017)
		Bacilllus firmus								
		Micrococcus luteus								
		I. limosus								
		Pantoea sp.								
		Klebsiella sp.								
		Bacillus ginsengihumi								
		Streptomyces costaricanus								
		Cellulosimicrobium cellulans								

sp., *A. fumigatus*, and *Bacillus subtilis*. Protease production can be done from *A. niger*, *Pseudomonas aeruginosa*, *Aspergillus awamori*, *Aspergillus oryzae*, *Bacillus* sp., *A. fumigatuss*, *Bacillus cereus*, *C. tropicalis*, *Thermus* sp., *Thermoactinomyces* sp. Lipase production can be done from *P. aeruginosa*, *A. niger.* Many other types of enzymes can be produced from the microorganisms such as xylanase, laccase, inulinase, and keratinase.

- **Production of detergents:** Microorganisms are the sources that are able to produce detergents such as surfactin, sophorolipids, peptidolipids, rhamnolipids act as bioactive compounds. *Pleurotus djamor* can produce surfactin and rhamnolipids, many other like *Bacillus amyloliquefaciens*, *B. subtilis* can be utilized for the surfactin production. *P. aeruginosa* used for the rhamnolipids production.
- **Production of biopesticides:** SSF process is more apt for the biopesticide production as the spores are resistant and stable in stress conditions. Commonly fungi and bacteria are used such as *Coniothyrium minitans* against the parasite, *Sclerotinia sclerotiorum*, *Beauveria bassiana* against larvae of *Musca domestica.*
- **Production of bioethanol:** Yeast and fungi are commonly used for the bioethanol production such as *Saccharomyces cerevisiae*, *A. niger*, *Aspergillus variabilis*, *T. reesei*, *Clostridium phytofermentans*, and *Zymomonas mobilis* (Abu Yazid et al., 2017) (Fig. 13.3).

13.8 CONCLUSION

Based on the recent studies, SSF has shown remarkable advancement for resolving various issues related to biotechnological applications. It has been observed that various approaches are implied for the accurate effect of fermenters including bioreactor modeling and designs, automated and upgraded controlling systems, physical and chemical parameters, etc. Moreover, it is also revealed that there are various factors that need to be taken into account such as aeration, oxygen transfer, agitation, temperature, moisture content, and humidity for the stable and cost-effective outcome of the processes. Therefore, for the large-scale production of important biomolecules various optimization parameters are required to be maintained for desired product formation such as pH, temperature, nitrogen source, carbon source, media composition, agitation speed, incubation time, etc. It is also observed that the SSF is preferred over SmF because of its low capital cost, less water requirement, small size of fermenters, low cost of horticultural

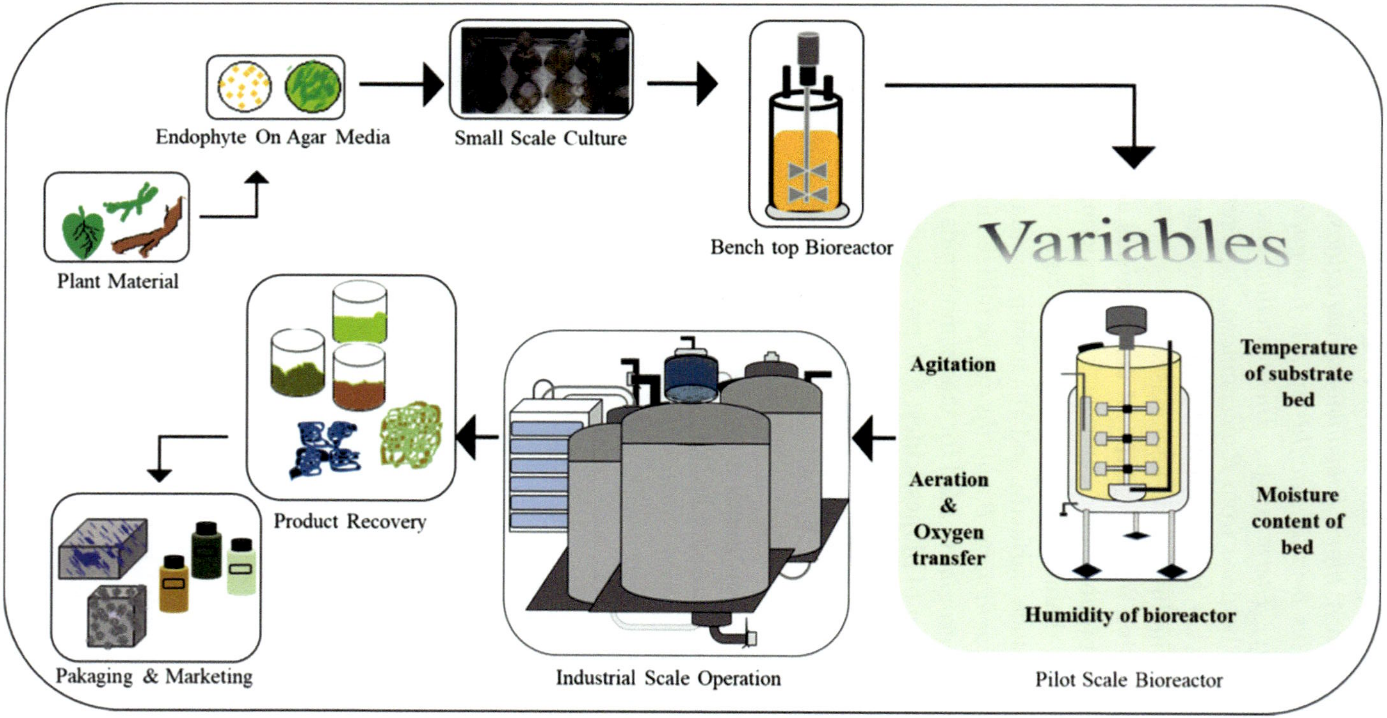

FIGURE 13.3 Outlay of scale-up strategy for bioactive compounds.

and industrial residual matter, less mixing. To conclude, SSF is suggested to be the best and most feasible strategy for the scale-up of bioactive secondary metabolites that are generated from microbial sources.

KEYWORDS

- **endophytes**
- **fermentation**
- **bioreactors**
- **scale-up**
- **physio-chemical parameters**
- **secondary metabolites**

REFERENCES

Abu Yazid, N.; Barrena, R.; Komilis, D.; Sánchez, A. Solid-State Fermentation As a Novel Paradigm for Organic Waste Valorization: A Review. *Sustainability* **2017,** *9* (2), 224.

Adi, D.; Oduro, I.; Simpson, B. K. Biological and Microbial Technologies for the Transformation of Fruits and Vegetable Wastes. In *Byproducts from Agriculture and Fisheries: Adding Value for Food, Feed, Pharma and Fuels*, 2019; p. 403.

Ahmad, Z. S.; Munaim, M. S. A. Response Surface Methodology Based Optimization of Sorbitol Production Via Solid State Fermentation Process. *Eng. Agric. Environ. Food* **2019,** *12* (2), 150–154.

Ali, H. K. Q.; Zulkali, M. M. D. Design Aspects of Bioreactors for Solid-State Fermentation: A Review. *Chem. Biochem. Eng. Q.* **2011,** *25* (2), 255–266.

Ali, N.; Aljuwaya, T.; Al-Dahhan, M. Evaluating the New Mechanistic Scale-Up Methodology of Gas-Solid Spouted Beds Using Gamma Ray Computed Tomography (CT). *Exp. Therm. Fluid Sci.* **2019,** *104*, 186–198.

Arora, J.; Ramawat, K. G. An Introduction to Endophytes. In *Endophytes: Biology and Biotechnology*, 2017; Springer; pp. 1–23.

Arora, S.; Richa, R.; Ghosh, S. Bioreactors in Solid State Fermentation Technology: Design, Applications and Engineering Aspects. *J. Biotechnol.* **2018,** *269*, 16–34.

Ashok, A.; Doriya, K.; Mohan Rao, D. R.; Kumar, D. S. Design of Solid State Bioreactor for Industrial Applications: An Overview to Conventional Bioreactors. *Biocatal. Agric. Biotechnol.* **2017,** *9*, 11–18.

Behera, S. S.; Ray, R. C. Solid State Fermentation for Production of Microbial Cellulases: Recent Advances and Improvement Strategies. *Int. J. Biol. Macromol.* **2016,** *86*, 656–669.

Castro, A. M.; Leda, R. C.; Freire, D. M. G. Performance of a Fixed-Bed Solid-State Fermentation Bioreactor with Forced Aeration for the Production of Hydrolases by *Aspergillus awamori*. *Biochem. Eng. J.* **2015,** *93*, 303–308.

Chakraborty, S.; Yadav, G.; Saini, J. K.; Kuhad, R. C. Comparative Study of Cellulase Production Using Submerged and Solid-State Fermentation. In *New and Future Developments in Microbial Biotechnology and Bioengineering*, 2019; Elsevier; pp. 99–113.

Chakravarty, K.; Gaur, S. Fungal Endophytes As Novel Sources of Anticancer Compounds. In *Anticancer Plants: Natural Products and Biotechnological Implements*, 2018; Springer; pp. 1–18.

Chhipa, H.; Deshmukh, S. K. Fungal Endophytes: Rising Tools in Sustainable Agriculture Production. In *Endophytes and Secondary Metabolites*, 2019; pp. 631–655.

Chithra, S.; Jasim, B.; Sachidanandan, P.; et al. Piperine Production by Endophytic Fungus *Colletotrichum gloeosporioides* Isolated from *Piper nigrum*. *Phytomedicine* **2014,** *21* (4), 534–540.

Crafack, M.; Keul, H.; Eskildsen, C. E.; et al. Impact of Starter Cultures and Fermentation Techniques on the Volatile Aroma and Sensory Profile of Chocolate. *Food Res. Int.* **2014,** *63*, 306–316.

Darabzadeh, N.; Hamidi-Esfahani, Z.; Hejazi, P. Optimization of Cellulase Production Under Solid-State Fermentation by a New Mutant Strain of *Trichoderma reesei*. *Food Sci. Nutr.* **2019,** *7* (2), 572–578.

Durand, A. Bioreactor Designs for Solid State Fermentation. *Biochem. Eng. J.* **2003,** *13* (2–3), 113–125.

Foong, C. W.; Janaun, J.; Krishnaiah, K.; Prabhakar, A. Effect of Superficial Air Velocity on Solid State Fermentation of Palm Kernel Cake in a Lab Scale Fermenter Using Locally Isolated Fungal Strain. *Ind. Crops Prod.* **2009a,** *30* (1), 114–118.

Foong, C. W.; Krishnaiah, K.; Janaun, J.; Krishnaiah, K.; Prabhakar, A. Heat and Mass Transfer Studies of Palm Kernel Cake (PKC) in Fluidized Bed Fermenter. *Ind. Crops Prod.* **2009b,** *30* (2), 227–234.

Ganaie, M. A.; Soni, H.; Naikoo, G. A.; et al. Screening of Low Cost Agricultural Wastes to Maximize the Fructosyltransferase Production and Its Applicability in Generation of Fructooligosaccharides by Solid State Fermentation. *Int. Biodeterior. Biodegrad.* **2017,** *118*, 19–26.

Garcia-Galindo, I.; Gómez-García, R.; Palácios-Ponce, S.; et al. New Features and Properties of Microbial Cellulases Required for Bioconversion of Agro-Industrial Wastes. In *Enzymes in Food Biotechnology*, 2019; Elsevier; pp. 535–550.

Gautam, A. K.; Avasthi, S. Fungal Endophytes: Potential Biocontrol Agents in Agriculture. In *Role of Plant Growth Promoting Microorganisms in Sustainable Agriculture and Nanotechnology*, 2019; Elsevier; pp. 241–283.

Greenfield, M.; Pareja, R.; Ortiz, V.; et al. A Novel Method to Scale Up Fungal Endophyte Isolations. *Biocontrol. Sci. Technol.* **2015,** *25* (10), 1208–1212.

Hölker, U.; Höfer, M.; Lenz, J. Biotechnological Advantages of Laboratory-Scale Solid-State Fermentation with Fungi. *Appl. Microbiol. Biotechnol.* **2004,** *64* (2), 175–186.

Huerta-Ochoa, S.; Castillo-Araiza, C. O.; Quijano, G. *Advances and Applications of Partitioning Bioreactors*, 2019; Academic Press; Vol. 54.

Jang, H. D.; Yang, S. S. Polyunsaturated Fatty Acids Production with a Solid-State Column Reactor. *Biores. Technol.* **2008,** *99* (14), 6181–6189.

Khan, R.; Naqvi, S. T. Q.; Fatima, N.; Muhammad, S. A. Study of Antidiabetic Activities of Endophytic Fungi Isolated from Plants. *Pure Appl. Biol.* **2019,** *8* (2), 1287–1295.

Kjer, J.; Debbab, A.; Aly, A. H.; Proksch, P. Methods for Isolation of Marine-Derived Endophytic Fungi and Their Bioactive Secondary Products. *Nat. Protoc.* **2010,** *5* (3), 479.

Krishania, M.; Sindhu, R.; Binod, P.; et al. Design of Bioreactors in Solid-State Fermentation. In *Current Developments in Biotechnology and Bioengineering*, 2018; Elsevier; pp. 83–96.

Krishna, C. Solid-State Fermentation Systems—An Overview. *Crit. Rev. Biotechnol.* **2005,** *25* (1–2), 1–30.

Kumar, S.; Katiyar, N.; Ingle, P.; Negi, S. Use of Evolutionary Operation (EVOP) Factorial Design Technique to Develop a Bioprocess Using Grease Waste As a Substrate for Lipase Production. *Biores. Technol.* **2011,** *102* (7), 4909–4912.

Lüth, P.; Eiben, U. Solid-State Fermenter and Method for Solid-State Fermentation. Google Patents, 2003.

Manan, M. A.; Webb, C. Design Aspects of Solid State Fermentation as Applied to Microbial Bioprocessing. *J. Appl. Biotechnol. Bioeng.* **2017,** *4* (1), 91.

Manan, M. A.; Webb, C. Control Strategies with Variable Air Arrangements, Forcefully Aerated in Single Circular Tray Solid State Bioreactors with Modified Gompertz Model and Analysis of a Distributed Parameter Gas Balance. *Biotechnol. Biotechnol. Equip.* **2018,** *32* (6), 1455–1467.

Martins, S.; Mussatto, S. I.; Martínez-Avila, G.; et al. Bioactive Phenolic Compounds: Production and Extraction by Solid-State Fermentation. A Review. *Biotechnol. Adv.* **2011,** *29* (3), 365–373.

Marzo, C.; Díaz, A. B.; Caro, I.; Blandino, A. Valorization of Agro-Industrial Wastes to Produce Hydrolytic Enzymes by Fungal Solid-State Fermentation. *Waste Manage. Res.* **2019,** *37* (2), 149–156.

Mishra, R.; Kushveer, J. S.; Revanthbabu, P.; Sarma, V. Endophytic Fungi and Their Enzymatic. In *Advances in Endophytic Fungal Research: Present Status and Future Challenges*, 2019; p. 283.

Mitchell, D. A.; Cunha, L. E. N.; Machado, A. V. L.; et al. A Model-Based Investigation of the Potential Advantages of Multi-Layer Packed Beds in Solid-State Fermentation. *Biochem. Eng. J.* **2010,** *48* (2), 195–203.

Mitchell, D. A.; Krieger, N. Solid-State Cultivation Bioreactors. In *Essentials in Fermentation Technology*, 2019; Springer; pp. 105–133.

Mitchell, D. A.; Krieger, N.; Berovič, M.; Luz, L. F. L. Group IVa: Continuously-Mixed, Forcefully-Aerated Bioreactors. In *Solid-State Fermentation Bioreactors*, 2006; Springer; pp. 115–128.

Moharram, A. M.; Zohri, A. A.; Seddek, N. H. L-Asparaginase Production by Endophytic Fungi Isolated from *Withania somnifera* in Egypt. *SS Int. J. Multidiscip. Res.* **2016,** *2*, 30–40.

Naik, B. S. Developments in Taxol Production Through Endophytic Fungal Biotechnology: A Review. *Orient. Pharm. Exp. Med.* **2019,** *19* (1), 1–13.

Nava, I.; Favela-Torres, E.; Saucedo-Castañeda, G. Effect of Mixing on the Solid-State Fermentation of Coffee Pulp with *Aspergillus tamarii*. *Food Technol. Biotechnol.* **2011,** *49* (3), 391.

nee'Nigam, P. S.; Pandey, A. Solid-State Fermentation Technology for Bioconversion of Biomass and Agricultural Residues. In *Biotechnology for Agro-Industrial Residues Utilisation*, 2009; Springer; pp. 197–221.

Nigam, P.; Singh, D. Solid-State (Substrate) Fermentation Systems and Their Applications in Biotechnology. *J. Basic Microbiol.* **1994,** *34* (6), 405–423.

Pena, L. C.; Jungklaus, G. H.; Savi, D. C.; et al. *Muscodor brasiliensis* sp. *nov*. Produces Volatile Organic Compounds with Activity Against *Penicillium digitatum*. *Microbiol. Res.* **2019,** *221*, 28–35.

Perez, C. L.; Casciatori, F. P.; Thoméo, J. C. Strategies for Scaling-Up Packed-Bed Bioreactors for Solid-State Fermentation: The Case of Cellulolytic Enzymes Production by a Thermophilic Fungus. *Chem. Eng. J.* **2019,** *361*, 1142–1151.

Pessoa, D. R.; Finkler, A. T. J.; Alex Vinícius Lopes Machado, A. V. L.; et al. CFD Simulation of a Packed-Bed Solid-State Fermentation Bioreactor. *Appl. Math. Model.* **2019,** *70*, 439–458.

Prabhakar, A.; Krishnaiah, K.; Janaun, J.; Bono, A. An Overview of Engineering Aspects of Solid State Fermentation. *Malays. J. Microbiol.* **2005,** *1* (2), 10–16.

Raghavarao, K. S. M. S.; Ranganathan, T. V.; Karanth, N. G. Some Engineering Aspects of Solid-State Fermentation. *Biochem. Eng. J.* **2003,** *13* (2–3), 127–135.

Rajamanikyam, M.; Vadlapudi, V.; Upadhyayula, S. M. Endophytic Fungi as Novel Resources of Natural Therapeutics. *Braz. Arch. Biol. Technol.* **2017,** *60*.

Rana, K. L.; Kour, D.; Sheikh, I.; et al. Biodiversity of Endophytic Fungi from Diverse Niches and Their Biotechnological Applications. In *Advances in Endophytic Fungal Research*, 2019; Springer; pp. 105–144.

Ranjbar, S.; Hejazi, P. Modeling and Validating *Pseudomonas aeruginosa* Kinetic Parameters Based on Simultaneous Effect of Bed Temperature and Moisture Content Using Lignocellulosic Substrate in Packed-Bed Bioreactor. *Food Bioprod. Process* **2019,** *117*, 51–63.

Rashid, S.; Charles, T. C.; Glick, B. R. Isolation and Characterization of New Plant Growth-Promoting Bacterial Endophytes. *Appl. Soil Ecol.* **2012,** *61*, 217–224.

Reisman, H. B. *Economic Analysis of Fermentation Processes*, 2019; CRC Press.

Robinson, T.; Nigam, P. Bioreactor Design for Protein Enrichment of Agricultural Residues by Solid State Fermentation. *Biochem. Eng. J.* **2003,** *13* (2–3), 197–203.

Rodríguez, C. S. Exploitation of Biological Wastes for the Production of Value-Added Products Under Solid-State Fermentation Conditions. *Biotechnol. J.: Healthcare Nutr. Technol.* **2008,** *3* (7), 859–870.

Roussos, S.; Raimbault, M.; Prebois, J. P.; Lonsane, B. K. Zymotis, a Large Scale Solid State Fermenter Design and Evaluation. *Appl. Biochem. Biotechnol.* **1993,** *42* (1), 37–52.

Rudakiya, D. M. Strategies to Improve Solid-State Fermentation Technology. In *New and Future Developments in Microbial Biotechnology and Bioengineering*, 2019; Elsevier; pp. 155–180.

Ruma, K.; Shailasree, S.; Sampath, K. K.; et al. Diversity of Fungal Endophytes from Two Endemic Tree Species *Artocarpus hirsutus* Lam. and *Vateria indica* Linn. of Western Ghats, India. *World J. Agric. Sci.* **2011,** *7* (5), 577–582.

Sala, A.; Barrena, R.; Artola, A.; Sánchez, A. Current Developments in the Production of Fungal Biological Control Agents by Solid-State Fermentation Using Organic Solid Waste. *Crit. Rev. Environ. Sci. Technol.* **2019,** *49* (8), 655–694.

Sangsurasak, P.; Mitchell, D. A. Validation of a Model Describing Two-Dimensional Heat Transfer During Solid-State Fermentation in Packed Bed Bioreactors. *Biotechnol. Bioeng.* **1998,** *60* (6), 739–749.

Shah, S.; Shrestha, R.; Maharjan, S.; et al. Isolation and Characterization of Plant Growth-Promoting Endophytic Fungi from the Roots of *Dendrobium moniliforme*. *Plants* **2019,** *8* (1), 5.

Sindhu, R.; Pandey, A.; Binod, P. Solid-State Fermentation for the Production of Poly (Hydroxyalkanoates). *Chem. Biochem. Eng. Q.* **2015,** *29* (2), 173–181.

Subbalaxmi, S.; Murty, V. R. Process Optimization for Tannase Production by *Bacillus gottheilii* M2S2 on Inert Polyurethane Foam Support. *Biocatal. Agric. Biotechnol.* **2016,** *7*, 48–55.

Suryanarayan, S. Current Industrial Practice in Solid State Fermentations for Secondary Metabolite Production: The Biocon India Experience. *Biochem. Eng. J.* **2003,** *13* (2–3), 189–195.

Syed, S.; Riyaz-Ul-Hassan, S.; Johri, S. A Novel Cellulase from an Endophyte, *Penicillium* sp. NFCCI 2862. *Am. J. Microbiol. Res.* **2013,** *1* (4), 84–91.

Umsza-Guez, M. A.; Díaz, A. B.; et al. Xylanase Production by *Aspergillus awamori* Under Solid State Fermentation Conditions on Tomato Pomace. *Braz. J. Microbiol.* **2011,** *42* (4), 1585–1597.

Valta, K.; Papadaskalopoulou, C.; Dimarogona, M.; Topakas, E. Bioethanol from Waste–Prospects and Challenges of Current and Emerging Technologies. In *Byproducts from Agriculture and Fisheries: Adding Value for Food, Feed, Pharma and Fuels*, 2019; p. 421.

von Meien, O. F.; Luz Jr, L. F. L.; Mitchell, D. A.; et al. Control Strategies for Intermittently Mixed, Forcefully Aerated Solid-State Fermentation Bioreactors Based on the Analysis of a Distributed Parameter Model. *Chem. Eng. Sci.* **2004,** *59* (21), 4493–4504.

Xu, X.; Yu, Y.; Shi, Y. Evaluation of Inert and Organic Carriers for *Verticillium lecanii* Spore Production in Solid-State Fermentation. *Biotechnol. Lett.* **2011,** *33* (4), 763–768.

Zaferanloo, B.; Bhattacharjee, S.; Ghorbani, M. M.; et al. Amylase Production by *Preussia minima*, a Fungus of Endophytic Origin: Optimization of Fermentation Conditions and Analysis of Fungal Secretome by LC-MS. *BMC Microbiol.* **2014,** *14* (1), 55.

CHAPTER 14

Orchids and Mycorrhizal Endophytes: A Hand-in-Glove Relationship

MADHVI KANCHAN[1], JAGDEEP VERMA[2], SAYEEDA KOUSAR BHATTI[3], KRANTI THAKUR[4], KUSUM[5], ANAND SAGAR[6], and JASPREET K. SEMBI[1*]

[1]*Department of Botany, Panjab University, Chandigarh 160014, India*

[2]*Department of Botany, Government College, Rajgarh 173101, Himachal Pradesh, India*

[3]*Department of Botany, Shoolini University of Biotechnology & Management Sciences, Solan 173212, Himachal Pradesh, India*

[4]*Department of Botany, Shoolini Institute of Life Sciences & Business Management, Solan 173212, Himachal Pradesh, India*

[5]*Department of Botany, St. Bede's College, Navbahar, Shimla 171002, Himachal Pradesh, India*

[6]*Department of Biosciences, Himachal Pradesh University, Shimla 171005, Himachal Pradesh, India*

**Corresponding author. E-mail: jaspreet.sembi@pu.ac.in*

ABSTRACT

Orchids depend on mycorrhizal fungi for nutrition, especially during early developmental stages. Some orchid species are specific in their interactions, while others have a variety of fungal associations. Orchid mycorrhiza belongs to at least five major taxonomic groups such as Glomeromycota, Basidiomycota, Ascomycota, Agaricomycetes and Sordariomycetes fungi. The terrestrial orchids are more dependent on orchid mycorrhizal fungi (OMF) interactions for the nutritional requirement in comparison to epiphytic orchids because their protocorms become photosynthetic at early life stages. Various omics approaches are employed to understand the complexity of OMF interaction, which indicate that fungus augments orchid development through regulation of various transcription factors (DMI, NSP, WRKY, GRAS, SWEET, CCaMK, ENODL, TPP etc.), involved in plant growth and development. In addition to this, tissue culture studies involving symbiotic seed germination and further development in the presence of the specific mycorrhizal partner, promotes seed germination and robustness of the seedlings. The studies on orchid mycorrhizal associations provide a conceptual framework to understand the mechanisms of selection of fungal partner, establishment of the symbiotic association, nutritional aspects, and ecological adaptations. The present chapter provides an outline on possible physiological, molecular and ecological approaches involved in the study of OMF interactions.

14.1 INTRODUCTION

It is believed that almost all land plants are, to some extent, engaged in symbiotic relationships with mycorrhizal fungi (Dickie et al., 2015). These mutually beneficial interactions act as important drivers of global plant biogeographical patterns (Delavaux et al., 2019). The interactions are constantly evolving where the mycorrhizal fungi have gradually widened their biotrophic capabilities to take advantage of their hosts for food and protection while the hosts have developed strategies to accommodate the fungal associates (Genre et al., 2020). Such relationship is quite crucial in Orchidaceae, where right from the germination of seed to the establishment seedlings, all the processes are positively correlated with successful mycorrhizal associations under natural conditions.

Orchids are well known worldwide for their unique and long-lasting flowers, and immense therapeutic properties. Presently, there are more than 28,000 species recorded across the globe (Govaerts et al., 2020). Opting chiefly for an epiphytic life mode, and the presence of velamen tissue in roots, and labellum (lip), gynostegium (column), and compound pollens in flowers, are some of the important characteristic features, which make them different from other plants. They represent the pinnacle of plant evolution but still depend upon suitable fungi and pollinators to complete their life cycle. Orchid seeds are the smallest in plant kingdom and are produced in large numbers. These seeds lack necessary nutritional reserves to sustain its own germination, and this inability forms the basis of various orchid–mycorrhizal interactions (Rasmussen and Rasmussen, 2009). The fungus aids in germination of these microscopic, nonendospermic seeds by providing the requisite nutrients. This association is so important that the abundance and distribution of mycorrhizal fungi act as a key factor affecting orchid population dynamics (McCormick et al., 2019). According to McCormick and Jacquemyn (2014), orchid mycorrhizal fungi (OMF) not only drive the local abundance and dynamics of individual orchid populations but also influence the coexistence and the regional distribution of various orchid species.

Association of fungi with orchid roots was first observed by Reissek (1847), and Frank (1885) proposed the term mycorrhiza, for this association. Wahrlich (1886) and Janse (1897) later confirmed the occurrence of mycorrhizal fungi in orchids. Further research carried out by Bernard (1903, 1904) and Burgeff (1936, 1943, 1959) demonstrated that orchid seeds cannot germinate without these fungal association. Orchids remain associated with fungal mycelia at least at some stage of their life cycle, but the requirement is critical during the early stages of their development when ambient nutritional resources are scarce (Harley, 1963; Jacquemyn et al., 2012).

14.2 MYCORRHIZAL ASSOCIATION VARIES WITH ORCHID LIFE MODE

Mycorrhizal association is more prominent in terrestrial orchids (Rasmussen, 1995; Sathiyadash et al., 2020; Phillips et al., 2020) as the mycorrhizae aid in survival of ground growing taxa under comparatively harsher conditions by making them better adapted to their habitats (Burgeff, 1959; Rasmussen, 1995; Swarts and Dixon, 2009; Smith et al., 2010). The leafless mycoheterotrophs, which prefer to grow in moist and humus-rich habitats, possess comparatively stronger mycorrhizal obligation, which is usually lifelong (Vij and Sharma, 1983; McKendrick et al., 2000; Smith and Read, 2008; Martos et al., 2009; Merckx, 2013). Such orchids involve a wider variety of mycorrhizal fungi belonging to Glomeromycota, Basidiomycota, and Ascomycota, as well as some saprobic taxa of Agaricomycetes (like *Hydropus*, *Gymnopus*, *Marasmiellus*) and Sordariomycetes fungi (like *Clonostachys*, *Resinicium*), and also exhibit higher degree of specificity with respect to their association (Furman and Trappe, 1971; Richardson et al., 1993; Taylor and Bruns, 1997; McKendrick et al., 2000; Taylor et al., 2002; Tsujita et al., 2009; Dearnaley et al., 2012). According to Merckx (2013), the retention of an entirely mycoheterotrophic state, where the plant remains totally dependent on the fungus even at maturity, has also evolved sporadically across Orchidaceae. Interestingly, under certain circumstances, the orchid seedlings also get carbon nutrition via ectomycorrhizal fungi which connect them with the roots of some neighboring autotrophic plants (Zelmer and Currah, 1995; Taylor and Burns, 1997; McKendrick et al., 2000; Selosse and Roy, 2009; Rasmussen et al., 2015) thereby making them behave as epiparasites (exploitative association) on the later (Taylor and Burns, 1997; Cullings et al., 1996; Bidartondo and Bruns, 2001; Ogura-Tsujita et al., 2009).

In epiphytic orchids, on the other hand, the mycorrhizal dependency is rather less as their protocorms become photosynthetic at early life stages (Hadley, 1982; Vij and Sharma, 1983; Arditti, 1992; Dearnaley, 2007; Manoharachary and Tilak, 2015). These orchids can partially meet their mineral nutrition from dust, organic debris, and stemflow along the host bark (Arditti, 1992; Zettler et al., 2011; Rasmussen et al., 2015) and, therefore, generally have facultative mycorrhizal associations (Goh et al., 1992; Richardson et al., 1993; Rasmussen, 2002; Benzing, 2004; Motomura et al., 2008; Zotz and Winkler, 2013). Comparatively lesser occurrence of pelotons in many adult epiphytic orchids has led some researchers to question their importance for plant nutrition, especially during their adulthood (Bayman et al., 2002). According to Rasmussen et al. (2015) and Phillips et al. (2020), our understanding about the mycorrhizal ecology of tropical epiphytic and lithophytic orchids is quite limited. Interestingly, habitat-driven mycorrhizal associations have also been indicated by Oja et al. (2015) and Ruibal et al. (2017) while investigating *Neottia ovata* and *Chiloglottis* populations growing in varied substrates and locations.

14.3 ORCHID MYCORRHIZAL FUNGI (OMF) HELP THE HOST IN MANIFOLD WAYS

Lack of chlorophyll and failure to utilize the available nutrient reserves make orchid seeds completely dependent upon their fungal associates for nutrition (Rasmussen and Rasmussen, 2009). OMF augment carbohydrate nutrition by breaking down the complex organic compounds in the soil/ substrate and facilitates their subsequent release in the orchid host (Rasmussen, 1995; Smith and Read, 1997; Dearnaley, 2007; Mehra et al., 2016). This includes carbon (Smith, 1967; Alexander and Hadley, 1985; Trudell et al., 2003; Cameron et al., 2006, 2008; Bougoure et al., 2010; Rasmussen et al., 2015; Mehra et al., 2016), phosphorus (Alexander et al., 1984; Smith and Read, 1997; Cameron et al., 2007; Nurfadilah et al., 2013; Zhao et al., 2014), nitrogen (Burgeff, 1936; Barrosso et al., 1986; Rasmussen, 1995; Smith and Read, 1997; Trudell et al., 2003; Cameron et al., 2006; Bougoure et al., 2010; Nurfadilah et al., 2013; Ding et al., 2014; Zhao et al., 2014; Rasmussen et al., 2015), water

(Yoder et al., 2000; Rasmussen and Whingham, 2002; Chang, 2007; Ding et al., 2014), and vitamins (Harvais and Pekkala, 1975; Barroso et al., 1986; Rasmussen, 1995; Rasmussen, 2002; Selosse, 2014). This supplementation is mainly facilitated by lysis of the fungal hyphae that form pelotons inside host tissue (Fochi et al., 2017; Phillips et al., 2020). Vitamins produced by the mycorrhizal fungus are reported to augment not only seed germination but also subsequent seedling growth (Burgeff, 1959; Hadley and Ong, 1978; Arditti et al., 1990; Rasmussen and Rasmussen, 2009) and aid in breaking down of pectin, cellulose, lignin, and metal phosphates (Perombelon and Hadley, 1965; Hadley, 1969; Jacobs et al., 2002; Nurfadilah et al., 2013; Manoharachary and Tilak, 2015). According to Gebauer et al. (2016), some photosynthetic orchids remain partial mycoheterotrophs and retain an ability to acquire nutrition from fungal partner even during adulthood. It is also worth mentioning that the endophytic fungi, which generally live peacefully within the hosts, may also behave as facultative pathogen under certain circumstances (Aly et al., 2010).

The fungal partner associated with any orchid also thrives in a heterotrophic mode. There are evidences to support that orchids may contribute carbon to the fungal associate (Cameron et al., 2006, 2008; Ogura-Tsujita et al., 2009; Nurfadilah et al., 2013; Liebel et al., 2015; Fochi et al., 2017; Phillips et al., 2020). The orchid mycorrhizae (OM) represent unusual symbioses; it is believed that during initial colonization process in the young, nonphotosynthetic orchid host, the fungus provides both organic and inorganic nutrition to the plant but receive nothing in return. However, it has been observed in some adult photosynthetic terrestrial orchids that there is some export of sugars from plants to the fungus also (Cameron et al., 2006), suggesting thereby that association is symbiotic. OMF are therefore thought to represent a true mutualism in both the early and mature stages of plant growth and development. However, the carbon dependence of fungus on orchid is not obligatory and it can grow independently and is not necessarily codistributed with the orchid host (Phillips et al., 2020).

14.4 HOST PREFERENCES OF THE MYCORRHIZAL FUNGUS

According to Sathiyadash et al. (2020), there are enough evidences to believe that OMF are not host-specific, but there are also indications to suggest a possible existence of a specific physiological compatibility in orchid–fungal relationship. This specificity is also augmented with reports on better seed germination when inoculated with mycorrhizal fungal isolated from the parent plant or from individuals belonging to the same species (Bernard, 1904; Burgeff, 1936; Clements, 1988; Smreciu and Currah, 1989; Masuhara and Katsuya, 1994; Phillips et al., 2011; Zhao et al., 2014; Herrera et al., 2016; Bhatti et al., 2016, 2017). On the other hand, there are reports suggesting that the fungal requirement is not very much specific and the seeds are capable of germination with fungi isolated even from other orchid species (Arditti et al., 1990; Johnson, 1994; Rasmussen, 2002; Taylor et al., 2002; Vij et al., 2002; Shefferson et al., 2007; Dearnaley, 2007; Roche et al., 2010; Salifah et al., 2011; Zi et al., 2014; Rasmussen et al., 2015).

14.5 ESTABLISHMENT OF MYCORRHIZAL ASSOCIATION

Orchid seeds are penetrated by the fungus through their general surface, embryonic rhizoids or from the micropylar end (Rasmussen, 2002; Sazak and Ozdener, 2006; Salifah et al., 2011; Hossain et al., 2013a; Chen et al., 2014; Kohler et al., 2015; Bhatti et al., 2016). In the case of seedlings, the fungal penetration may be through root hairs (Burgeff, 1936; Peterson and Farquhar, 1994; Senthilkumar, 2001; Muthukumar et al., 2011; Sathiyadash et al., 2012), or directly through root epiblema cells (Burges, 1939), or even through both of these structures (Vij and Sharma, 1983; Vij et al., 1985; Kaliamoorthy, 2007; Sathiyadash et al., 2012; Eswaranpillai et al., 2015). After initial penetration, the fungus usually reaches the cortical region through thin-walled passage cells of exodermis (Esnault et al., 1994) where it forms pelotons, intracellular coils, or hyphal aggregates (Peterson and Farquhar, 1994; Balachandar et al., 2019). Cortical cells, therefore, represent the sites of fungal digestion (Vij and Sharma, 1983). Disappearance of starch grains during fungal digestion and their reappearance immediately after the completion of process has been reported in this root zone (Rasmussen, 1995; Sathiyadash et al., 2012; Manoharachary and Tilak, 2015). The digestion process is triggered by the release of enzymes (peroxidases and phosphatases) resulting from increased ionic concentration of the cell sap, or due to the presence of some fungistatic compounds (Vij and Sharma, 1983; Kumar and Krishnamurthy, 1999), and is completed in a controlled manner (Rasmussen and Whingham, 2002). The formation and digestion of pelotons occurs throughout the year. Such a type of repeated cycles of peloton formation and digestion represents tolypophagus type of fungal digestion (Burgeff, 1959; Hadley, 1982; Rasmussen and Whingham, 2002; Sathiyadash et al., 2012). Rasmussen (1995) observed another kind of digestion (phytophagous) in mycoheterotrophic orchids where the hyphal tips get lysed immediately while entering the cortical cells of the host. This facilitates the release of hyphal cell contents, including soluble carbohydrates, essential ions, and water into the host cells (Smith, 1966; Alexander et al., 1984; Alexander and Hadley, 1985; Manoharachary and Tilak, 2015; Mehra et al., 2016).

It has been reported that roots experience maximum colonization by the fungus, but the fungus is also found associated with protocorms (Hayakawa et al., 1999; Zettler et al., 2005; Zi et al., 2014; Khamchatra et al., 2016), rhizomes (Harley, 1959; Yagame et al., 2008; Sudheep and Sridhar, 2012), tubers, corms, pseudobulbs (Harley, 1959; Rasmussen, 2002; Kaliamoorthy, 2007; Tondello et al., 2012), and even leaves (Arnold and Lutzoni, 2007). Moreover, the degree of fungal spread in root cortical cells varies with respect to habit, habitat, and life cycle stages of orchids as well as season of the year (Arditti, 1992; Goh et al., 1992; Rasmussen, 2002; Cameron et al., 2006, 2008; Bertolini et al., 2014).

Orchids are known to form mycorrhizal associations with phylogenetically and ecologically diverse fungi. They mainly belong to Basidiomycetes (*Ceratobasidium*, *Ceratorhiza*, *Epulorhiza*, *Mycena*, *Rhizoctonia*, *Sebacina*, *Thanatephorus, Tulasnella,* etc.) and Ascomycetes (*Alternaria*, *Bionectria*, *Cladosporium*, *Cochliobolus*, *Fusarium*, *Trichoderma*, *Xylaria,* etc.) (Warcup and Talbot, 1966; Williams, 1985; Rasmussen, 2002; Chen et al., 2011; Hossain et al., 2013a,b; Ma et al., 2015; Bhatti et al., 2016; Ruibal et al., 2017). *Rhizoctonia*-like fungi (Ceratobasidiaceae, Tulasnellaceae) are quite common both in epiphytic and ground growing orchids (Zettler et al., 2011). Species belonging to genera *Armillaria*, *Corticium*, *Fomes*, *Hypochnus*, *Marasmius*, and *Xerotus* have also been reported to develop mycorrhizal associations with orchids (Currah and Sherburne, 1992).

14.6 ISOLATION AND CHARACTERIZATION OF MYCORRHIZAL FUNGUS

Scientists have generally employed root sections of varied thickness (up to 2.0 cm) for isolating mycorrhizal fungi in orchids (Bernard, 1903; Curtis, 1939; Sazak and Ozdener, 2006; Siddiquee et al., 2010; Hossain et al., 2013b; Ding et al., 2014; Zhao et al., 2014; Ma et al., 2015; Bhatti et al., 2016). However, a few root cortical cells having fungal colonization or even a single peloton have also been used for this purpose (Currah et al., 1987; Kristiansen et al., 2001; Athipunyakom et al., 2004; Chen et al., 2011; Khamchatra et al., 2016). Root sections/pelotons are cultured on different artificial media such as potato dextrose agar (PDA), cornmeal agar (CMA), oatmeal agar (OMA), and water agar under in vitro conditions (Currah et al., 1987; Ding et al., 2014; Zhao et al., 2014). Recently, Zettler and Corey (2018) summarized various methods to isolate and identify peloton-forming fungi in the *Rhizoctonia* complex.

The identification of mycorrhizal fungi has always been challenging because many species lack distinguishing morphological characters. This problem is even more acute in the case of many OMF that mostly belong to the phylum Basidiomycota. Earlier, the isolated fungi were studied using morphological characteristics only (Warcup and Talbot, 1967; Rasmussen, 1995; Hossain et al., 2013a; Ding et al., 2014; Zhao et al., 2014; Ma et al., 2015). But the rapid advancement of various biotechnological tools has helped overcoming the bottlenecks associated with the traditional identification methods; it is now largely done by using modern molecular techniques (Taylor and Bruns, 1997; Kristiansen et al., 2001; Shefferson et al., 2007; Tao et al., 2008; Sawmya et al., 2013; Ding et al., 2014; Zhao et al., 2014; Ma et al., 2015; Sathiyadash et al., 2020). Sequencing the internal transcribed spacer (ITS) regions of ribosomal RNA gene is one of such methods which has proved useful in delimiting even various strains belonging to same species (Taylor and Burns, 1997; Kristiansen et al., 2001; Rasmussen, 2002; Ding et al., 2014; Pereira et al., 2015a,b).

14.7 MOLECULAR BASIS OF ORCHID MYCORRHIZAL ASSOCIATION

The mycorrhizal fungi provide carbohydrates and other nutrients until heterotrophic protocorms (nonphotosynthetic) develop into photosynthetic seedlings (Dearnaley et al., 2016). The above strategy is termed mycoheterotrophy or OM (Suetsugu et al., 2017). These are different from the arbuscular mycorrhizae (AM) in the mechanism of nutrient exchange. In AM, the carbon exchange is unidirectional, while in OM, there is bidirectional flow of carbon between plants and fungus and flow of nitrogen or phosphorus from fungus to plants. The fungus stimulates orchid development through upregulation of cell cycle proteins, purine recycling, ribosome biogenesis, energy metabolism, and secretion in the plant (Valadares et al., 2014). Earlier studies for determination of the fungal symbionts mainly involved morphological identification and analysis of in vitro isolated strains. The molecular mechanism of OM is the least understood among mycorrhizal symbiotic associations (Zhao et al., 2014). Since the pioneering works by Bernard (1899), a number of researches for understanding the diversity and specificity of OMF have been undertaken but the expression and interaction of genes during the orchid mycorrhizal associations remains unclear.

In order to understand the dynamics of gene expression in OM relationships, genomic, transcriptomic, and proteomic studies have been done to unravel the key molecular events during the plant–fungus interactions (Li et al., 2012; Perotto et al., 2014; Valadares et al., 2014; Zhao et al., 2014). During this exploration of the molecular basis of the symbiotic association in orchids, a number of genes have been reported to be involved in the perception and transduction of microbial signals during root colonization and nutrient exchange. These genes have been referred to as common symbiosis genes (CSGs). Some of the common CSG genes reported in orchids are *nodulin-like* genes, *calcium-dependent protein kinase* (*CDPKs*) genes, GRAS family transcription factor *NSP1*, auxin efflux facilitator gene, *PIN1* and *SWEET* gene family (sugar transporters) (Zhao et al., 2013; Perotto et al., 2014; Suetsugu et al., 2017; Miura et al., 2018). In a terrestrial orchid *Bletilla striata*, the highest expression of CSGs genes is reported in root cells (Miura et al., 2018). In this study, the role of a *calcium- and calmodulin-dependent protein kinase* gene (*BsCCaMK*) in AM symbiosis has been studied by cross-complementation assay with *Lotus japonicas ccamk-3* mutant with *BsCCaMK* and indicates its role in calcium mediated signaling during orchid–fungal interactions. The *CCaMK* gene is very important gene and plays a vital role in actinorhizal symbiosis and is also studied in *Oryza sativa*, where it rescued defective nodulation phenotype of *ccamk* mutant, which lead to the formation of a mature nodule (Banba et al., 2008). The *calcium- and calmodulin-dependent protein kinase* (*CCaMK*) gene is the member of family of serine/threonine kinases and responds to many environmental stresses and is rapidly activated upon exposure to biotic and abiotic stresses. It has been reported in *Oncidium sphacelatum* that calcium signaling is the heart of OM symbiosis signal transduction and leads to the accumulation of differential proteins such as calmodulin (a core component of the calcium signal transduction pathway) and inositol-5-phophatase (involved in IP3 hydrolysis) (Chen et al., 2008; Yang and Poovaiah, 2003; Valadares et al., 2014). *Calcium-dependent protein kinases* (*CDPK*s) comprise a large gene family as reported in *Arabidopsis* and rice (Harmon et al., 2001; Asano et al., 2005). In orchids, two *CDPK* genes, *CDPK1* and *CDPK32* were identified in symbiotically germinated *Dendrobium officinale* seeds using suppression subtractive hybridization (SSH) cDNA library and were predicted to have an important role in *D. officinale—Sebacina* sp. symbiotic association. Similarly, two homologues of *CDPK* genes were found to be upregulated in roots during Ca^{2+} spiking in the nucleus and perinuclear region upon cocultivation with different mycorrhizal fungi in *Cymbidium hybridum* (Zhao et al., 2014). *CDPK1* has also been predicted to be transcriptionally activated in response to low temperature, wounding, and pathogen infection in *Phalaenopsis amabilis* (Tsai et al., 2007). Similarly, *CDPK1* gene has also been identified from *D. officinale* roots infected by an OM fungus, *Mycena* sp. and reported to be accumulated in roots after one month of fungal infection indicating the role of this gene in symbiosis between *D. officinale* and *Mycena* sp. (Zhang et al., 2012). In *Gastrodia elata*, the highest expression of genes responsible for the expression of pathogenesis-/wound-related proteins, peroxidases, and serine/threonine-protein kinases is reported in late-stage protocorms, which signifies that these genes have a role in fungal colonization that triggers the defense responses (He et al., 2017; Zeng et al., 2018).

Another important pathogen-responsive gene family is *WRKY*. In *D. officinale*, *DoWRKY* genes were reported to be involved in multiple biological mechanisms, out of which many *DoWRKY* genes were differentially expressed between symbiotic- and asymbiotic-germinated seeds indicating that *DoWRKYs* might be involved in promoting in vitro symbiotic germination of seeds with *Tulasnella*-like fungi (Wang et al., 2018). In *Arabidopsis thaliana*, a *WRKY* gene has been reported to induce a response to pathogen infection (Zheng et al., 2006). *WRKY* gene family is one of the largest family of transcriptional regulators in plants, with two *WRKY* conserved domains with N-terminus WRKYGQK motif and the C-terminus zinc-binding motif (Wang et al., 2018).

The nodulin-like factors belonging to the Phycocyanin (PC) protein family are the ancient plant blue copper-binding proteins (BCPs) and show binding with single-type I copper atoms and acts as an electron transporters. Phytocyanin (PC) genes have been reported in *A. thaliana, O. sativa*, and *Brassica rapa* (Mashiguchi et al., 2009; Ma et al., 2011; Li et al., 2013). This family is divided into four subfamilies, uclacyanins, stellacyanins, plantacyanins, and early nodulin-like proteins (ENODLs) (Mashiguchi et al., 2004). In *Phalaenopsis equestris*, PC gene family has been reported to have a role in orchid–mycorrhizal associations (Xu et al., 2017).

Studies have also been done on developing a model system to predict the involvement of various genes in orchid–fungi association, in *Serapias vomeracea* orchid colonized by the *Rhizoctonia*-like fungus, *Tulasnella calospora* (Balestrini et al., 2014; Perotto et al., 2014). The expression analysis in mycorrhizal and nonmycorrhizal protocorm tissues shows that nodulin-like protein (ENODL) containing a plastocyanin-like domain expressed only in protocorm cells containing intracellular fungal hyphae (Balestrini et al., 2014; Perotto et al., 2014). This study suggests that *ENODL* genes play important roles not only in the nodulation process, but also in symbiotic development. In another study, laser microdissection method has been applied for the identification of genes involved in AM in *S. vomeracea*, to detect transcripts corresponding to fungal and plant nutrient transporters with same functions (e.g., amino acid and ammonium transporters) and *ENODL* genes in peloton-containing cells (Fochi et al., 2017).

Sugars act as vital primary metabolites, nutrients, and signal molecules in plants during orchid–fungus interactions. An important group of genes which are predicted to be involved in these processes are the *SWEET* (sugar transporter-like proteins) genes. On the basis of phylogenetic analyses, SWEET protein family is divided into four clades, Clades I and II include AtSWEET1-8 which mainly act as glucose transporters (Chen et al., 2010), Clade III consists of AtSWEET9-15 which are the sucrose and glucose transporters, and AtSWEET16-17 is of clade IV mainly acts as sucrose, glucose, and fructose transporters (Klemens et al., 2013; Han and Jiang, 2015). The SWEET proteins are predicted to have seven transmembrane helices (7-TM) with two MtN3/saliva domains (Chen et al., 2012; Yuan et al., 2014). A number of *SWEET* genes have been identified in *D. officinale* and *P. equestris* (Wang et al., 2018). *SvNod9* gene, which encodes a predicted sugar transporter of the SWEET gene family, has been reported in orchid mycorrhizal symbiosis between protocorms of *S. vomeracea* and the fungus *T. calospora* (Perotto et al., 2014). The comparison of the transcriptomes of *Epipogium aphyllum* and *Neottia nidus-avis* with other mycoheterotrophic orchids indicates that these plants have highly upregulated trehalose and trehalose-6-P phosphatases (TPP), which showed that fungi provides trehalose to plants (Lallemand et al., 2019; Jakalski et al., 2020). Along with this, the presence of *SWEET* transporter orthologues is also reported in these mycoheterotrophic orchids. The presence of SWEET gene transporters is also reported in achlorophyllous mutants of mixotrophic orchid *Epipactis helleborine* (Jakalski et al., 2020). However, a detailed analysis of the *SWEET* gene family in orchids is yet to be explored.

Another type of CSGs genes reported in mycoheterotrophic symbiosis is the *GRAS*-domain transcription factor gene. The name GRAS is derived from first three members: GIBBERELLIC-ACID INSENSITIVE (GAI), REPRESSOR of GAI (RGA), and SCARECROW (SCR). This gene family consist of a various distinct domain at C-terminal region such as leucine heptad repeat I (LHR I), VHIID, leucine heptad repeat II (LHR II), PFYRE and SAW motif, out of which VHIID and PFYRE motifs are found highly conserved and LHR motif play significant in protein–protein interactions (Tian et al., 2004). *GRAS* genes have been reported from *Dendrobium catenatum* (Zeng et al., 2019). It is reported in *C. hybridum* that NSP1 and NSP2 (Nodulation signaling pathway 1) transcription factors of GRAS gene family are essential for rhizobial infection and induction of cortical cell divisions for formation nodule primordium (Yang and Poovaiah, 2003). The *DIM2* (*Does Not Make Infections2*) gene encodes a receptor-like kinase, which is necessary for root endosymbiosis and consequently, activates downstream transcription regulators (CDPK1, CDPK2, NSP1, and NSP2), which govern the expression of mycorrhization genes. In *C. hybridum*, Ca^{2+} spiking in the nucleus and perinuclear region of root hair cells digital analysis revealed the presence of transcripts encoding a homologue of *DMI2* nodulation receptor kinase and one cyclic nucleotide-gated ion channel protein (Yang and Poovaiah, 2003; Zhao et al., 2014). Many genes reported in mycoheterotrophic interaction are related to nitrogen transport (Fochi et al., 2017) (Table 14.1).

TABLE 14.1 Genes **Involved in Orchid Mycorrhizae (OM) Associations**

Name of the gene	Orchid species	References
DMI2 (Does Not Make Infections2)	*Cymbidium hybridum*	Zhao et al. (2015)
NSP1/NSP2 (Nodulation Signaling pathway)	*C. hybridum*	Zhao et al. (2015)
WRKY	*Dendrobium officinale*	Wang et al. (2018)
GRAS	*C. hybridum*	Zhao et al. (2015)
	Dendrobium catenatum	Zeng et al. (2019)
SWEET	*D. officinale*	Wang et al. (2018)
	Phalaenopsis equestris	Perotto et al. (2014)
	Serapias vomeracea	Jakalsk et al. (2020)
	Epipogium aphyllum	
	Neottia nidus-avis	
	Epipactis helleborine	
CCaMK (Calcium- and calmodulin-dependent protein kinase)	*Bletilla striata*	Zhang et al. (2012)
	D. officinale	Balestrini et al. (2014)
	Oncidium sphacelatum	Valadares et al. (2014)
	Phalaenopsis amabilis	Tsai et al. (2007)
	C. hybridum	Zhao et al. (2015)
ENODL(Nodulin-like protein)	*P. equestris,*	Xu et al. (2017)
	S. vomeracea	Balestrini et al. (2014)
TPP (Trehalose and Trehalose-6-P phosphatase)	*E. aphyllum*	Jakalsk et al. (2020)
	N. nidus-avis	

Root transcriptome analysis of partially mycoheterotrophic orchid *E. helleborine* and its mutant (mycoheterotrophic/achlorophyllous) suggest that during mycorrhizal association genes encoding glycerol 3-phosphate acyltransferase and subtilisin are upregulated, and genes related to gibberellin synthesis are downregulated (Suetsugu et al., 2017; Takeda et al., 2015). Studies on comparative transcriptomic and proteomic of asymbiotic or symbiotic seed germination in *D. officinale* show that proteins related to "carbohydrate metabolism" and "post-translational modification" have the highest expression at the seedling stage following the protocorm stage and germination in the case of symbiotic seed germination (Chen et al., 2017).

14.8 NONSYMBIOTIC FUNGAL ASSOCIATES IN ORCHIDS

A number of non–mycorrhizal endophytic fungi (i.e. *Alternaria*, *Cercospora*, *Lasiodiplodia*, *Phyllosticta*) have also been reported from orchids (Tao et al., 2008; Salifah et al., 2011; Pecoraro et al., 2013; Sawmya et al., 2013; Ma et al., 2015). It is also possible that some surface contaminating (*Aspergillus*, *Penicillium*, *Cladosporium*, etc.) or soil dwelling fungi (*Trichoderma*, *Verticillium*, etc.) could had mistakenly identified as orchid mycorrhizal endophytes (Kasmir et al., 2011; Salifah et al., 2011).

14.9 MYCORRHIZAE IN SEED GERMINATION AND CONSERVATION

Orchids are inherently slow growers. The conventional methods of their multiplication through cuttings, divisions, kiekis, pseudobulbs, back bulbs, tubers, etc. usually prove inadequate to meet their demand for reasons related to their science and commerce. Therefore, propagation by means of symbiotic and asymbiotic seed culture has emerged out as an important technique to mass multiply them. It is now a well-established fact that the OMF not only promote seed germination but also stimulate the growth and development of protocorms and seedlings. The mycorrhizal fungi have been used to induce seed germination both under natural (in situ) and laboratory (in vitro) conditions (Rasmussen, 1995; Bhatti et al., 2016), and they contribute to better growth and development (Dutta et al., 2014). The orchid–mycorrhizal associations are more important in leafless orchids like *Galeola falconeri* as their absence may result in shrinkage of orchid populations (Das and Khumbongmayu, 2006). Therefore, researches on OMF are of notable importance not only in orchid commercialization but also in their conservation (Dearnaley et al., 2012; Riofrio et al., 2013).

Symbiotically raised seedlings have been reported to be better adapted to the environmental fluctuations as compared to those produced by asymbiotic culture (Guimaraes et al., 2013). The fungal endophytes in such associations ensure better accessibility of orchid plants to the soil resources and help them becoming tolerant to the environmental stresses (Rasmussen, 1995). Seed germination, growth, and ecological fitness of several orchids have been promoted by inoculation with a range of OMF belonging mainly to genera *Ceratobasidium*, *Ceratorhiza*, *Epulorhiza*, *Mycena*, *Rhizoctonia*, *Mortierella*, *Scytalidium*, *Tulasnella*, etc. (Powell and Arditti, 1975; Hayakawa et al., 1999; Qui-Xia et al., 2014; Zhao et al., 2014; Bhatti et al., 2016, 2017). According to Zettler and Hofer (1997) the long-term survival of orchids depends upon their efficient association with appropriate fungal partner for seedling recruitment and plant nutritional support. OMF can influence the size of plant populations and the distribution of OMF is potentially a critical factor in determining the distribution and population dynamics in orchids (McCormick et al., 2019). As the fungal support facilitates high germination rates and yield comparatively vigorous seedlings, it can play an important role in the rehabilitation of threatened orchid species in their natural (Phillips et al., 2020) or natural-like habitat.

14.10 CONCLUSIONS

The indispensability of the mycorrhizal fungus for orchid seed germination and growth has been well documented. OMF facilitate the host in a number of ways, including supply of carbon and other nutrients. Various advancements in the field of orchid–fungi interactions have led to the better understanding of the evolutionary patterns and population dynamics of the fungus as well as the host. Isolation and molecular characterization of the fungal partner has paved way for better understanding of various pathways involved in mycoheterotrophy, including carbohydrate and nitrogen metabolism and pathogen responses. The present review elucidates this hand-in-glove relationship in the orchid mycorrhizal associations.

KEYWORDS

- **orchid mycorrhizal fungi**
- **protocorms**
- **symbiotic germination**
- **OMF interactions**
- **Glomeromycota**

REFERENCES

Alexander, C.; Alexander, I. J.; Hadley, G. Phosphate Uptake in Relation to Mycorrhizal Infection. *New Phytol.* **1984,** *97*, 401–411.

Alexander, C.; Hadley, G. Carbon Movement between Host and Mycorrhizal Endophyte during the Development of the Orchid *Goodyera repens*. *New Phytol.* **1985,** *101*, 657–665.

Aly, A. H.; Debbab, A.; Kjer, J.; Proksch, P. Fungal Endophytes from Higher Plants: A Prolific Source of Phytochemicals and Other Bioactive Natural Products. *Fungal Diversity* **2010,** *41*, 1–16.

Arditti, J. *Fundamentals of Orchids Biology*; John Wiley and Sons: New York, 1992.

Arditti, J.; Ernst, R.; Yam, T. W.; Glabe, C. The Contribution to Orchid Mycorrhizal Fungi to Seed Germination: A Speculative Review. *Lindleyana* **1990,** *5*, 249–255.

Arnold, A. E.; Lutzoni, F. Diversity and Host Range of Foliar Fungal Endophytes: Are Tropical Leaves Biodiversity Hotspots. *Ecol. Soc. Am.* **2007,** *88* (3), 541–549

Asano, T.; Tanaka, N.; Yang, G.; Hayashi, N.; Komatsu, S. Genome-Wide Identification of the Rice Calcium-Dependent Protein Kinase and Its Closely Related Kinase Gene Families: Comprehensive Analysis of the CDPKs Gene Family in Rice. *Plant Cell Physiol.* **2005,** *46* (2), 356–366.

Athipunyakom, P.; Manoch, L.; Piluek, C. Isolation and Identification of Mycorrhizal Fungi from Eleven Terrestrial Orchids. *Kasetsart J. Nat. Sci.* **2004,** *38*, 216–228.

Balachandar, M.; Ravi, R. K.; Ranjithamani, A.; Muthukumar, T. Comparative Vegetative Anatomy and Mycorrhizal Morphology of Three South Indian *Luisia* species (Orchidaceae) with the Note on Their Epiphytic Adaptations. *Flora* **2019,** *251*, 39–61.

Balestrini, R.; Lumini, E. Focus on Mycorrhizal Symbioses. *Appl. Soil Ecol.* **2018,** *123*, 299–304.

Balestrini, R.; Nerva, L.; Sillo, F.; Girlanda, M.; Perotto, S. Plant and Fungal Gene Expression in Mycorrhizal Protocorms of the Orchid *Serapias vomeracea* colonized by *Tulasnella calospora*. *Plant Signal. Behav.* **2014,** *9* (11): e977707.

Banba, M.; Gutjahr, C.; Miyao, A.; Hirochika, H.; Paszkowski, U.; Kouchi, H.; Imaizumi-Anraku, H. Divergence of Evolutionary Ways Among Common Sym Genes: CASTOR and CCaMK Show Functional Conservation between Two Symbiosis Systems and Constitute the Root of a Common Signaling Pathway. *Plant Cell Physiol.* **2008,** *49*: 1659–1671.

Barrosso, J.; Neves, H. C.; Pais, M. S. S. Production of Indole-3-Ethanol and IAA by Mycorrhizal Fungus of *Ophrys lutea* (Orchidaceae). *New Phytol.* **1986,** *103*, 745–750.

Bayman, P.; Eduardo, J.; Gonzalez, J.; Fumero, J.; Raymond, I. Are Fungi Necessary. How Fungicides Affect Growth and Survival of the Orchid *Lepanthes rupestris* in the field. *J. Ecol.* **2002,** *90*, 1002–1008.

Benzing, D. H. Vascular Epiphytes. In *Forest Canopies*, 2nd ed.; Elsevier Academic Press, 2004; pp 175–211.

Bernard, J. M. Seasonal Changes in Standing Crop and Primary Production in a Sedge Wetland and an Adjacent Dry Old-Field in Central Minnesota. *Ecology* **1974,** *55*, 350–359.

Bernard, N. L'evolution dans la symbiose: les orchidees et leurs champignons commensaux. *Ann. Sci. Nat. Bot.* **1909,** 1–196.

Bernard, N. *La germination des orchidees*; Comptes Rendus de l'Academie des Sciences: Paris, 1903; Vol. 137, pp 483–485.

Bernard, N. Recherches expérimentales sur les Orchidées. *Rev. Gén. Bot.* **1904,** *16*, 405–451.

Bernard, N. Seances Sur la germination du *Neottia nidus-avis*. *Acad. Sci.* **1899,** *128*, 1253–1255.

Bertolini, V.; Cruz-Blasi, J.; Damon, A.; Valle Mora, J. F. Seasonality and Mycorrhizal Colonization in Three Species of Epiphytic Orchids in Southeast Mexico. *Acta Bot. Brasilica* **2014,** *28* (4), 512–518.

Bhatti, S. K.; Verma, J.; Sembi, J. K.; Kumar, A. Mycobiont Mediated *in vitro* Seed Germination of an Endangered 'Fox-Tail' Orchid, *Rhynchostylis retusa* (L) Blume. *J. Pure Appl. Microbiol.* **2016,** *10* (1), 663–670.

Bhatti, S. K.; Verma, J.; Sembi, J. K.; Pathak, P. Symbiotic Seed Germination of *Aerides multiflora* Roxb. – A Study In Vitro. *J. Orchid Soc. India* **2017,** *31*, 85–91.

Bidartondo, M. I.; Bruns, T. D. Extreme Specificity in Epi-Parasitic Monotropoideae (Ericaceae): Widespread Phylogenetic and Geographical Structure. *Mol. Ecol.* **2001,** *10*, 2285–2295.

Bougoure, J. J.; Ludwig, M.; Brundrett, M.; Grierson, P. Identity and Specificity of the Fungi Forming Mycorrhizas with the Rare Myco-Heterotrophic Orchid *Rhizanthella gardneri*. *Mycol. Res.* **2010,** *113*, 1097–1106.

Burgeff, H. *Die samenkeimung der Orchideen;* Gustav Fisher: Jena, 1936.

Burgeff, H. Mycorrhiza of Orchids. In *The Orchids: A Scientific Survey*; Withner, C. L., Ed.; Ronald Press: New York, 1959; pp 361–395.

Burgeff, H. Problematik der Mycorhiza. *Naturwissenschaften* **1943,** *31*, 558–567.

Burges, A. The Defensive Mechanism in Orchid Mycorrhizas. *New Phytol.* **1939,** *38*, 273–283.

Cameron, D. D.; Johnson, I.; Leake, J. R.; Read, D. J. Mycorrhizal Acquisition of Inorganic Phosphorus by the Green-Leaved Terrestrial Orchid *Goodyera repens*. *Ann. Bot.* **2007,** *99*, 831–834.

Cameron, D. D.; Johnson, L.; Read, D. J.; Leake, J. R. Giving and Receiving: Measuring the Carbon Cost of Mycorrhizas in the Green Orchid, *Goodyera repens*. *New Phytol.* **2008,** *180*, 176–184.

Cameron, D. D.; Leake, J. R.; Read, D. J. Mutualistic Mycorrhiza in Orchids: Evidence from Plant-Fungus Carbon and Nitrogen Transfers in the Green-Leaved Terrestrial Orchid *Goodyera repens*. *New Phytol.* **2006,** *171*, 405–416.

Chang, D. C. N. The Screening of Orchid Mycorrhizal Fungi (OMF) and Their Applications. In *Orchid Biotechnology*; Chen, W. H., Chen, H. H., Eds.; World Scientific Publishing Co Pvt Ltd.: Hong Kong, 2007; pp 77–78.

Chen, J. H.; Wang, S. S.; Liu, Y.; Li, Y.; Guo, S. X. Ultrastructure of Symbiotic Germination of the Orchid *Dendrobium officinale* with Its Mycobiont, *Sebacina* sp. *Aust. J. Bot.* **2014,** *62* (3), 229–234.

Chen, J.; Hu, K. X.; Hou, X. Q.; Guo, S. X. Endophytic Fungi Assemblages from 10 *Dendrobium* Medicinal Plants (Orchidaceae). *World J. Microbiol. Biotechnol.* **2011,** *27*, 1009–1016.

Chen, J.; Liu, S. S.; Kohler, A.; Yan, B.; Luo, H. M.; Chen, X. M.; Guo, S. X. iTRAQ and RNA-Seq Analyses Provide New Insights into Regulation Mechanism of Symbiotic Germination of *Dendrobium officinale* Seeds (Orchidaceae). *J. Proteome Res.* **2017,** *16*, 2174–2187.

Chen, L. Q.; Hou, B. H.; Lalonde, S.; Takanaga, H.; Hartung, M. L.; Qu, X. Q.; Guo, W. J.; Kim, J. G.; Underwood, W.; Chaudhuri, B.; Chermak, D. Sugar Transporters for Intercellular Exchange and Nutrition of Pathogens. *Nature* **2010,** *468* (7323), 527–532.

Chen, L. Q.; Qu, X. Q.; Hou, B. H.; Sosso, D.; Osorio, S.; Fernie, A. R.; Frommer, W. B. Sucrose Efflux Mediated by SWEET Proteins as a Key Step for Phloem Transport. *Science* **2012,** *335* (6065), 207–211.

Chen, X.; Lin, W. H.; Wang, Y.; Luan, S.; Xue, H. W. An Inositol Polyphosphate 5-Phosphatase Functions in PHOTOTROPIN1 Signaling in *Arabidopsis* by Altering Cytosolic Ca^{2+}. *Plant Cell* **2008,** *20* (2), 353–366.

Clements, M. A. Orchid Mycorrhizal Associations. *Lindleyana* **1988,** *3*, 73–76.

Cullings, K. W.; Szaro, T. M.; Bruns, T. D. Evolution of Extreme Specialization within a Lineage of Ectomycorrhizal Epiparasites. *Nature* **1996,** *379*, 63–66.

Currah, R. S.; Sherburne, R. Septal Ultrastructure of Some Fungal Endophytes from Boreal Orchid Mycorrhizas. *Mycol. Res.* **1992,** *96* (7), 583–587.

Currah, R. S.; Sigler, L.; Hambleton, S. New Records and New Taxa of Fungi from the Mycorrhizae of Terrestrial Orchids of Alberta. *Can. J. Bot.* **1987,** *65*, 2473–2482.

Curtis, J. T. The Relation of Specificity of Orchid Mycorrhizal Fungi to the Problem of Symbiosis. *Am. J. Bot.* **1939,** *26*, 390–399.

Das, A. K.; Khumbongmayu, A. D. *Galeola falconeri* Hook. f., an Endangered Giant Saprophytic Orchid. *Curr. Sci.* **2006,** *91* (7), 871–873.

Dearnaley, J. D. W. Further Advances in Orchid Mycorrhizal Research. *Mycorrhiza* **2007,** *17*, 475–486.

Dearnaley, J. D.; Cameron, D. D. Nitrogen Transport in the Orchid Mycorrhizal Symbiosis-Further Evidence for a Mutualistic Association. *New Phytol.* **2016,** *213* (1), 10–12.

Dearnaley, J. D.; Martos, F.; Selosse, M. A. Orchid Mycorrhizas: Molecular Ecology, Physiology, Evolution and Conservation Aspects. In *Fungal associations*; Springer: Berlin, Heidelberg, 2012; pp 207–230.

Dearnaley, J.; Perotto, S.; Selosse, M. A. Structure and Development of Orchid Mycorrhizas. In *Molecular Mycorrhizal Symbiosis*, 2016; pp 63–86.

Dearnaley, JDW.; Cameron, D. D. Nitrogen Transport in the Orchid Mycorrhizal Symbiosis–Further Evidence for a Mutualistic Association. *New Phytol.* **2017,** 213.

Delavaux, C. S.; Weigelt, P.; Dawson, W.; et al. Mycorrhizal Fungi Influence Global Plant Biogeography. *Nat. Ecol. Evol.* **2019,** *3*, 424–429.

Dickie, I. A.; Alexander, I.; Lennon, S.; Opik, M.; Selosse, M.; van der Heijden, M. G. A.; Martin, F. M. Evolving Insights to Understanding Mycorrhizas. *New Phytol.* **2015,** *205*, 1369–1374.

Ding, R.; Chen, X.; Zhang, L.; Yu, X.; Qu, B.; Duan, R.; Xu, Y. F. Identity and Specificity of *Rhizoctonia*-Like Fungi from Different Populations of *Liparis japonica* (Orchidaceae) in Northeast China. *PLoS One* **2014,** *9* (8), 172–177.

Dutta, D.; Puzari, KC.; Gogoi, R.; Dutta, P. Endophytes: Exploitation as a Tool in Plant Protection. *Braz. Arch. Biol. Technol.* **2014,** *57* (5), 1–10.

Esnault, A. L.; Masuhara, G.; McGee, P. A. Involvement of the Exodermal Passage Cells in Mycorrhizal Infection of Some Orchids. *Mycol. Res.* **1994,** *98*, 672–676.

Eswaranpillai, U.; Raman, R.; Muthukumar, T. Morphology, Anatomy and Mycotrophy of Pseudobulb and Subterranean Organs in *Eulophia epidendraea* and *Malaxis acuminata* (Epidendroideae, Orchidaceae). *Flora* **2015,** *217*, 14–23.

Fochi, V.; Chitarra, W.; Kohler, A.; Voyron, S.; Singan, V. R.; Lindquist, E. A.; Barry, K. W.; Girlanda, M.; Grigoriev, I. V.; Martin, F.; Balestrini, R. Fungal and Plant Gene Expression in the *Tulasnella calospora–Serapias vomeracea* Symbiosis Provides Clues About Nitrogen Pathways in Orchid Mycorrhizas. *New Phytol.* **2017,** *213*: 365–379.

Frank, A. B. Uber die auf Wurzelsymbiose beruhende Ernahrung gewisser Baume durch Unterirdische Pilze. *Berichte der Deutschen Botanischen Gesellschaft (in German)* **1885,** *3*, 128–145.

Furman, T. E.; Trappe, J. M. Phylogeny and Ecology of Mycotrophic Achlorophyllous Angiosperms. *Q. Rev. Biol.* **1971,** *46*, 219–225.

Gebauer, G.; Preiss, K.; Gebauer, A. C. Partial Mycoheterotrophy is More Widespread Among Orchids Than Previously Assumed. *New Phytol.* **2016,** *211*, 11–15.

Genre, A.; Lanfranco, L.; Perotto, S.; et al. Unique and Common Traits in Mycorrhizal Symbioses. *Nat. Rev. Microbiol.* **2020,** *18*, 649–660.

Genre, A.; Russo, G. Does a Common Pathway Transduce Symbiotic Signals in Plant–Microbe Interactions. *Front. Plant Sci.* **2016,** *7*, 96.

Goh, C. J.; Sim, A. A.; Lim, G. Mycorrhizal association in some tropical orchids. *Lindleyana* **1992,** *7*, 13–17.

Govaerts, R.; Bernet, P.; Kratochvil, K.; Gerlach, G.; Carr, G.; et al. World Checklist of Orchidaceae; Facilitated by the Royal Botanic Gardens, Kew. Published on the Internet, 2020; http://wcsp.science.kew.org/ Retrieved 9 December 2020.

Guimaraes, F. A. R.; Pereira, M. C.; Felicio, C. S. Symbiotic Propagation of Seedlings of *Cyrtopodium glutiniferum* Raddi (Orchidaceae). *Acta Bot. Brasilica* **2013,** *27*, 590–596.

Hadley, G. Cellulose as a Carbon Source for Orchid Mycorrhiza. *New Phytol.* **1969,** *69* (4), 1015–1023.

Hadley, G. Orchid Mycorrhiza. In *Orchid Biology Reviews and Perspectives*; Arditti, J., Ed.; Cornell University Press: New York, 1982; Vol. 2, pp 83–118.

Hadley, G.; Ong, S. H. Nutritional Requirements of Orchid Endophytes. *New Phytol.* **1978,** *81* (3), 561–569.

Han, J.; Jiang, J. Genome-Wide Analysis of SWEET Gene Family in *Arabidopsis thaliana, Oryza sativa* and *Lycopersicum esculentum*. *Mol. Plant Breed.* **2015,** *13* (3), 581–588.

Harley, J. L. Mycorrhiza. In *The Ecology of Soil–Borne Plant Pathogens.* Berkley, 1963; pp 218–229.

Harley, J. L. *The Biology of Mycorrhiza*. Leonard Hill Books: London, 1959.

Harmon, A. C.; Gribskov, M.; Gubrium, E.; Harper, J. F. The CDPK Superfamily of Protein Kinases. *New Phytol.* **2001,** *151* (1), 175–183.

Harvais, G.; Pekkala, D. Vitamin Production by a Fungus Symbiotic with Orchids. *Can. J. Bot.* **1975,** *53*, 156–163.

Hayakawa, S.; Yukari, U.; Akira, O. Identification of Symbiotic *Rhizoctonias* from Naturally Occurring Protocorms and Roots of *Dactylorhiza aristata* (Orchidaceae). *J. Faculty Agric.* **1999,** *69* (2), 129–141.

He, F.; Sheng, M.; Tang, M. Effects of *Rhizophagus irregularis* on Photosynthesis and Antioxidative Enzymatic System in *Robinia pseudoacacia* L. Under Drought Stress. *Front. Plant Sci.* **2017,** *8*, 183.

Herrera, H.; Valadares, R.; Contreras, D.; Bashan, Y.; Arriagada, C. Mycorrhizal Compatibility and Symbiotic Seed Germination of Orchids from the Coastal Range and Andes in South Central Chile. *Mycorrhiza* **2016,** *27* (3), 175–188.

Hossain, M. M.; Rahi, P.; Gulati, A.; Sharma, M. Improved *Ex Vitro* Survival of Asymbiotically Raised Seedlings of *Cymbidium* Using Mycorrhizal Fungi Isolated from Distant Orchid Taxa. *Sci. Hortic.* **2013b,** *159*, 109–112.

Hossain, M.; Kant, M. R.; Pham, T. V.; Winarto, B.; Zeng, S.; Silva, J. A. T. The Application of Biotechnology to Orchids. *Crit. Rev. Plant Sci.* **2013a,** *32*, 69–139.

Jacobs, H.; Bosewell, G. P.; Harper, F. A.; Ritz, K.; Davidson, F. A.; Gadd, G. M. Solubilization of Metal Phosphates by *Rhizoctonia solani. Mycol. Res.* **2002,** *106*, 1468–1479.

Jacquemyn, H.; Deja, A.; De-hert, K.; Bailatore, B. C.; Lievens, B. Variation in Mycorrhizal Associations with Tulasnelloid Fungi Among Populations of Five *Dactylorhiza* Species. *PLoS One* **2012,** *7* (8), e42212.

Jakalski, M.; Minasiewicz, J.; Caius, J.; May, M.; Selosse, MA.; Delannoy, E. The Genomic Impact of Mycoheterotrophy: Targeted Gene Losses but Extensive Expression Reprogramming. *BioRxiv* **2020.**

Janse, J. M. Les endophytes radicaux de quelques plantes Javanaise. *Annales du Jar din Botanique de Buitenzorg* **1897,** *14*, 53–112.

Johnson, S. R. Symbiotic Seed Germination in Relation to Potential Naturalization Ability of *Bletilla striata* (Orchidaceae). *Lindleyana* **1994,** *9* (2), 99–101.

Kaliamoorthy, S. Pattern of Mycorrhizal Infection in the Roots of *Aerides maculosum* Lindl. and *Calanthe triplicata* (Willem.) Ames. *Mycorrhiza News* **2007,** *19*, 14–18.

Kasmir, J.; Senthilkumar, S. R.; Britto, S. J. L.; Raj, J. M. Identification of Fungal Endophytes from Orchidaceae Members based on nrITS (Internal Transcribed Spacer) Region. *Int. Res. J. Biotechnol.* **2011,** *2* (6), 139–144.

Khamchatra, N.; Dixon, K. W.; Tantiwiwat, S. Symbiotic Seed Germination of an Endangered Epiphytic Slipper Orchid, *Paphiopedilum villosum* (Lindl.) Stein. from Thailand. *S. Afr. J. Bot.* **2016,** *104*, 76–81.

Klemens, P. A.; Patzke, K.; Deitmer, J.; Spinner, L.; Le Hir, R.; Bellini, C.; Bedu, M.; Chardon, F.; Krapp, A.; Neuhaus, H. E. Overexpression of the Vacuolar Sugar Carrier AtSWEET16 Modifies Germination, Growth, and Stress Tolerance in *Arabidopsis*. *Plant Physiol.* **2013,** *163* (3), 1338–1352.

Kohler, A.; Kuo, A.; Nagy, L. G. Convergent Losses of Decay Mechanisms and Rapid Turnover of Symbiosis Genes in Mycorrhizal Mutualists. *Nat. Genet.* **2015,** *47*, 410–500.

Kristiansen, K. A.; Freudenstein, J. V.; Rasmussen, F. N.; Rasmussen, H. N. Molecular Identification of Mycorrhizal Fungi in *Neuwiedia veratrifolia* (Orchidaceae). *Mol. Phylogenet. Evol.* **2004,** *33* (2): 251–258.

Kristiansen, K. A.; Taylor, L.; Kjoller, R.; Rasmussen, H. N.; Rosendah, S. Identification of Mycorrhizal Fungi from Single Pelotons of *Dactylorhiza majalis* (Orchidaceae) Using Single-Strand Conformation Polymorphism and Mitochondrial Ribosomal Large Subunit DNA Sequences. *Mol. Ecol.* **2001,** *10*, 2089–2093.

Lallemand, F.; Martin-Magniette, M. L.; Gilard, F.; Gakière, B.; Launay-Avon, A.; Delannoy, É.; Selosse, M. A. *In situ* Transcriptomic and Metabolomic Study of the Loss of Photosynthesis in the Leaves of Mixotrophic Plants Exploiting Fungi. *Plant J.* **2019,** *98*.

Li, H. Y.; Wei, D. Q.; Shen, M.; Zhou, Z. P. Endophytes and Their Role in Phytoremediation. *Fungal Diversity* **2012,** *54*, 11–18.

Li, J.; Gao, G.; Zhang, T.; Wu, X. The Putative Phytocyanin Genes in Chinese Cabbage (*Brassica rapa* L.): Genome-Wide Identification, Classification and Expression Analysis. *Mol. Genet. Genomics* **2013,** *288* (1–2), 1–20.

Liebel, H. T.; Bidartondo, M. I.; Gebauer, G. Are Carbon and Nitrogen Exchange between Fungi and the Orchid *Goodyera repens* Affected by Irradiance. *Ann. Bot.* **2015,** *115* (2), 251–261.

Ma, H.; Zhao, H.; Liu, Z.; Zhao, J. The Phytocyanin Gene Family in Rice (*Oryza sativa* L.): Genome-Wide Identification, Classification and Transcriptional Analysis. *PLoS One* **2011,** *6* (10), e25184.

Ma, X.; Kang, J.; Nontachaiyapoom, S.; Wen, T.; Hyde, K. D. Non-Mycorrhizal Endophytic Fungi from Orchids. *Curr. Sci.* **2015,** *109* (1), 72–87.

Manoharachary, C.; Tilak, K. V. B. R. Orchid Mycorrhizae: The Forgotten Biological World. *J. Indian Bot. Soc.* **2015,** *94* (3,4), 153–159.

Martos, F.; Dulormne, M.; Pailler, T.; Bonfante, P.; Faccio, A.; Fournel, J.; Dubois, M. P.; Selosse, M. A. Independent Recruitment of Saprotrophic Fungi as Mycorrhizal Partners by Tropical Achlorophyllous Orchids. *New Phytol.* **2009,** *184*, 668–681.

Mashiguchi, K.; Asami, T.; Suzuki, Y. Genome-Wide Identification, Structure and Expression Studies, and Mutant Collection of 22 Early Nodulin-Like Protein Genes in *Arabidopsis. Biosci. Biotechnol. Biochem.* **2009,** *73* (11), 2452–2459.

Mashiguchi, K.; Yamaguchi, I.; Suzuki, Y. Isolation and Identification of Glycosylphosphatidylinositol-Anchored Arabinogalactan Proteins and Novel β-glucosyl Yariv-Reactive Proteins from Seeds of Rice (*Oryza sativa*). *Plant Cell Physiol.* **2004,** *45* (12), 1817–1829.

Masuhara, G.; Katsuya, K. *In Situ* and *In Vitro* Specificity between *Rhizoctonia* sp. and *Spiranthes sinensis* (Persoon) Ames. var. *amoena* (M. Bieberstein) Hara (Orchidaceae). *New Phytol.* **1994,** *127*, 711–718.

McCormick, M. K.; Jacquemyn, H. What Constrains the Distribution of Orchid Populations?. *New Phytol.* **2014,** *202*: 392–400.

McCormick, M. K.; Whigham, D. F.; Canchani-Viruet, A. Mycorrhizal Fungi Affect Orchid Distribution and Population Dynamics. *New Phytol.* **2019,** *219*: 1207–1215.

McKendrick, S. L.; Leake, J. R.; Taylor, D. L.; Read, D. J. Symbiotic Germination and Development of Myco-Heterotrophic Plants in Nature: Transfer of Carbon from Ectomycorrhizal *Salix repens* and *Betula pendula* to the Orchid *Corallorhiza trifida* through Shared Hyphal Connections. *New Phytol.* **2000,** *145*, 539–548.

Mehra, S. P.; Morrison, D.; Coates, F.; Lawrie, A. C. Differences in Carbon Source Utilisation by Orchid Mycorrhizal Fungi from Common and Endangered Species of *Caladenia* (Orchidaceae). *Mycorrhiza* **2016,** *27* (2), 95–108.

Merckx, V. S. F. T. Mycoheterotrophy: An Introduction. In *Mycoheterotrophy: The Biology of Plants Living on Fungi*; Merckx, V. S. F. T., Ed.; Springer: Berlin, Germany, 2013; pp 1–17.

Miura, C.; Yamaguchi, K.; Miyahara, R.; Yamamoto, T.; Fuji, M.; Yagame, T.; Imaizumi-Anraku, H.; Yamato, M.; Shigenobu, S.; Kaminaka, H. The Mycoheterotrophic Symbiosis between Orchids and Mycorrhizal Fungi Possesses Major Components Shared with Mutualistic Plant-Mycorrhizal Symbioses. *Mol. Plant-Microbe Interact.* **2018,** *31* (10), 1032–1047.

Motomura, H.; Ueno, O.; Kagawa, A.; Yukawa, T. Carbon Isotope Ratios and the Variation in the Diurnal Pattern of Malate Accumulation in Aerial Roots of CAM Species of *Phalaenopsis* (Orchidaceae), *Photosynthetica* **2008,** *46*, 531–536.

Muthukumar, T.; Uma, E.; Karthikeyan, A.; Sathiyadash, K.; Jaison, S.; Priyadharsini, P.; Chongtham, I.; Muniappan, V. Morphology, Anatomy and Mycorrhizae in Subterranean Parts of *Zeuxine gracilis* (Orchidaceae). *An. Biol.* **2011,** *33*, 127–134.

Nurfadilah, S.; Swarts, N. D.; Dixon, K. W.; Lambers, H.; Merritt, D. J. Variation in Nutrient-Acquisition Patterns by Mycorrhizal Fungi of Rare and Common Orchids Explains Diversification in a Global Biodiversity Hotspot. *Ann. Bot.* **2013,** *111* (6), 1233–1241.

Ogura-Tsujita, Y.; Gebauer, G.; Hashimoto, T.; Umata, H.; Yukawa, T. Evidence for Novel and Specialized Mycorrhizal Parasitism: The Orchid *Gastrodia confuse* Gains Carbon from Saprotrophic *Mycena*. *Proc. R. Soc. London B: Biol. Sci.* **2009,** *276*, 761–768.

Oja, J.; Kohout, P.; Tedersoo, L.; Kull, T.; Koljalg, U. Temporal Patterns of Orchid Mycorrhizal Fungi in Meadows and Forests as Revealed by 454 Pyrosequencing. *New Phytol.* **2015,** *205*, 1608–1618.

Pecoraro, L.; Girlanda, M.; Kull, T.; Perini, C.; Perotto, S. Fungi from the Roots of the Terrestrial Photosynthetic Orchid *Himantoglossum adriaticum*. *Plant Ecol. Evol.* **2013,** *146* (2), 145–152.

Pereira, G.; Albornoz, V.; Munoz-Tapia, L.; Romero, C.; Atala, C. Asymbiotic Germination of *Bipinnula fimbriata* (Orchidaceae) Seeds in Different Culture Media. *Seed Sci. Technol.* **2015b,** *43* (3), 1–11.

Pereira, M. C.; Rocha, D. I.; Veloso, T. G. R.; Pereira, O. G.; Francino, D. M. T.; Meira, R. M. S.; Kasuya, M. C. Characterization of Seed Germination and Protocorm Development of *Cyrtopodium glutiniferum* (Orchidaceae) Promoted by Mycorrhizal Fungi *Epulorhiza* sp. *Acta Bot. Brasilica* **2015a,** *29* (4), 567–574.

Perombelon, M.; Hadley, G. Production of Pectic Enzymes by Pathogenic and Symbiotic *Rhizoctonia* Strains. *New Phytol.* **1965,** *64*, 144–151.

Perotto, S.; Rodda, M.; Benetti, A.; Sillo, F.; Ercole, E.; Rodda, M.; Balestrini, R. Gene Expression in Mycorrhizal Orchid Protocorms Suggests a Friendly Plant–Fungus Relationship. *Planta* **2014,** *239* (6), 1337–1349.

Peterson, R. L.; Farquhar, M. L. Mycorrhizas Integrated Development between Roots and Fungi. *Mycologia* **1994,** *86*, 311–326.

Phillips, R. D.; Barrett, M. D.; Dixon, K. W.; Hopper, S. D. Do Mycorrhizal Symbioses Cause Rarity in Orchids. *J. Ecol.* **2011,** *99*, 585–569.

Phillips, R. D.; Reiter, N.; Peakall, R. Orchid Conservation: From Theory to Practice. *Ann. Bot.* **2020,** *126*: 345–362.

Powell, K. B.; Arditti, J. Growth Requirements of *Rhizoctonia repens* M32. *Mycopathology* **1975,** *55*, 163–167.

Qui-Xia, W.; Ning, Y.; Da-Gan, J. I.; Shu-Yan, L. I.; JiangMiao, H. U.; Hong, H. U. Mycorrhizal Fungi Promote Growth and Nitrogen Utilization by *Dendrobium nobile* (Orchidaceae). *Plant Diversity Res.* **2014,** *36*: 321–330.

Rasmussen, H. N. Recent Developments in the Study of Orchid Mycorrhiza. *Plant Soil* **2002,** *224*, 149–163.

Rasmussen, H. N. *Terrestrial Orchids: From Seeds to Mycotrophic Plants*. Cambridge University Press: Cambridge, UK, 1995.

Rasmussen, H. N.; Dixon, K. W.; Jersakova, J.; Tesitelova, T. Viewpoint: Part of a Highlight on Orchid Biology Germination and Seedling Establishment in Orchids: A Complex of Requirements. *Ann. Bot.* **2015,** *116*, 391–402.

Rasmussen, H. N.; Whingham, D. F. Phenology of Roots and Mycorrhiza in Orchid Species Differing in Phototrophic Strategy. *New Phytol.* **2002,** *154*, 797–807.

Rasmussen, H.; Rasmussen, F. Orchid Mycorrhiza: Implications of a Mycophagous Life Style. *Oikos* **2009,** *118*, 334–345.

Reissek, S. Über endophyten der Pflanzenzelle: eine gesetzmässige, den Samenfaden oder beweglichen Spiralfasern analoge Erscheinung. *Naturwiss. Abh.* **1847,** 1.

Richardson, K. A.; Currah, R. S.; Hambleton, S. Basidiomycetous Endophytes from the Roots of Neotropical Epiphytic Orchidaceae. *Lindleyana* **1993,** *8*, 127–137.

Riofrio, M. L.; Cruz, D.; Torres, E.; Cruz, M.; Iriondo, J. M.; Suarez, J. P. Mycorrhizal Preferences and Fine Spatial Structure of the Epiphytic Orchid *Epidendrum rhopalostele*. *Am. J. Bot.* **2013,** *100* (12), 2339–2348.

Roche, S.; Carter, R. J.; Peakall, R.; Smith, L. M.; Whitehead, M. R.; Linde, C. C. A Narrow Group of Monophyletic *Tulasnella* (Tulasnellaceae) Symbiont Lineages are Associated with Multiple Species of *Chiloglottis* (Orchidaceae): Implications for Orchid Diversity. *Am. J. Bot.* **2010,** *97*, 1313–1327.

Ruibal, M. P.; Triponez, Y.; Smith, L. M.; Peakall, R.; Linde, C. C. Population Structure of an Orchid Mycorrhizal Fungus with Genus-Wide Specificity. *Sci. Rep.* **2017,** *7*, 5613.

Salifah, H. A. B.; Muskhazli, M.; Rusea, G.; Nithiyaa, P. Variation in Mycorrhizal Specificity for *In Vitro* Symbiotic Seed Germination of *Grammatophyllum speciosum* Blume. *Sains Malaysiana* **2011,** *40*, 451–455.

Sarsaiya, S.; Jain, A.; Jia, Q.; Fan, X.; Shu, F.; Chen, Z.; Chen, J. Molecular Identification of Endophytic Fungi and Their Pathogenicity Evaluation against *Dendrobium nobile* and *Dendrobium officinale*. *Int. J. Mol. Sci.* **2020,** *21* (1), 316.

Sathiyadash, K.; Muthukumar, T.; Karthikeyan, V.; Rajendran, K. Orchid Mycorrhizal Fungi: Structure, Function, and Diversity. In *Orchid Biology: Recent Trends & Challenges*; Khasim, S., Hegde, S., González-Arnao, M., Thammasiri, K., Eds.; Springer: Singapore, 2020.

Sathiyadash, K.; Muthukumar, T.; Uma, E.; Rama, P. R. Mycorrhizal Association and Morphology in Orchids. *J. Plant Interac.* **2012,** *7*, 238–247.

Sawmya, K.; Vasudevan, T. G.; Mural, T. S. Fungal Endophytes from Two Orchid Species Pointer towards Organ Specificity. *Czech Mycol.* **2013,** *65* (1), 89–101.

Sazak, A.; Ozdener, Y. Symbiotic and Asymbiotic Germination of Endangered *Spiranthes spiralis* (L.) Chevall. and *Dactylorhiza osmanica* (KI.) Soo var. *Osmanica* (endemic). *Pak. J. Biol. Sci.* **2006,** *9*, 222–228.

Schreiber, L.; Franke, R. B. *Endodermis and Exodermis in Roots*; John Wiley and Sons: New York, 2001.

Selosse, M. A. The Latest News from Biological Interactions in Orchids: In Love, Head to Toe. *New Phytol.* **2014,** *202* (2), 347–350.

Selosse, M. A.; Roy, M. Green Plants that Feed on Fungi: Facts and Questions about Mixotrophy. *Trends Plant. Sci.* **2009,** *14*, 64–70.

Senthilkumar, S. Problems and Prospectus of Orchid Mycorrhizal Research. *J. Orchid Soc. India* **2001,** *15*, 23–32.

Shefferson, R. P.; Taylor, D. L.; Weiss, M.; Garnica, S.; McCormick, M. K.; Adams, S.; Gray, H. M.; McFarland, J. W.; Kull, T.; Tali, K.; Yukawa, T. The Evolutionary History of Mycorrhizal Specificity among Lady's Slipper Orchids. *Evolution* **2007,** *61*, 1380–1390.

Siddiquee, S.; Yusuf, U. K.; Zainudin, N. Morphological and Molecular Detection of *Fusarium chlamydosporum* from Root Endophytes of *Dendrobium crumenatum*. *Afr. J. Biotechnol.* **2010,** *9*, 26–30.

Smith, S. E. Carbohydrate Translocation in Orchid Mycorrhizas. *New Phytol.* **1967,** *66*, 371–378.

Smith, S. E. Physiology and Ecology of Orchid Mycorrhizal Fungi with Reference to Seedling Nutrition. *New Phytol.* **1966,** *65*, 488–499.

Smith, S. E.; Read, D. J. *Mycorrhizal Symbiosis*, 2nd ed.; Academic Press: San Diego, 1997.
Smith, S. E.; Read, D. J. *Mycorrhizal Symbiosis*; Academic Press and Elsevier: London, UK, 2008.
Smith, Z. F.; James, E. A.; McLean, C. B. Mycorrhizal Specificity of *Diuris fragrantissima* (Orchidaceae) and Persistence in a Reintroduced Population. *Aust. J. Bot.* **2010,** *58*, 97–106.
Smreciu, E. A.; Currah, R. S. Symbiotic Germination of Seeds of Terrestrial Orchids of North America and Europe. *Lindleyana* **1989,** *4*, 6–15.
Sudheep, N. M.; Sridhar, R. K. Non-Mycorrhizal Fungal Endophytes in Two Orchids of Kaiga Forest (Western Ghats). *India J. For. Res.* **2012,** *23* (3), 453–460.
Suetsugu, K.; Yamato, M.; Miura, C.; Yamaguchi, K.; Takahashi, K.; Ida, Y.; Kaminaka, H. Comparison of Green and Albino Individuals of the Partially Mycoheterotrophic Orchid *Epipactis helleborine* on Molecular Identities of Mycorrhizal Fungi, Nutritional Modes and Gene Expression in Mycorrhizal Roots. *Mol. Ecol.* **2017,** *26* (6), 1652–1669.
Swarts, N. D; Dixon, K. W. Terrestrial Orchid Conservation in the Age of Extinction. *Ann. Bot.* **2009,** *104* (3), 543–556.
Takeda, N.; Handa, Y.; Tsuzuki, S.; Kojima, M.; Sakakibara, H.; Kawaguchi, M. Gibberellins Interfere with Symbiosis Signaling and Gene Expression and Alter Colonization by Arbuscular Mycorrhizal Fungi in *Lotus japonicus*. *Plant Physiol.* **2015,** *167* (2), 545–557.
Tao, G.; Liu, Z. Y.; Hyde, K. D.; Liu, X. Z.; Yu, Z. N. Whole rDNA Analysis Reveals Novel and Endophytic Fungi in *Bletilla ochracea* (Orchidaceae). *Fungal Diversity* **2008,** *33*, 101–122.
Taylor, D. L.; Bruns, T. D. Independent, Specialized Invasion of Ectomycorrhizal Mutualism by Two Nonphotosynthetic Orchids. *Proc. Nat. Acad. Sci. U.S.A.* **1997,** *94*, 4510–4515.
Taylor, D. L.; Bruns, T. D.; Leake, J. R.; Read, D. J. Mycorrhizal Specificity and Function in Myco-Heterotrophic Plants. In *Mycorrhizal Ecology*; Van der, M. G., Heijden, A., Sanders, I., Eds.; Springer-Verlag: Berlin, 2002; pp 157–160.
Tian, C.; Wan, P.; Sun, S.; Li, J.; Chen, M. Genome-Wide Analysis of the GRAS Gene Family in Rice and *Arabidopsis*. *Plant Mol. Biol.* **2004,** *54* (4), 519–532.
Tondello, E. V.; Villani, M.; Baldan, B.; Squartini, A. Fungi Associated with the Southern Eurasian orchid *Spiranthes spiralis* (L.) Chevall. *Fungal Biol.* **2012,** *116* (4), 543–549.
Trudell, S. A.; Rygiewicz, P. T.; Edmonds, R. L. Nitrogen and Carbon Stable Isotope Abundances Support the Mycoheterotrophic Nature and Host Specificity of Certain Achlorophyllous Plants. *New Phytol.* **2003,** *160*, 391–401.
Tsai, T. M.; Chen, Y. R.; Kao, T. W.; Tsay, W. S.; Wu, C. P.; Huang, D. D.; Huang, H. J. PaCDPK1, a Gene Encoding Calcium-Dependent Protein Kinase from Orchid, *Phalaenopsis amabilis*, is Induced by Cold, Wounding, and Pathogen Challenge. *Plant Cell Rep.* **2007,** *26* (10), 1899–1908.
Tsujita, Y. O.; Gebauer, G.; Hashimoto, T.; Umata, H.; Yukawa, T. Evidence for Novel and Specialized Mycorrhizal Parasitism: The Orchid *Gastrodia confuse* Gains Carbon from Saprotrophic *Mycena*. *Proc. Biol. Sci.* **2009,** *276*, 761–767.
Valadares, R. B. S.; Perotto, S.; Santos, E. C.; Lambais, M. R. Proteome Changes in *Oncidium sphacelatum* (Orchidaceae) at Different Trophic Stages of Symbiotic Germination. *Mycorrhiza* **2014,** *24* (5), 349–360.
Vij, S. P.; Lakhanpal, T. N.; Gupta, A. Orchidoid Mycorrhiza and Techniques to Investigate. In *Techniques in Mycorrhizal Studies*; Springer: Dordrecht, 2002; pp 385–434.
Vij, S. P.; Sharma, M. Mycorrhizal Associations in North Indian Orchidaceae. A Morphological Study. *Bibl. Mycol.* **1983,** *91*, 467–503.

Vij, S.; Sharma, P. M.; Datta, S. S. Mycorrhizal Endophyte of *Spiranthes lancea* (Sw.) Baker White Flowered Taxon (Orchidaceae). *J. Indian Bot. Soc.* **1985,** *64*, 175–179.

Wahrlich, W. *Beitrag zur kenntniss der orchideenwurzelpilze*; Druck von Breitkopf & Härtel, 1886.

Wang, T.; Song, Z.; Wei, L.; Li, L. Molecular Characterization and Expression Analysis of WRKY Family Genes in *Dendrobium officinale*. *Genes Genomics* **2018,** *40* (3), 265–279.

Wang, W.; Shi, J.; Xie, Q.; Jiang, Y.; Yu, N.; Wang, E. Nutrient Exchange and Regulation in Arbuscular Mycorrhizal Symbiosis. *Mol. Plant* **2017,** *10*, 1147–1158.

Warcup, J. H.; Talbot, P. H. B. Perfect States of *Rhizoctonias* Associated with Orchids. *New Phytol.* **1967,** *66*, 631–641.

Warcup, J. H.; Talbot, P. H. B. Perfect States of Some *Rhizoctonias*. *Trans. Br. Mycol. Soc.* **1966,** *49*, 427–435.

Williams, P. G. Orchidaceous *Rhizoctonias* in Pot Cultures of Vesicular-Arbuscular Mycorrhizal Fungi. *Can. J. Bot.* **1985,** *63*, 1329–1333.

Xu, L.; Wang, X. J.; Wang, T.; Li, L. B. Genome-Wide Identification, Classification, and Expression Analysis of the Phytocyanin Gene Family in *Phalaenopsis equestris*. *Biol. Plant.* **2017,** *61* (3), 445–452.

Yagame, T.; Yamato, M.; Mii, M.; Suzuki, A.; Iwase, K. Developmental Processes of Achlorophyllous Orchid *Epipogium roseum* from Seed Germination to Flowering Under Symbiotic Cultivation with Mycorrhizal Fungus. *J. Plant Res.* **2008,** *120*, 229–236.

Yang, T.; Poovaiah, B. W. Calcium/Calmodulin-Mediated Signal Network in Plants. *Trends Plant. Sci.* **2003,** *8* (10), 505–512.

Yoder, J. A.; Zettler, L.; Wand, S. L. Water Requirements of Terrestrial and Epiphytic Orchid Seeds and Seedlings, and Evidence for Water Uptake by Means of Mycotrophy. *Plant Sci. Limerick* **2000,** *156*, 145–150.

Yuan, M.; Zhao, J.; Huang, R.; Li, X.; Xiao, J.; Wang, S. Rice MtN3/saliva/SWEET Gene Family: Evolution, Expression Profiling, and Sugar Transport. *J. Integr. Plant Biol.* **2014,** *56* (6), 559–570.

Zelmer, C. D.; Currah, R. S. *Ceratorhiza pernacatena* and *Epulorhiza calendulina* spp.: Mycorrhizal Fungi of Terrestrial Orchids. *Can. J. Bot.* **1995,** *73*, 1981–1985.

Zeng, X.; Li, Y.; Ling, H.; Chen, J.; Guo, S. Revealing Proteins Associated with Symbiotic Germination of *Gastrodia elata* by Proteomic Analysis. *Bot. Stud.* **2018,** *59* (1), 8.

Zeng, X.; Ling, H.; Chen, X.; Guo, S. Genome-Wide Identification, Phylogeny and Function Analysis of GRAS Gene Family in *Dendrobium catenatum* (Orchidaceae). *Gene* **2019,** *705*, 5–15.

Zettler, L. W.; Corey, L. L. Orchid Mycorrhizal Fungi: Isolation and Identification Techniques. In Orchid Propagation: From Laboratories to Greenhouses—Methods and Protocols. Springer Protocols Handbooks; Lee, Y. I., Yeung, E. T., Eds.; Humana Press: New York, NY, 2018.

Zettler, L. W.; Corey, L. L.; Jacks, A. L.; Gruender, L. T.; Lopez, A. M. *Tulasnella irregularis* (Basidiomycota: Tulasnellaceae) from Roots of *Encyclia tampensis* in South Florida, and Confirmation of Its Mycorrhizal Significance through Symbiotic Seed Germination. *Lankesteriana* **2005,** *13*, 119–128.

Zettler, L. W.; Corey, L. L.; Richardson, L. W.; Ross, A. Y.; Moller-Jacobes, I. Protocorms of an Epiphytic Orchid (*Epidendrum amphistomum*) Recovered In Situ, and Subsequent Identification of Associated Mycorrhizal Fungi Using Molecular Markers. *Eur. J. Environ. Sci.* **2011,** *1*, 108–114.

Zettler, L. W.; Hofer, C. J. Sensitivity of *Spiranthes odorata* Seeds to Light During *In Vitro* Symbiotic Seed Germination. *Lindleyana* **1997,** *12* (1), 26–29.

Zhang, G.; Zhao, M.; Li, B.; Song, C.; Zhang, D.; Guo, S. Cloning and Expression Analysis of a Calcium-Dependent Protein Kinase Gene in *Dendrobium officinale* in Response to Mycorrhizal Fungal Infection. *Acta Pharm. Sin.* **2012,** *47* (11), 1548–1554.

Zhao, M. M.; Zhang, G.; Zhang, D. W.; Hsiao, Y. Y.; Guo, S. X. ESTs Analysis Reveals Putative Genes Involved in Symbiotic Seed Germination in *Dendrobium officinale. PLoS One* **2013,** *8* (8), e72705.

Zhao, X. L.; Yang, J.; Liu, S.; Chen, C. L.; Zhu, H. Y.; Cao, J. X. The Colonization Patterns of Different Fungi on Roots of *Cymbidium hybridum* Plantlets and Their Respective Inoculation Effects on Growth and Nutrients Uptake of Orchid Plantlets. *World J. Microbiol. Biotechnol.* **2014**.

Zheng, Z.; Qamar, S. A.; Chen, Z.; Mengiste, T. *Arabidopsis* WRKY33 Transcription Factor Is Required for Resistance to Necrotrophic Fungal Pathogens. *Plant J.* **2006,** *48* (4), 592–605.

Zi, X. M.; Sheng, C. L.; Goodale, U. M.; Shao, S. C.; Gao, J. Y. *In Situ* Seed Baiting to Isolate Germination-Enhancing Fungi for an Epiphytic Orchid, *Dendrobium aphyllum* (Orchidaceae). *Mycorrhiza* **2014,** *24* (7), 487–499.

Zimmer, K.; Hynson, N. A.; Gebauer, G.; Allen, E. B.; Allen, M. F.; Read, D. J. Wide Geographical and Ecological Distribution of Nitrogen and Carbon Gains from Fungi in Pyroloids and Monotropoids (Ericaceae) and in Orchids. *New Phytol.* **2007,** *175*, 166–175.

Zotz, G.; Winkler, U. Aerial Roots of Epiphytic Orchids: The Velamen Radicum and Its Role in Water and Nutrient Uptake. *Oecologia* **2013,** *171*, 733–441.

Index

F

G

H

I

L

M

For Product Safety Concerns and Information please contact our
EU representative GPSR@taylorandfrancis.com Taylor & Francis
Verlag GmbH, Kaufingerstraße 24, 80331 München, Germany